ECOLOGY OF
FRESH WATERS

MAN AND MEDIUM

ECOLOGY OF FRESH WATERS
MAN AND MEDIUM

BRIAN MOSS

Reader in Environmental Sciences
University of East Anglia
Norwich

Second Edition

OXFORD
BLACKWELL SCIENTIFIC PUBLICATIONS
LONDON   EDINBURGH   BOSTON
MELBOURNE   PARIS   BERLIN   VIENNA

© 1980, 1988 by
Blackwell Scientific Publications
Editorial offices:
Osney Mead, Oxford OX2 0EL
25 John Street, London WC1N 2BL
23 Ainslie Place, Edinburgh EH3 6AJ
238 Main Street, Cambridge
  Massachusetts 02142, USA
54 University Street, Carlton
  Victoria 3053, Australia

Other Editorial Offices:
Librairie Arnette SA
2, rue Casimir-Delavigne
75006 Paris
France

Blackwell Wissenschafts-Verlag GmbH
Düsseldorfer Str. 38
D-10707 Berlin
Germany

Blackwell MZV
Feldgasse 13
A-1238 Wien
Austria

First published 1980
Second edition 1988
Reprinted 1989, 1991, 1992, 1993 (twice)

Typeset in Great Britain
by William Clowes Limited, Beccles and
London

Printed and bound in Great Britain by
Redwood Books, Trowbridge, Wiltshire

DISTRIBUTORS
Marston Book Services Ltd
PO Box 87
Oxford OX2 0DT
(*Orders:* Tel: 0865 791155
        Fax: 0865 791927
        Telex: 837515)

USA
Blackwell Scientific Publications, Inc.
238 Main Street
Cambridge, MA 02142
(*Orders:* Tel: 800 759–6102
        617 876–7000)

Canada
Oxford University Press
70 Wynford Drive
Don Mills
Ontario M3C 1J9
(*Orders:* Tel: 416 441–2941)

Australia
Blackwell Scientific Publications Pty Ltd
54 University Street
Carlton, Victoria 3053
(*Orders:* Tel: 03 347–5552)

British Library
Cataloguing in Publication Data
Moss, Brian
  Ecology of fresh waters.—2nd ed.
  1. Freshwater ecosystems
  I. Title
  574.5'2632

  ISBN 0-632-01642-6

Library of Congress
Cataloging-in-Publication Data
Moss, Brian
  Ecology of fresh waters.
  Bibliography: p.
  Includes indexes.
  1. Freshwater ecology.   I. Title
  QH541.5.F7M67 1988
  574.5'2632   88-6128

  ISBN 0-632-01642-6

FOR MY WIFE JOYCE
for her loyalty and love

FOR ANGHARAD BRONMAI
who is just discovering (I think) that
wetness is wonderful

FOR RICHARD STRAUSS
for his Four Last Songs and to

JESSYE NORMAN
for her singing of them

AND FOR ALDO LEOPOLD
whom I believe had real wisdom

'The shallow-minded modern who has lost his rootage in the land assumes
that he has already discovered what is important: it is such who prates of
empires, political or economic, that will last a thousand years. It is only the
scholar who appreciates that all history consists of successive excursions
from a single starting-point, to which man returns again and again to organise
yet another search for a durable scale of values. It is only the scholar who
understands why the raw wilderness gives definition and meaning to the
human enterprise.'

ALDO LEOPOLD · A Sandy County Almanac, 1949

Contents

Preface and Acknowledgements xiii

1   Introduction. The Pongolo Floodplain 1
1.1   Introduction 1
1.2   The Pongolo floodplain ecosystem 2
    1.2.1   Cycles of change on the floodplain 4
    1.2.2   Problems of the human population 5
    1.2.3   The Pongolapoort dam 7
    1.2.4   Future management of the floodplain 8

2   On Living in Water 10
2.1   Properties of water 11
    2.1.1   Physical properties 11
    2.1.2   Water as a solvent 14
    2.1.3   Solubility of non-ionic compounds 16
2.2   Land and water habitats and the evolution of organisms 17
    2.2.1   Physiological problems of living in water 18
    2.2.2   Brackish and freshwater invertebrates 19
    2.2.3   Osmotic relationships in vertebrates 20
    2.2.4   Colonization of fresh waters from the land 22
    2.2.5   Respiration in water 23
    2.2.6   Invertebrate air breathers in fresh waters 24
    2.2.7   Conclusions on physiology 25
2.3   Time and fleetingness—a fundamental difference between freshwater and marine ecosystems 25

3   From Atmosphere to Stream—the Chemical Birth of Fresh Waters 32
3.1   Dissolving of atmospheric gases and the acidity of rain 33
3.2   Contribution of sea spray to rain 34
3.3   Atmospheric pollution 35
3.4   The composition of water draining from the catchments 36
    3.4.1   A chemical catalogue for run-off waters 37
    3.4.2   Rock weathering 39
    3.4.3   Weathering of sedimentary rocks 42

3.5   Effects of soil development and vegetation on the chemistry of drainage
        waters 44
        3.5.1   Nitrogen fixation 45
        3.5.2   Storage in the plant biomass 45
        3.5.3   Vegetation and the supply of suspended silt and dissolved and
                suspended organic matter to drainage waters 50
        3.5.4   Dissolved organic matter 51
3.6   Effects of human activities on the composition of drainage waters 51
        3.6.1   Agriculture 51
        3.6.2   Lowland agriculture 52
        3.6.3   Settlement 55
        3.6.4   Industry 57
        3.6.5   Industrial atmospheric sources 58

4   Upland Streams and Rivers 60
4.1   Introduction 60
4.2   Upland streams—three general questions 62
        4.2.1   The erosive stream habitat and survival in it 62
        4.2.2   Adaptation to moving water 64
4.3   Sources of food and energy flow in upland streams 65
        4.3.1   Hot spring streams 65
        4.3.2   Bear Brook and other streams in wooded catchments 68
        4.3.3   Mechanics of processing of organic matter in woodland streams 70
        4.3.4   The shredders 71
        4.3.5   Collectors, scrapers and carnivores 72
        4.3.6   New Zealand streams 73
        4.3.7   Fish in upland streams 75
        4.3.8   The Atlantic salmon 77
        4.3.9   Animal production in streams 78
4.4   Stream communities 81
        4.4.1   Do distinct stream communities exist? 82
        4.4.2   Competition in structuring stream communities 86
4.5   Upland streams and human activities 87
        4.5.1   River blindness 87
        4.5.2   Game fisheries 88
4.6   Alterations to upland streams by human activities 90
        4.6.1   Acidification 91
        4.6.2   Changes in land use 93
        4.6.3   Physical alteration of the streams 95
        4.6.4   River regulating reservoirs 96
        4.6.5   Alterations by man of the fish community 98

5   Lowland Rivers and their Floodplains 100
5.1   Submerged plants 101
5.2   Growth of submerged plants 103
5.3   Methods of measuring the primary productivity of submerged plants 105
        5.3.1   Whole community methods 106
        5.3.2   Enclosure methods 107

5.4 Submerged plants and the river ecosystem 109
    5.4.1 Plant bed management in rivers 111
5.5 Further downstream—swamps and floodplains 114
    5.5.1 Productivity of swamps and floodplain grasslands 114
    5.5.2 Swamp soils and the fate of the high primary production 116
    5.5.3 Oxygen supply and soil chemistry in swamps 117
    5.5.4 Emergent plants and flooded soils 120
5.6 Swamp and marsh animals 122
    5.6.1 Whitefish and blackfish 125
5.7 Human societies of floodplains 125
    5.7.1 The Marsh Arabs 125
    5.7.2 The Nuer of Southern Sudan 127
5.8 Floodplain fisheries 130
5.9 Modification of floodplain ecosystems 134
    5.9.1 Wetland values 134
    5.9.2 Swamps and nutrient retention 134
    5.9.3 Floodplain swamps and human diseases 136
5.10 Drainage and other alterations to floodplain ecosystems 139
    5.10.1 The Florida Everglades 139
    5.10.2 Drainage and river management in temperate regions 144
5.11 Lowland river channels 145
    5.11.1 Pollution by organic matter 146
    5.11.2 Sewage treatment 147
    5.11.3 Pollution monitoring 149
    5.11.4 Current problems of river pollution 152
    5.11 5 Heavy metals 152
    5.11.6 Problems in pollution management 154

6 Lakes, Pools and Other Standing Waters—some Basic Features of their Productivity 156
6.1 Exorheic lakes 157
6.2 The essential features and parts of a lake 159
    6.2.1 Light availability 160
    6.2.2 The euphotic zone 163
    6.2.3 Thermal stratification and the structure of water masses 164
    6.2.4 Key nutrients 168
    6.2.5 Nutrient 'limitation' 170
    6.2.6 How the total phosphorus concentration of a lake is established 171
    6.2.7 Consequences of thermal stratification for water chemistry 174
    6.2.8 Loss of phosphorus to the sediment 177
    6.2.9 Sediment and the oxidized microzone 178
    6.2.10 Aquatic plant communities and the shapes of basins 180
6.3 General models of lake production 182
    6.3.1 Brylinsky's synthesis 182
    6.3.2 Schindler's analysis 184
    6.3.3 Models incorporating other features 185

6.4   Eutrophication and acidification—changes in the production of
       phytoplankton in lakes 185
       6.4.1   Eutrophication 186
       6.4.2   Solving the eutrophication problem 188
       6.4.3   What are the present supplies of phosphorus and do they contribute
               equally? 190
       6.4.4   Relationship of the phosphorus concentration to the algal crop 190
       6.4.5   Methods available for reducing total phosphorus concentrations 191
       6.4.6   In-lake methods 194
       6.4.7   Complications for phosphorus control 194
6.5   Acidification 197
6.6   Variations on the theme—other standing waters 198
       6.6.1   Rainwater pools 199
       6.6.2   Meromictic lakes 201
       6.6.3   Endorheic lakes 202

7   The Plankton and Fish Communities of the Open Water 205
7.1   The structure of the plankton community 205
7.2   Phytoplankton 205
       7.2.1   Photosynthesis and growth of phytoplankton 211
       7.2.2   Net production and growth 213
       7.2.3   Nutrient uptake and growth rates of phytoplankton 214
       7.2.4   Distribution of freshwater phytoplankton 216
       7.2.5   The desmid plankton 217
       7.2.6   Mixing, stratification and washout 218
       7.2.7   Blue-green algal blooms 219
       7.2.8   Phytoplankton communities and drinking water 221
7.3   Microconsumers of the phytoplankton—bacteria and protozoa 221
7.4   Protozoa and fungi 223
7.5   Zooplankton 223
       7.5.1   Grazing 226
       7.5.2   Feeding and grazing rates of zooplankton 229
7.6   Fish in the open water community 231
       7.6.1   Predation on the zooplankton and fish production 231
       7.6.2   The small mouth yellowfish in Lake Le Roux 232
       7.6.3   Predation by fish 234
       7.6.4   Predation and the composition of zooplankton communities 236
       7.6.5   Predator avoidance by the zooplankton 239
       7.6.6   Consensus 240
7.7   Functioning of the open water community 241
       7.7.1   Cycling of phosphorus in the plankton 241
       7.7.2   The nitrogen cycle in the plankton 244
7.8   Seasonal changes in the plankton 246
       7.8.1   Mechanisms underlying algal periodicity 249
7.9   Practical applications of plankton biology—treatment of eutrophication by
       biomanipulation 251

8   The Edges and Bottoms of Lakes and their Communities 257
8.1   A variety of habitats 257
8.2   Submerged plant communities in lakes 259
        8.2.1   Microbial and animal communities in plant beds 261
        8.2.2   Epiphytic algae 263
        8.2.3   Invertebrates 265
8.3   Competition between submerged plants and phytoplankton 269
        8.3.1   Consequences of loss of aquatic plants 271
8.4   Bare rocks and sandy littoral habitats 272
        8.4.1   Distribution of triclads in the British Isles 272
        8.4.2   Specialization in the rocky littoral 276
        8.4.3   Sandy shores 278
8.5   Relationships between the littoral zone and the open water 279
8.6   The profundal benthos 281
        8.6.1   Biology of selected benthic invertebrates 282
                Chironomus anthracinus 282
                Ilyodrilus hammoniensis and *Pisidium casertanum* 285
                Carnivorous benthos—*Chaoborus* and *Procladius* 286
        8.6.2   What the sediment-living invertebrates really eat 287
8.7   Influence of the open water community on the profundal benthos 288

9   Fish Production and Fisheries in Lakes 295
9.1   Some basic fish biology 296
        9.1.1   Eggs 297
        9.1.2   Feeding 298
        9.1.3   Breeding 301
9.2   Choice of fish for a fishery 303
9.3   Measurement of fish production 304
        9.3.1   Growth measurement 306
9.4   Commercial fisheries 307
9.5   The North Buvuma Island fishery 309
        9.5.1   Estimation of t_b, F_b and M_b for the Buvuma *Oreochromis* fishery 310
9.6   Approximate methods for yield assessment 314
9.7   Changes in fisheries 315
        9.7.1   The North American Great Lakes 316
        9.7.2   The East African Great Lakes 319
9.8   Fish culture 323

10   The Birth, Development and Passing of Lakes 326
10.1   Introduction 326
10.2   Man-made lakes 327
        10.2.1   Fisheries in new tropical lakes 330
        10.2.2   Effects downstream of the new lake 332
        10.2.3   New tropical lakes and human populations 333
        10.2.4   Man-made tropical lakes, the balance of pros and cons 333
10.3   The development of lake ecosystems 335
        10.3.1   Dating the sediment 336

10.3.2 Radiometric techniques 338
10.3.3 Non-radiometric methods 338
10.4 Sources of information in sediments 339
10.4.1 Chemistry 339
10.4.2 Fossils 341
10.4.3 Diatom remains 342
10.4.4 Pollen 344
10.4.5 General problems of interpretation of evidence from sediment cores 344
10.5 Examples of lake development 346
10.5.1 Blea Tarn, English Lake District 346
10.5.2 Esthwaite 348
10.5.3 Pickerel Lake 351
10.5.4 Lago di Monterosi 353
10.6 Filling in of shallow lakes 356
10.6.1 Tarn Moss, Malham 356
10.7 Consensus—natural eutrophication 359
10.8 Finale 361

Further Reading 365

References 367

Index 399

Preface and Acknowledgements

This edition has been almost completely rewritten. Some sections have survived almost intact from the first, but I felt that a different format and a wider scope were necessary to meet the growing changes both in freshwater ecosystems and myself. The subtitle reflects a recognition that human activities are increasingly the main architects of freshwater ecosystem structure and function. The book is still intended for undergraduates and beginning postgraduates wishing to gain a wide view before digging themselves into the isolated pits of their sub-specialities. I hope it will also have wider appeal.

My debts to those whom I acknowledged in the first edition—Charles Sinker and Frank Round—remain, but I realize also the profound influence particularly of the older members and those now dead or retired of the Freshwater Biological Association at Windermere. They seem to have had a powerful knack of unearthing interesting and fundamental stories.

The literature list has expanded from the first edition, but comprises still only a small fraction of the available literature. Usually there are several possible papers to hand (and probably several times more not immediately available) to illustrate any point one wishes to make, so I hope those excluded will not be offended by this random element. Realization of it makes a splendid nonsense of citation indexes and the like used by those who can't to judge those who can.

Finally, I am very grateful to Mrs. Barbara Slade who cheerfully typed a first draft half as long again as the final version and did not mind the major surgery I exercised upon it.

BRIAN MOSS

1
Introduction
The Pongolo Floodplain

1.1 INTRODUCTION

Water is all to us. From space the planet earth looks blue because of the optical properties of the water in the ocean and atmosphere. This water defines a thin, moist film, only about 18 km thick from the bottoms of the deepest ocean trenches to the top of the habitable atmosphere, which harbours the biosphere. The planetary crust itself is brittle and prone to crack and move as it floats on a plastic mass of inner rock, so the biosphere rests like a delicate moist film on the surface of an egg and may be just as ephemeral. Yet there is virtually no activity of any living organism that can go on indefinitely without the presence of uncombined water. The reactions that sustain life would be too slow in a solid system, too uncontrolled in a gaseous one. Only in a liquid matrix can they be satisfactorily achieved; and water is the only common liquid on Earth.

This remarkable liquid is continually distilled by the sun from the ocean. It enters the atmosphere, is moved by the winds and, condensing to droplets or crystals of ice, picks up chemical substances produced in the air or released to it by man and other organisms. As rain or ice it is delivered to the land surfaces, a dilute but complex solution, there to be further altered by the rocks, soils, vegetation and man before it drains directly, or through the ground, to streams and rivers. Now it becomes both the architecture of the ecosystem and transporter of raw materials. Its properties dominate and mould the living organisms that can live in it; it brings nutrients and bears downstream the chemical products of the communities it has served. From the turbulent upland streams it passes, chemically yet more complex, to sluggish reaches where riverine communities, swamps, and the beginnings of a drifting plankton take advantage of the nutrient salts and organic silt it has borne from upstream. It may pass from river to estuary or, its silt left in some delta, may be delayed in a lake. There it clears to allow sufficient light to penetrate to the microscopic algae of the plankton to sustain a food web in the open water. Eventually, though some may pass back to the air at every stage, its full cycle ends in the ocean, there to leave the salts it has carried from the land, itself soon, or after some thousands or more of years, to return again to the atmosphere.

1

This book is about water and the freshwater ecosystems it defines. It is also about people, for fresh water is among the most used of the world's resources. No settlement can be far from a source of it and many of the world's civilizations have taken their names from the rivers which served them—Nile, Indus and Mesopotamia, which means the land between the rivers (Tigris and Euphrates). Some of the most impressive structures of engineering harness the fresh water supply—the dams at Kariba on the Zambezi, Akosombo on the Volta and Aswan on the Nile, whilst for much of mankind the basic foods are from fresh water—rice and inland fish. In turn, this closeness to water has inflicted diseases, carried by water or animals that live in it—malaria, schistosomiasis (bilharzia), filariasis, infectious hepatitis, polio and cholera. Some of these infect more than 200 million people at any time. Death rates of infants less than a year old average 14% in the world's 43 poorest countries, and owe much to diarrhoea which might be prevented by a reliable water supply and improved hygiene. In the rich countries such diseases have been almost removed but the fresh waters must be used for many purposes—water supply, disposal of sewage and industrial waste, fisheries, recreation, irrigation and navigation—that are often mutually incompatible. Because the quality of the water is determined also by the uses made of the catchment from which it drains, there is almost no human activity which does not have some effect on freshwater ecosystems.

There was a time, though it was some millenia ago, when an account of freshwater ecosystems might have sensibly described them separately from the activities of man. We may yearn for pristine wilderness, serving rivers and lakes unsullied by any human hand, but there are few. There is no longer a pure ecology and an applied ecology. There is simply an ecology which accommodates the role of man no less than it does of algae, snails, fish or crocodiles. On such a basis this book is written.

It follows the passage of water, and the organisms in and around it, from rain falling on the catchments to the lakes and lowland rivers. It deals with general ideas and principles, often through the medium of examples. Freshwater ecologists are called upon (though not often enough) to advise communities and governments on how fresh waters should be managed in particular cases. Advice based on an understanding of how the ecosystems work may often conflict with aspirations of others who wish to use the water for their own purposes. A study of the Pongolo floodplain in Natal may serve as an introduction to illustrate how specific problems of management are bound with the general principles of freshwater ecology.

1.2 THE PONGOLO FLOODPLAIN ECOSYSTEM

Near the village of Mkuzi in northern Natal, Rider Haggard, the local stationmaster, wrote his novel *King Solomon's Mines*. Itshaneni, the 'Ghost Mountain' of the book rises above the village and from its foot, to the north, stretches a dry, flat land. To the west, small tributaries rising in the Lebombo mountains feed the Pongolo river which crosses the land to join the Usutu river some 150 km from Itshaneni. Figure 1.1 shows the geography of the area.

The river is permanent but swells in summer when much of the 500–600 mm of rain falls. It dwindles in winter to a flow only a tenth or so of the peak. In the wetter summer the Pongolo river floods, dropping its coarser sediments close to the channel as banks or levees over which the water spills to gather in shallow lakes called pans (for their flat bottoms). There are more than 90 pans, accounting for much of the floodplain. The climate is subtropical and the much greater area of the Makatini Flats which surround the

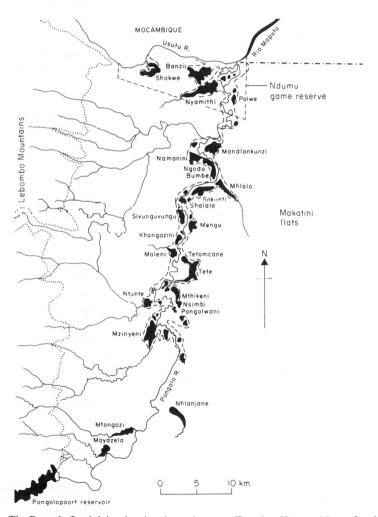

Fig. 1.1. The Pongolo floodplain, showing the major pans. (Based on Heeg and Breen [340].)

floodplain is arid; winds are strong and evaporation rates high. The Pongolo river floodplain is a narrow, green ribbon in a grey-brown, thorny landscape. Not suprisingly much of the region's population, originally of the Tembe-Thonga tribe, later assimilated by the Zulus, clusters in the floodplain, which now has about 13 000 people.

1.2.1 Cycles of change on the floodplain

These people, and the other life of the floodplain, depend on the rhythmic flood and recession of the river and the pans [340]. In spring, water levels are at their lowest and those pans still retaining water after the dry winter are cut off from the river. Their water is cool and clear, for the river-borne silt has long settled. Great beds of submerged plants, (*Potamogeton crispus*), the curly pondweed, are at their densest, and flocks of white-faced duck (*Dendrocygna viduata*) up-end in the water to reach the energy-rich, compact shoots, or turions, by which the plants will last the summer. Fifty or more species of fish live in the pans; most have broad diets which include plants and are tolerant of both lake and river regimes. These fish are caught by several traditional methods. One of them, the isiFonya, has a line of people, each wielding a basket, moving across the pan and scooping the fish as they become concentrated at one end. The co-operation needed for this seems to have importance in preserving the social structures of the villages. The fishery provides much protein and the peoples of the plain are among the best fed black Africans in Natal. To the north, in a section of the floodplain (Ndumu) reserved for game, the fertile pans support large numbers of Nile crocodile and about 200 hippopotami. The list of visiting and resident bird species is impressively long, for the Pongolo plain lies at the junction of the ranges of both tropical and sub-tropical species.

With summer coming and with increasing water temperature, the curly pondweed in the pans starts to die back. Scarce nutrients—nitrogen and phosphorus compounds—are translocated to the turions, incorporated into animals, particularly snails which can readily digest the decaying vegetation, or returned to the sediments in animal faeces or through chemical absorption. These nutrients are then bound and conserved within the pans before they might be washed out by the floodwaters. The warming water also stimulates the development of gonads and the production of ripe sex cells in many of the fish species. The waters by this time have increased slightly in salinity. Chlorides and other salts have been sucked out of the underlying salt-laden groundwater by evaporation from the surface of the floodplain. The local rock contains much salt from its origin in a Cretaceous sea.

From October or November the summer floods begin; the river brings nutrient-rich silt to the floodplain, water levels rise in the pans and accumulated salt is washed out and down river. The rising water covers land at the edges of the pans, quickening the growth of grasses, particularly

Cynodon dactylon; the fish spawn. When they have hatched and exhausted the yolk in their sacs, the fish larvae feed richly among the new grass, on organic detritus washed in from the river, and on the invertebrates which have thrived in the beds of decaying pondweed. The weed cannot grow in the warm, turbid water [689] but the supply of energy it provided through photosynthesis in the winter is now matched by rich river-silt brought in from the soils of the catchment area, and boosted by the dung of hippopotami which feed on land by night and rest in the water by day. Pans which were previously dry are now full and fish move through channels among them, making full use of the water area for spawning and growth; fishing methods change to those which use fences and baskets fixed across the channels.

In autumn (from about March), the flood ends and evaporation starts to lower the levels of the pans. The *Cynodon* grasslands at the pan margins reach full production and the people move cattle to graze on them, redoubling their supply of milk and meat. Silt settles, the pan waters clear, and as winter begins the turions of the pondweed germinate in the cooler water. The Tembe cultivate areas previously flooded and now enriched with silt, for maize, groundnuts, pumpkins, sorghum, tomatoes, tubers and sugar cane while the soils are still moist. The duck return from migration and the annual cycle is completed (Fig. 1.2). It illustrates the complex linkages to be found between natural environmental cycles, the ecology of plants and animals, and traditional small-scale human societies. But all is not utopian.

1.2.2 Problems of the human population

The Tembe suffer from two severe water-borne diseases: malaria and bilharzia (schistosomiasis). Malaria is a protozoan disease of the blood, carried by some mosquito species whose larvae live in water. The larvae are not common in the main pans, where they are readily eaten by fish, but thrive in small, isolated pools too temporary for fish to survive in them. Bilharzia is caused by mating pairs of trematode worms which live in the veins around the human bladder or gut. Fluid accumulates and there is painful swelling and general weakness. Sufferers release eggs of the worms in urine or faeces (depending on the schistosome species) into the river or pans. Over 80 % of children of the floodplain are infected by urinary schistosomiasis (*Schistosoma haematobium*) and about 4 % by the intestinal form (*S. mansoni*). The eggs hatch into larvae (miracidia) which infect particular snail species (*Bulinus globosus, Biomphalaria pfeifferi*). These snails are well adapted to the floodplain, becoming dormant in dried mud then active again as the pans fill. In the snails the parasite transforms to a cercarium, another larva, which is released to swim in the water and burrow into the exposed flesh of its eventual human host.

The human population is also fast increasing; there was a rise of 13.8 % on the floodplain between 1960 and 1970. Many young men and some women

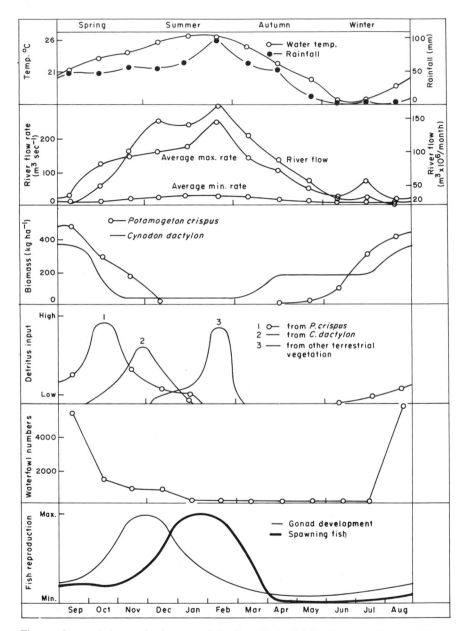

Fig. 1.2. Seasonal changes in the pans of the Pongolo floodplain. (Based on Heeg and Breen [340]).

need to move away for part of the year to find work and the male population has fallen relative to that of the women. Inevitably this causes social problems. Numbers of cattle, and of goats, which are particularly destructive grazers, have increased so that the seasonal *Cynodon* grasslands as well as the neighbouring drier flats are over-grazed. Stock theft is a new problem.

1.2.3 The Pongolapoort dam

In the 1970s the government of South Africa built the Pongolapoort dam, some 70 m high, across the gorge where the Pongolo river cuts through the Lebombo mountains (see Fig. 1.1). The intention was to store the river flood water for irrigation of a large area of the Makatini flats. Frost-free winters and the favourable soils make the area attractive for cash crops—citrus fruits, winter wheat and vegetables, sugar cane, cotton, rice, groundnuts, tobacco, maize, beans, sorghum, soya beans and cassava—if water can be provided; and the yield and quality of the Pongolo water are suitable.

A major irrigation scheme would ultimately increase the wealth of the area and perhaps provide increased local employment and better health care. Substantial diversion of the flood waters would also disrupt the linked ecology and traditional societies of the floodplain. Releases of water from the dam to the river would, if experience elsewhere is repeated, be determined by the needs of irrigation, be small, and might bear no seasonal relationship to the needs of the floodplain ecosystem. The present fish productivity of the pans, the natural irrigation of subsistence farmlands close to the river, and the winter *Cynodon* pastures would all be lost.

It is never easy to know what a human population wants its future to be; decisions must be made without the benefit of hindsight. The Tembe-Thonga people might wish to preserve their traditional society. Almost certainly it offers values that at least compete with those that would be transferred from the developed world along with an irrigation scheme and a cash economy. On the other hand the social problems presently felt from the population increase, and a possible control of malaria and bilharzia might easily persuade the younger people to accommodate a new way of life, and increasingly so as those brought up in the traditional society age and die.

The eventual future, however, is not the present problem which faces freshwater ecologists, sociologists, engineers and the others concerned with the Pongolo floodplain. The problem is that it will take many years to establish an irrigation scheme capable of supporting all the floodplain population, even is this is wanted, whilst destruction of the present subsistence based on the annual flooding can be (and almost was) immediate. The problem is to cater for the interim decade or more of social transition whilst also conserving as much as possible of the wildlife value of the floodplain. What minimum water flow is needed to support the floodplain system whilst allowing as much water as possible to be diverted to the irrigation scheme?

1.2.4 Future management of the floodplain

Initially the new lake behind the Pongolapoort dam filled more slowly than anticipated because of lower than usual rainfall. In the summer of 1969–70 only 0.6 m ^{3} s $^{-1}$ of water was released from the dam to the river, compared with the uninterrupted flow of perhaps 250 m ^{3} s $^{-1}$ in an average year. Immediately there was a scarcity of drinking water, so the people were forced to dig in the drying river bed for supplies, which became contaminated; typhoid followed. Then it was seen that there were sound reasons for conserving the floodplain ecosystem, and in the making of a management plan the findings of ecologists have been crucial.

An early suggestion was that ground water might be pumped into the river at the right time of the year, but the supply is not sufficient and is too saline. The water has up to 24 g l $^{-1}$ of salts, whilst the pondweed in the pans will not grow at more than 5 g l $^{-1}$ and many of the invertebrates die at lower values. Water seeping through a future irrigation system could also not be used for it, too, would become salty.

Two practicable schemes were put forward. The first allows release of sufficient water in summer just to flush out and refill the pans once. It would need about 126 × 10^6 m^3 of water from an annual total averaging 1000 × 10^6 m^3. The cost of this water would be a little under 1 % per year of the total cost of the development (dam, channels, irrigation scheme) but would take 30% from the annual potential profit of the scheme. This was felt to be too much, though the profit was not calculated to allow for many of the costs which cannot readily be given cash values (social disruption; health problems arising from a change from a balanced diet to one more dependent on carbohydrate crops; wildlife losses).

A second proposal would use less water more efficiently. Inflatable weirs would be moved progressively downriver, diverting the released water in turn to the pans without 'losing' any by passage directly downriver. The cost would only be a third of that of the first scheme (10.5% of the proportional profit). The snags of either of these schemes, particularly the latter, cannot be known fully in advance. Both schemes represent attempts to salvage something from the *fait accompli* of the building of a very expensive dam.

The natural assets of the floodplain could be developed to cover the cost of the water needed to maintain it. The fishery will bear expansion and some fish-farming is possible; a sport fishery and encouragement of modest tourism will at least cover the costs of the second scheme and at best the first (Table 1.1). These measures will, however, bring some additional problems. Tourist development would mean control of bilharzia by use of chemicals to kill snails; this may disrupt the mechanisms by which scarce nutrients are

Table 1.1. Cost–benefit analysis summarizing costs to agricultural irrigation scheme and the benefits realizable from conservation of the Pongolo floodplain ecosystem using inflatable weirs or limited free release of water from the dam. (From Heeg and Breen [340])

Cost (as loss of profit to agriculture, 1982 prices, million rand per year)			Benefits (million rand per year)	
Limited free release	Inflatable weirs		Minimum	Possible maximum
		Fisheries	0.2	0.68
		Subsistence agriculture	0.6	2.5
		Tourist development	0.025	0.09
1.575	0.525	Total	0.825	3.27

conserved in the pans by snail-grazing on the senescing pondweed. Of course it is desirable to control bilharzia anyway, but preferably by proper sanitation which would take time to provide; tourists wealthy enough to bring sufficient income would demand immediate measures.

In deciding the best course of action, one possible view is that any disruption of the floodplain system is bound to be harmful—the system having been produced through natural selection of its components to be the most stable one under the prevailing conditions. Alternatively, a system normally prone to a year-to-year change in river flows must also have a built-in resilience to such unpredictability which should serve it well as its flows are disrupted by man. Indeed the releases of water to it must not be completely regular. The survival of the pondweed in the pans depends on both turions and on seeds; the seeds are produced in years when the pans dry out early and rest in the sediment as a hedge against possible death of all the turions by a severe drought. Absolutely regular flooding, undoubtedly an arrangement convenient to the dam managers, could lead to complete dependence on turion formation and exhaustion of the stored bank of seeds.

The current proposals [1] are for unrestricted flooding (the first scheme) but on specific dates (November and February) and represent a move towards the best possible compromise over a difficult problem.

2

On Living in Water

Earth has much water: a continuous ocean covers over seven-tenths of its surface. The fresh waters, in contrast, are divided and small, both individually and in total volume (Table 2.1). There is nonetheless a large annual turnover—addition and loss, inflow and outflow—from this 'pool'. On average the surface fresh waters are naturally replaced in days or months compared with residence times of thousands of years for the oceans. A few, very large, deep lakes, like Baikal in the USSR and Malawi and Tanganyika in central Africa, have long residence times like the ocean but most fresh waters are transitory. This is the first main difference between the seas and the fresh waters.

There is another, more obvious, difference. The ocean is salty (about 35 g l^{-1} of salts), the fresh waters (on average less than 0.1 g l^{-1}) are not. Water flows into the ocean but leaves only by evaporation: whilst some salts are precipitated, those dissolved in the inflowing water have been concentrated over time. In fresh waters the outflow is mostly as liquid rather than as vapour, and although there is some concentration of salts by evaporation, particularly in very arid areas (for example parts of California, East Africa and Australia), there is, in general, little opportunity for salt accumulation.

Table 2.1. Distribution of water on Earth, and the time taken for complete replacement (residence or renewal time)* of each category

Category	Water volume* (km^3)	Fraction of total (%)	Renewal time†
Atmosphere	1.3×10^4	0.001	7–11 days
River channels	1.2×10^3	0.0001	7 days
Freshwater lakes	1.2×10^6	0.009	330 days
Saline lakes and inland seas	1.0×10^5	0.008	1–4 years
Soil water	6.6×10^4	0.005	
Ground water	8.2×10^6	0.62	60–300 years
Ice caps and glaciers	2.9×10^7	2.15	12 000 years
Ocean	1.3×10^9	97.2	300–11 000 years

* From Leopold [466].
† Estimates vary—values taken from Gregory and Walling [302] and Ward [840]. The residence time is defined as the volume of the water body concerned divided by the volume added to it in a given time. The dimensions of this are thus length³ divided by length³ time⁻¹.

These two distinctions of residence time and saltiness are matched by a third—the large contrast in the variety of life that lives in fresh waters and in oceans. Rather more phyla (56) (the next largest category below kingdoms in the classification scheme commonly used) are present in the ocean than in fresh waters, where there are 41. No phyla are confined to fresh waters but 15 phyla are found only in the ocean. Although full data are more difficult to obtain, this probably applies more to the total numbers of species present. For example, of the nearly 19 000 species of fish, just under 7000 are found in fresh waters; 18 of the 46 orders of fish are confined to the sea, and only half as many to fresh water [588].

For largely historical reasons the ecological aspects of the water cycle have been somewhat rigidly divided between marine biologists and freshwater biologists (with estuarine biologists apparently a sub-culture of the former). Yet water, whether fresh or salt, has so many shared features as a medium in which to live and is so different from air that its properties and the similar problems of living in it should be emphasized as much as the differences between the fresh water and the ocean. Chapter 2 explores first the similarities of living in, and then the reasons behind the differences in species diversity between fresh waters and the sea.

2.1 PROPERTIES OF WATER

2.1.1 Physical properties

For all its familiarity, water is a quite remarkable substance. It is the hydride of oxygen, an element which lies in the Periodic Table above sulphur and selenium and between fluorine and nitrogen. Elements within vertical groups and along the rows in the Table usually have graded series of properties, but H_2O is clearly anomalous compared with the related H_2S and H_2Se or NH_3 and HF (Table 2.2). It remains a liquid at earth-surface temperatures, whereas its position in the Periodic Table suggests that it should be a gas. It also freezes at a much higher temperature than would be expected.

Liquids are not so easy to describe as are solids or gases. The molecules of solids are held in a more or less constant relationship with one another; individual molecules vibrate, but about a fixed position. On the other hand the molecules of gases move fast and randomly and have no structural

Table 2.2. Comparative properties of the hydrides of oxygen (water), and of those elements close to it in the Periodic Table. The upper value is the melting point (°C) and the lower value the boiling point (°C) in each case. (From Hutchinson [379])

CH_4	NH_3	H_2O	H_2S	H_2Se	HF
-182.6	-77.7	0	-82.9	-64.0	-83
-161.4	-33.4	$+100$	-59.6	-42.0	-19.4

relationship with each other. Liquids can be seen either as highly condensed gases, in which the random molecular movements are dominant though the molecules are closer to one another, or as disturbed solids in which the ordering of the molecules is much reduced but still present and important. The results of recent X-ray and neutron diffraction techniques emphasize the latter view—and liquid water has more structure than most. Some of the crystal structure of ice is preserved in the liquid. It is this that makes water odd compared with hydrogen sulphide and hydrogen selenide, so what is the reason for it?

The water molecule has two hydrogen atoms held at an angle of $104° 27'$ and at distances of nearly 0.1 nm from the oxygen atom. The hydrogen atoms are held at about 0.15 nm from each other. The molecule is covalently bonded, electrons being shared between the hydrogens and the oxygen giving overall electrical neutrality, but the oxygen nucleus has greater affinity for electrons than that of the hydrogen. There is thus, on average, a slight displacement of negative charge towards the oxygen, which leaves a slight positive charge on the hydrogen atoms. Consequently if, under suitable conditions of temperature and pressure, the molecules are brought close together, there is an attraction between the slightly positive hydrogen on one molecule, and the slightly negative oxygen on the other, which links them together with a 'hydrogen bond'. The angles at which oxygen and hydrogen are held in the water molecule, coupled with this hydrogen bonding, result in a crystal structure for ice which is based on tetrahedra. Oxygen is at the centre of each tetrahedron, surrounded by four hydrogen atoms, two covalently- and two hydrogen-bonded. Such a crystal is quite open, compared with those of other substances which often have twelve neighbours (as opposed to four in ice) packed around each molecule. Ice consequently has a low density. It floats on liquid water and, by forming an insulating layer at the surface of water bodies, often prevents them from freezing solid and thus killing fish and other organisms.

The hydrogen bonding in ice is quite strong because the displacement of negative charge towards the oxygen atom is powerful. The temperature at which melting of ice takes place—a measure of the energy needed to begin breaking down the hydrogen-bonded structure—is thus relatively high (see Table 2.2), compared with H_2S and H_2Se, where the charge displacement in the molecules is small.

The latent heat of fusion—the energy needed to convert the solid to the liquid once melting begins—is, however, low for water compared with many other substances. This suggests that much of the orderly structure in ice is not destroyed on melting. More of the structure is destroyed as temperature increases but it is not until the relatively high boiling point is reached (see Table 2.2) that all of it is lost. The high latent heat of evaporation—the energy needed to break down the crystal structure entirely, which is over six times higher than the latent heat of fusion—reflects this. For organisms living in water this tenacity of structure is important in giving water a high specific heat. It takes much gain or loss of heat to change its temperature, for temperature is a measure of the amount of free movement of molecules. Hence temperature ranges and fluctuations in water are muted. An aquatic habitat might experience a range of about 25°C or less, whereas continental land habitats can have ranges of air temperature between winter and summer of more than twice this.

A further property of water, related to its retention of structure as a liquid, is its maximum density at 3.94°C (under standard pressure). As ice melts, the collapse of some of the molecules into the parts of the structure that are retained, leads to a partial filling of the open crystal and an increase in density. This might be likened to a partly demolished building where rubble fills the still-standing shell. A second process, of movement further apart of the released molecules, also occurs and tends to decrease the density: the rubble is carted away. The former process predominates up to 3.94°C and the latter at higher temperatures. The density differences around the peak (Fig. 2.1) are small but this characteristic leads to a layering in winter

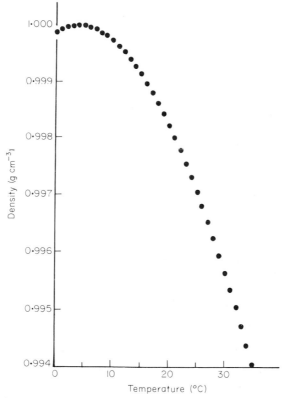

Fig. 2.1. Graph of density of pure water, under one standard atmosphere pressure, in relation to temperature. The peak density is reached at 3.94°C.

of the water in deeper basins such that the colder layers ($< 3.94°C$) are at the top and the warmer ($3.94°C$) layer is at the bottom, insulated by the lighter colder layers from further cooling and freezing. At higher temperatures the less dense, warm water may float on cooler water leading again to an isolation of the deeper layers from the atmosphere (see Chapter 6).

2.1.2 **Water as a solvent**

The charge displacement, which gives some properties of a charged ion to an essentially non-ionic substance, brings water its versatile properties as a solvent. The act of dissolving needs a chemical attraction between the solvent and the solute. A charged or ionic solvent cannot then dissolve a completely neutral solute nor a neutral one a charged solute. Because of the separation of charge in its molecule, water acts as a charged solvent and will attack ionic crystals, such as salts, and bring them into solution.

The extent to which such ions are dissolved depends on their attraction to water molecules. In turn this depends on their own charge or valency (the number of electrons by which the outer orbit of the atom exceeds or falls short of the number in the outer orbit of the closest related inert gas) and on the size of the ion. The attraction increases with the charge (either positive or negative) but decreases with the size of the ion. This is because the charge is weakened by being spread over a greater surface area of the ion. If Z is the charge on and r the radius of the ion, the affinity for water can be measured as the ionic potential Z/r. The efficiency of water as a solvent does not simply increase with increasing ionic potential of the solute however. Ions with Z/r less than 3.0 or greater than 12.0 are readily dissolved but those with Z/r between these values tend to be precipitated.

Ions with an ionic potential less than 3.0 are cations derived from metals (Fig. 2.2). The charge is sufficient for attraction to the water molecules and to bind the ions with water (the results being called hydrated cations) in solution. The ions of sodium, potassium, and calcium are good examples.

Ionic radius is not a simple function of atomic weight and some heavier elements have quite small ionic radii despite a high weight and a high charge. In these, which have an ionic potential between 3.0 and 12.0, the charge is sufficient to attract the oxygen atom in water so close that the binding between the oxygen and one of the hydrogens in the water molecule is weakened. A hydrogen ion is then ejected into solution and a metal hydroxide is formed. Such hydroxides have little surplus charge left for the attraction of water molecules, and they precipitate. Aluminium and silica are good examples. So are iron and manganese, though these transition metals have several valency states and can also behave as the first group.

In the third case, elements with ionic potentials greater than 12.0, the charge attraction for the oxygen in water is so great that both the hydrogens are ejected into solution as H^+ (where they bond with water molecules to

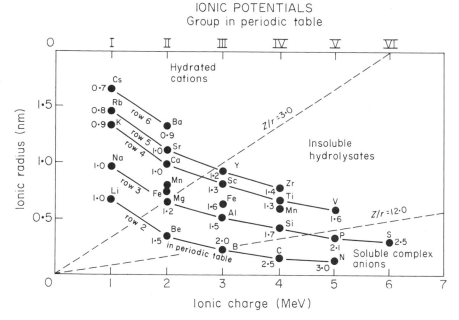

Fig. 2.2. Relationship between ionic charge (Z) and ionic radius (r) of several elements. Lines connect elements in the same row of the Periodic Table. Values of the ionic potential (Z/r) distinguish elements which behave as hydrated cations, ($Z/r < 3.0$), insoluble hydroxides ($Z/r > 3.0$, < 1.20), and soluble complex ions ($Z/r < 12.0$). (From Raiswell *et al.* [658])

form H_3O^+, hydrated hydrogen ions, and remain in solution) and an oxy-anion is formed. Such oxy-anions are called complex, for they involve two elements, and the ionic potential of the complex ion (as opposed to those of its elemental components) is reduced to a value at which it attracts water molecules and remains in solution. Nitrate, carbonate and sulphate are good examples, with the position of phosphate (see Fig. 2.2) bringing it close to the borderline between the soluble oxy-anion and the insoluble middle group.

The charge properties of water thus interact with those of other elements to bring ions into solution to varying extents, and the inorganic composition of natural water, where the highly soluble Na^+, K^+, Mg^{++}, Ca^{++}, HCO_3^-, SO_4^{--} and Cl^- are normally major components, reflects this. The final composition, however, also reflects the availability in the surface of the Earth's crust of elements for potential solution (see Chapter 3). Because of this almost none of the more soluble elements will approach saturation, except in conditions of very high evaporation. And in mixtures of ions the affinity of one to another may overcome the attraction of either to water and lead to precipitation. For example, both calcium and carbonate are highly soluble each in the absence of the other, but readily precipitate as $CaCO_3$ if mixed together. As a result of such processes, natural waters are dilute ionic solutions, more so for fresh waters than the sea, which results from concentration by evaporation.

2.1.3 Solubility of non-ionic compounds

Water is a charged compound and so substances whose molecules have no charge displacement will not dissolve in it to any extent. Many organic compounds, for example aliphatic (straight chain) and aromatic (benzene-ring) hydrocarbons are thus not soluble. But a wide variety of organic compounds do have slightly charged (polar) groups in them, for example hydroxyl (OH^-), amino (NH_2^-) and sulphide (S^{2-}) and so have some affinity for and hence solubility in water.

In a sense all substances will dissolve to some extent in that if a source of them is held in contact with water, there will be some diffusion, though the equilibrium will lie heavily weighted towards the source, for non-polar compounds, as opposed to the reverse for polar ones. Most of the atmospheric gases fall into the non-polar group so solubilities of N_2, O_2 and the inert gases are low. They vary with temperature and pressure and with their concentrations in the atmosphere (Table 2.3).

Table 2.3. Solubilities of atmospheric gases in pure water under conditions of one atmosphere pressure of each gas and in equilibrium with atmospheric air ($21\%\ O_2$, $78\%\ N_2$, $0.03\%\ CO_2$) with a total pressure of one atmosphere. Both values are for a temperature of $15°C$

Gas	Under 1 atmosphere		With atmospheric air
	ml (STP) l^{-1}	mg l^{-1}	mg l^{-1}
Oxygen	34.1	46.2	9.7
Nitrogen	16.9	20.03	15.8
Carbon dioxide	1019.0	1897.0	0.57

The exception is carbon dioxide which is relatively soluble because it reacts with water, being itself polar, and exists in equilibrium with ions of carbon such as bicarbonate and carbonate in natural waters. Solubilities of gases also decrease with temperature (Fig. 2.3) because the increased molecular movement increases the possibility of escape from the liquid to the overlying vapour, whereas solubility of ions increases with temperature. This is because ionic solubility is a chemical reaction not a passive mixing.

One consequence of the differential solubilities of inorganic charged compounds and non-polar compounds (other than atmospheric gases) is that research in aquatic ecosystems has concentrated on the more readily analysable inorganic substances, because they are relatively abundant. This is to the detriment of the scarcer, non-polar or less polar organic compounds,

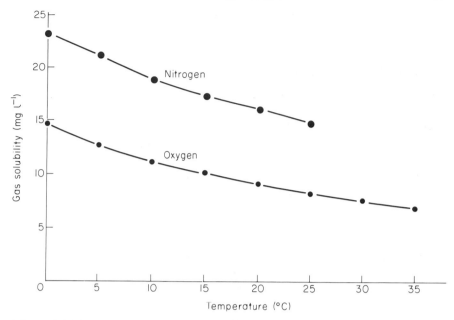

Fig. 2.3. Solubility of oxygen and nitrogen (mg l^{-1}) in water in equilibrium with the atmosphere.

which may nonetheless be extremely important biologically. Because living organisms are themselves composed largely of organic compounds, this consequence of the properties of water could have led to biases in understanding.

2.2 LAND AND WATER HABITATS AND THE EVOLUTION OF ORGANISMS

There seems little doubt that life evolved in water, but no certain evidence that the water was initially fresh or salt. The earliest fossils are of microorganisms—bacteria and other prokaryotic groups like the blue-green bacteria or Cyanophyta. These need access to small molecules of nutrients and gases and hence are permeable also to water. They must have at least a covering film of liquid water to prevent their drying out. Representatives of all of the earliest phyla, however, are found in both the sea and fresh waters. Some protistan families (Protista are eukaryote microorganisms, including algae and protozoa) are entirely marine, a comparable number entirely freshwater, and most have representatives, even fellow species in the same genus, in both habitats.

What does seem clear, however, is that multicellular animals evolved in the sea and the multicellular plants had at least three routes—one via the sea to brown seaweeds (Phaeophyta), a second via fresh waters from the blue-green algae to the now mostly marine red algae (Rhodophyta) and a third through the freshwater green algae (Chlorophyta) to the freshwater stoneworts (Charophyta) and the essentially land-living mosses and liverworts (Bryophyta) and vascular plants. Subsequently there was a secondary movement of some multicellular animal groups into fresh waters and onto the land and even a further movement of some land animals and land plants into fresh waters and back to the sea. This overall picture implies a considerable flux over time between the three main habitat types. The evidence for it depends largely on fossil and morphological comparisons and to a lesser extent on comparisons of the physiological modifications needed to exploit the three habitats. It seems that physiological adjustment has been, on the whole, rather readily and quickly achieved and that movement among fresh waters, sea water and the land has not been hampered by insuperable physiological problems. The lower diversity of organisms in fresh waters compared with the other two habitats is probably thus not particularly a result of fresh water being a more 'difficult' medium to colonize, but some discussion of the problems faced in each habitat is needed.

2.2.1 Physiological problems of living in water

Organisms must be open systems, allowing movement of materials like oxygen and wastes between themselves and the environment. In satisfying this need they are inevitably open also to movement of water and salts, whose amounts they must closely regulate, lest they dry out or burst or become too salty or too dilute for their enzymes to function.

Of the three main habitats, sea water is the most steady. Its salinity varies only between about 32 and 38 parts per thousand (except in very local instances). The proportional composition of the salts in it varies even less from place to place. Fresh waters vary much more (Table 2.4) in concentration and composition.

Marine invertebrates and wall-less microorganisms have an internal salt concentration (or osmoconcentration) slightly higher than that in sea water, or roughly 1000 osmolal. (Osmolality is a measure of the number of particles (ions or undissociated molecules) present per kg of water.) There is thus a small inward movement of water. The similar salt concentrations inside and outside the body mean that salts do not necessarily diffuse in or out of the body, but because the salt composition of the body fluids usually does differ from that of the sea, there are active (energy-requiring) mechanisms which exclude certain salts or pump in others. The salt concentrations inside the cells themselves may be much lower than those of the body fluids for sodium and potassium ions at seawater concentrations inhibit many enzymes. The cell contents must then be maintained at similar osmotic concentration to the body fluids or outside medium by production of amino acids and other soluble organic compounds.

Table 2.4. Concentrations of some of the most common ions in various natural waters. Values are given in millimoles per kilogram. A single set of values is representative for sea water, but fresh waters and inland saline waters vary greatly between themselves and even (for example L. Chilwa) between years, depending on water level

Ions	Fresh waters		Inland saline waters		
	Amazon streams*	Barton Broad† UK	Lake Chilwa‡ Malawi	Dead Sea§ Israel	Average sea water
Sodium	0.009	2.1	1.6–142	1955	475
Potassium	0.004	0.23		219	10.1
Calcium	0.001	2.8	0.24–0.95	481	10.3
Magnesium	0.002	0.46	0.04–0.74	2029	54.2
Chloride	0.031	2.4	0.74–88.0	7112	554
Sulphate	—	—	—	5.3	28.6
Bicarbonate	0.036	3.4	1.9–88.0	3.7	2.4

* Furch [247]; † Moss [563]; ‡ Morgan and Kalk [548]; § Steinhorn *et al.* [771];   Potts and Parry [650].

Excess salts enter the body in the food of an organism and by diffusion through permeable surfaces like the gills if there are relatively lower concentrations of particular ions in the body than in the medium, and must then be excreted. Many marine organisms exclude Na^+ and Cl^- ions through special cells in the gills and through production of small quantities of urine (which balances the osmotic inflow of water) into which unwanted Mg^{++} and SO_4^{--} ions have been pumped. The energy required to maintain the pumps which regulate the salt concentration is apparently small in comparison with the total energy needs of the organism.

2.2.2 Brackish and freshwater invertebrates

Representatives of several phyla of marine invertebrates have remained confined to the sea; others have adjusted to brackish conditions in estuaries (15–900 mosmol l^{-1}) and eventually to fresh waters (<15 mosmol l^{-1}). Estuarine brackish waters usually have a salt composition dominated by the seawater composition, but some inland brackish waters (see Table 2.4), produced by evaporation of fresh water or salt springs in confined areas, may have a very different ionic concentration and even a higher total salt concentration than sea water. They are called athallasic (not sea-like) and, if appropriate, hypersaline or brine waters).

Colonization of estuaries, linking rivers and the sea, by marine invertebrates has meant coping with both reduced and variable salt concentrations, as the mixture of fresh and salt water varies with river flow and state of the tide. Some such colonizers (e.g. the lugworm, *Arenicola*) avoid the problem by burrowing in the sediment which is generally infused with the denser almost full-strength sea water and whose composition daily changes only a little. Others close up protective shells at low concentrations (e.g. mussels, barnacles) or move down towards the sea (some crabs). Many, at least in larger bodies of brackish water with some stability in salt concentration, like the Baltic Sea, are osmoconformers.

Osmoconformity means they have reduced their internal osmoconcentration to close to that of the water by pumping out salts. This minimizes absorption of water by osmosis and consequent severe body swelling. As external salt concentrations are reduced further, such animals may begin to regulate their salt composition by active absorption and become osmoregulators (Fig. 2.4).

Such an ability was necessary for those animals which colonized fresh water, for an internal salt concentration as low as that of the external medium in fresh water would be insufficient to allow many enzymes to function. Freshwater invertebrates thus maintain, by active salt uptake, an internal concentration of between about 30 and 300 mosmol l^{-1}. This means that although most of the body may be relatively impermeable to both water and salt they continually take up water by osmosis, particularly through the gill surfaces which must be large because of the low oxygen concentration in water (see below), and permeable to small molecules. In turn they must eliminate this water as urine and in doing so inevitably lose some salts dissolved in the urine. The salt balance is maintained by active uptake through cells in the gills and, in some insects, through flat tail-fans or 'anal gills'. This uptake may be made less costly in energy by exchanging NH_4^+ ions, produced as waste by food metabolism, for Na^+ and HCO_3^- produced through respiration, for Cl^-.

The problems of coping with the more dilute fresh water (exclusion of water, maintenance of internal salt concentration) are certainly different from those of coping with sea water (regulating ionic composition at high salt level) but not very costly in energy [649]. Representatives of many phyla occur in both the sea and fresh water—bivalve and gastropod molluscs, crustaceans and annelid worms being particularly good examples—but it is in the vertebrates, particularly fish, that evidence for a comparatively easy movement between the habitats comes most readily.

2.2.3 Osmotic relationships in vertebrates

Only one small group of vertebrates, the hagfish, which is an early-evolved group of jawless fishes, has body fluids with both ionic and osmotic concentrations similar to those of sea water. The remaining jawless fish (the lampreys), the cartilaginous fish (sharks and rays), and the bony (or teleost) fish, all have ionic concentrations and compositions in the body fluids very different and much lower than those of sea water. The osmotic concentrations of the freshwater fishes of all groups are similar to those of freshwater invertebrates and the fish cope with the same problems of high water uptake and salt loss by generally similar mechanisms. The osmotic concentrations

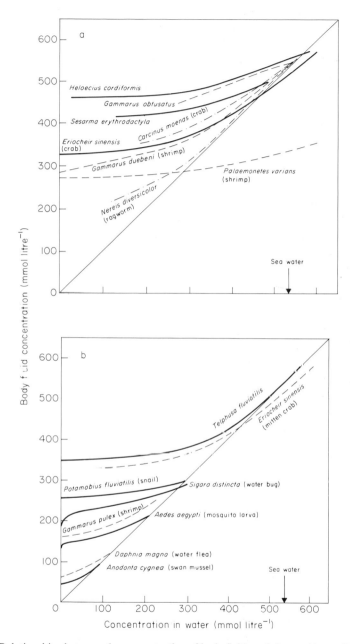

Fig. 2.4. Relationships between the concentration of body fluids and the outside medium in (a) various brackish water animals and (b) various freshwater animals. From Beadle [40]. Diagonal lines indicate equal concentrations in the body fluid and the medium. At high external concentrations, the species (if they survive) are osmoconformers. At low concentrations they regulate their internal concentration around an optimal level irrespective of that of the medium.

of the marine cartilaginous fish are about the same as those of sea water because large quantities of at least two organic compounds, urea and trimethylamine oxide are stored in the body fluids to add to the low osmoconcentration provided by ions alone. Although jawed cartilaginous fish are now scarce in fresh waters, they may have had their evolutionary origin there. On moving back to the sea they have found an alternative means of increasing their osmoconcentrations that is very different from that of their ultimate invertebrate ancestors. Amphibians are a largely freshwater group with low ionic osmoconcentration, but one species, a crab-eating frog of the mangrove swamps of south-east Asia, has colonized sea water, also by increasing the urea concentration in its blood to similar osmoconcentrations as sea water.

The bony fish appear to have had their origin also in fresh water, for the marine bony fish have a much lower osmoconcentration, as well as ionic concentration, than sea water. They thus tend to lose water by osmosis to the sea water and must compensate for this by drinking a large amount of sea water then excreting the excess salt through glands in the gills (Na^+, Cl^-), or in their urine (Mg^{++}, SO_4^{--}).

The low osmoconcentrations of the body fluids of both freshwater and marine bony fish allow a relatively ready movement of certain species between the sea and freshwater lakes and rivers. The migrations of lampreys, salmon and eels are well known but many inshore marine fish penetrate estuaries to the lower reaches of lowland rivers. Some cartilaginous fish also move between fresh waters and the sea by lowering their urea concentrations and quickening their urine production in fresh water.

Eels live most of their lives in fresh water but are called 'catadromic' because they move to the sea to breed. On moving into the sea the eel starts to lose water by osmosis, and may lose about 4% of its body water before it adjusts, by drinking sea water, to a new equilibrium, in 1 or 2 days. Young eels, moving to the rivers from the sea at first absorb water but soon compensate for this by increasing their urine flow.

2.2.4 Colonization of fresh waters from the land

Many freshwater organisms, judged from the presence of morphologically similar relatives in brackish water and the sea, had a marine origin and it would seem that problems of adjusting to changed salinity were relatively easily overcome. Some groups, perhaps many Protista and the bony fish, had their origins probably in fresh water and equally easily adjusted physiologically to a recolonization of the sea.

A second evolutionary pathway for the establishment of freshwater organisms was colonization from the land. The freshwater insects and the higher (vascular) plants seem to have come by this path. The freshwater vascular plants often still retain features characteristic of a land existence such as stomata and wind and terrestrial insect pollination. In both cases the diversity of species in fresh waters is much lower than it is on land. Thus as for the similar contrast between fresh water and the sea, the question of whether this arises from physiological difficulties posed by fresh water can be asked.

The main problem for land organisms is the threat of dehydration. They can be regarded as organisms adapted to a habitat where fresh water is scarce. In a sense all land animals represent envoys moving out to foreign parts from a home-base of fresh water to which they must return to drink. Land plants in turn, with their roots in moist soil are effectively freshwater organisms in a freshwater habitat (damp soil) always perilously close to drying out. The colonization of more conventional freshwater habitats should thus have been relatively easy for land organisms in that the commodity in greatest scarcity for them is present in unlimited supply. Many other advantages are also given by fresh water (Table 2.5). Water has a higher viscosity and density, eliminating the need for investment of much energy in supporting structures like wood or bone, and a high heat capacity, buffering temperature change. Carbon dioxide availability for photosynthesis is also much greater in water because of the dissolved reserve, as bicarbonate, which easily dissociates to form CO_2. But there are problems also. The oxygen content of water is relatively very low and the rate at which it diffuses to body surfaces is much lower in water than in air (Table 2.5). The great advantage of colonization of the land was access to a rich supply of atmospheric oxygen and the possibilities of greater manipulation of energy which this offered. A move to water was to accept the loss of this advantage.

Table 2.5. A comparison of certain characteristics of atmospheric air and fresh water. Values are given at standard temperature (273 K) and pressure (1 atmosphere (101.3 kPa)). (Modified from Schmidt-Nielsen [725])

Characteristic	Water	Air	Ratio water:air
Oxygen concentration (ml l^{-1})	7.0	209.0	1:30
Density (kg l^{-1})	1.000	0.0013	800:1
Dynamic viscosity (cP)	1.0	0.02	50:1
Heat capacity (cal l^{-1} (°C)$^{-1}$)	1000.0	0.31	3000:1
Diffusion coefficient (oxygen) cm^2 s^{-1}	2.5×10^{-5}	0.198	1:8000
(carbon dioxide) cm^2 s^{-1}	1.8×10^{-5}	0.155	1:9000

2.2.5 Respiration in water

The oxygen concentrations and the rate of oxygen diffusion in water are low. Aquatic animals must increase the surface area of the tissues (gills, which are evaginations and folds of the body surface, or lungs, which are similar invaginations) available for oxygen uptake, or both. Or they must maximize the rate of flow of oxygenated water past them, to maintain the highest possible concentration of oxygen at their surfaces. Alternatively aquatic animals can continue to breathe air. Those present in fresh waters (mostly insects) and the sea (mostly mammals and birds) which had land ancestors usually do this.

The smaller invertebrates may simply use their body surfaces as a gill as long as the path for diffusion into the centre of the body is no more than about a millimetre or so (e.g. in flatworms, water fleas, rotifers. Beyond that, more specialized folds of tissue, rich in a blood supply, are needed, and often some pumping mechanism which forces water past them. For example bivalve molluscs (mussels, clams, oysters) pump water in through tubes called siphons and may also use the fine particles in this water as a source of food. Fish force water over the gills by rhythmically filling the mouth with water and then reducing the mouth volume so that the water is forced over the gills then out through a slit or slits behind the head. When great activity is needed, they may simply keep the mouth open and allow the surge of water created by their movement forward to flood the gills. The drawback to gill respiration is that large areas of tissue permeable to almost any small molecule must be exposed so that the freshwater problems of osmotic water uptake and body salt dilution and the marine one of excess salt uptake are exacerbated.

2.2.6 Invertebrate air breathers in fresh waters

Few invertebrate groups have recolonized the sea after a land ancestry, but the insects have been widespread in invading fresh waters. Often they have complex life histories with juvenile larval or nymph stages. (Larvae look very different from the adult whereas nymphs resemble the adults in many though not all respects.) The aquatic juvenile stages may breathe water or air. The adults are usually air breathers and may live in water or on land. Many nymphs of the mayflies (Ephemeroptera), for example, have gills on their backs, which they fan in the water, whereas mosquito larvae suspended from the surface tension film at the surface absorb atmospheric air through a large pore or spiracle at their hind end.

Water beetle larvae have gills but the adults acquire an air bubble at the water surface which they hold under their wing covers when they dive. Through their spiracles they absorb oxygen from this bubble, and to maintain gaseous equilibrium, more oxygen diffuses into the bubble from the water, which is itself in equilibrium with the atmosphere. The bubble may thus give an oxygen supply for some time but eventually collapses. This is because as oxygen is absorbed by the animal, the percentage of nitrogen in the bubble temporarily increases. Nitrogen then diffuses to the water which is in equilibrium with the lesser proportion of nitrogen in the air. The bubble thus progressively gets smaller and must eventually be renewed at the surface.

Water spiders maintain a similar temporary reserve of air under a 'bell' of silk strung between the stems and leaves of aquatic plants. They return to this bell to breathe between hunting forays. Some insects maintain a permanent air bubble close to their bodies by supporting it with a very dense mat of very fine, water-repellent, hairs called a plastron. These prevent collapse, by temporary nitrogen loss, of the bubble and allow the animal to breathe atmospheric air indefinitely whilst underwater.

2.2.7 Conclusions on physiology

By and large therefore the colonization of fresh waters from the land has been readily achieved, primarily by invertebrates, whilst few significant problems have been met by mammals and birds colonizing fresh waters. They breathe atmospheric air and have therefore no particular osmotic or salt problems. Indeed more severe problems were involved for such vertebrates returning to the sea where salt ingress in food must be coped with. Yet there are many more, truly aquatic, marine warm-blooded animals than there are in fresh waters. Likewise the only reptiles able to spend their entire life history in water are sea snakes. The lower diversity of the freshwater biota compared with that of land or sea is again therefore unlikely to be a result of a physiologically difficult habitat.

2.3 TIME AND FLEETINGNESS—A FUNDAMENTAL DIFFERENCE BETWEEN FRESHWATER AND MARINE ECOSYSTEMS

Because the properties of water itself are shared and physiological problems of salinity tolerance have been so readily solved by organisms moving, both literally for some and in evolutionary terms for many, between fresh waters and the sea, what is the reason behind the relative lack of diversity in the communities of freshwater organisms? W. J. Sollas in 1884 thought that the flow of rivers might be a physical barrier to the delicate drifting larvae which characterize many marine animals [754]: in general freshwater animals do not have such larvae. But the copepods, a widespread group in freshwaters, do have larvae, called nauplii, to which the rivers cannot have been a barrier; and in contrast, the vigorously moving marine squids seem never to have colonized fresh waters. Others [522] have proposed that the more widely fluctuating temperatures of fresh waters compared with those of the sea might also have posed problems. Yet the diverse intertidal zone is seasonally

as variable in temperate regions as any lake, whilst the fluctuation in a large
equatorial lake is no greater than that at the tropical sea surface.

The idea of change in the habitat, extending to a longer time-scale, may,
however, be a profitable one to follow. Fresh waters have small volumes, and
are liable to be disturbed by climatic changes which, for periods of thousands
of years, bring drought to one area or ice to another. They are subject also to
geological upheavals which may alter drainage, emptying lakes and reversing
river flows. The ocean has not been constant either. Continents have moved
across the earth's surface altering the shapes of the ocean basins and the
polar glaciations have cooled the water. There has, nonetheless, always been
an ocean. A continuous sequence of sedimentary rocks laid down in the sea
testifies to at least one ever-present huge volume of water on a planet that for
over 4 billion years has never been completely dry.

For over one hundred thousand years until only 10–15 000 years ago most
of the northern hemisphere, whose land areas now bear extensive river
systems, and most of the world's natural freshwater lakes, was covered with
a polar ice-sheet several kilometres thick. The freshwater fauna previously
present would have been crushed; some, more active, members doubtless
moved towards the equator or were washed southwards in melt waters at the
edge of the sheet. Further back, 250 million years ago, the presently cool and
damp northern England was an arid dune-desert. In Africa few areas of the
continent have been undisturbed by volcanic and major geological movements
for less than 300 million years whilst whole mountain chains have risen over
the whole of western North America in the last 200 million years. Between
periods of change in one place there may have been periods—hundreds of
thousands or even millions of years—of relative stability but these are short
compared with the continuity of the ocean for four and a half thousand
million years.

The distinction between fresh waters and the ocean can then perhaps be
seen in terms of the predictability of these habitats. The ocean has not been
disrupted to the same extent as the fresh waters. It has probably changed
both chemically and physically over geological time but these changes have
been gradual on account of its bulk and the buffering properties, such as the
high specific heat, of water itself. Species have been able to differentiate
without violent interruption. Assuredly the rate of this process has been
different in different places—the coral reefs of tropical seas provide a more
stable habitat than the polar seas at the edge of a fluctuating ice sheet. But
the existence still of species such as the coelocanth (*Latimeria chalumnae*),
which has changed little from fossils recorded in the Devonian period, 300
million years ago, testifies to considerable stability.

The differentiation of species in fresh waters, however, may never have
been able to proceed for long without disruption and the discontinuity of
freshwater systems, each separated in its own drainage-basin, has hampered
movement and escape when catastrophe has struck. Almost all natural

temperate lakes were created about 10 000 years ago as the ice retreated, leaving a hummocky landscape and natural moraine dams across many valleys. These lakes have had to be colonized by freshwater organisms which found refuge in the warmer areas towards the Equator. There has been little time yet for more species to differentiate within them. That such a process does go on may be shown by a number of animal species which occur in lakes around the head of the Baltic sea. These lakes, just after the ice retreated and sea levels rose from the melt water, were continuous with the Baltic. Then the land, relieved of the pressure of the ice, rose relative to sea level, and the lakes became isolated and freshened by the rivers. A group of animals, apparently originally brackish water organisms, still present in the Baltic Sea, survived in them. In a series of lakes of different age of freshening, some of these animals show alterations in body shape relative to their ancestors, which may indicate a gradual evolutionary change. An example is that of *Limnocalanus macrurus* (Fig. 2.5). That change will not proceed very far, however, for the area is likely to be glaciated again some tens of thousands of years hence.

Away from the polar ice, fresh waters have still been disrupted so frequently as to prevent attainment of the diversity found in the sea. In the Miocene period some 25 million years ago, Africa was a continent of subdued relief with well-separated shallow, swampy basins. A series of volcanic eruptions and movement of the crustal plates, which underly the continent, created a watershed which separates east from west, and the rift valley which houses the basins of many of the east African lakes. More recently the wetter 'pluvial' periods, which bear some relationship to the warm, interglacial periods in higher latitudes, supported savannah and swamps in what is now the Sahara. There, rock paintings from only a few thousand years ago show hippopotami. A later drying out isolated Lake Turkana (L. Rudolf) from its once connection with the Nile and left many lakes reduced to remnant salty basins.

In these, for example Lake Chilwa [422] in Malawi, the unpredictability from year to year (for occasionally they dry out completely only to re-fill) perhaps mirrors the general unpredictability of inland waters. They support a limited biota of generalist species, unfussy in their requirements for existence and breeding and taking a range of food from algae to fish (Fig. 2.6). When Lake Chilwa dries out, for example, its fish move into the remaining trickling rivers or even into still, wet mud pools in the surrounding swamps.

In contrast, not far north of Lake Chilwa, Lake Malawi parallels the ocean. Among lakes it is long-lived, for its basin is very deep and although the water level has fluctuated as climate has changed, there has probably been a permanent water mass for perhaps a million years. In Lake Malawi and in the similarly old Lake Tanganyika and the Russian Lake Baikal there has been a progressive differentiation of very many specialist species dividing

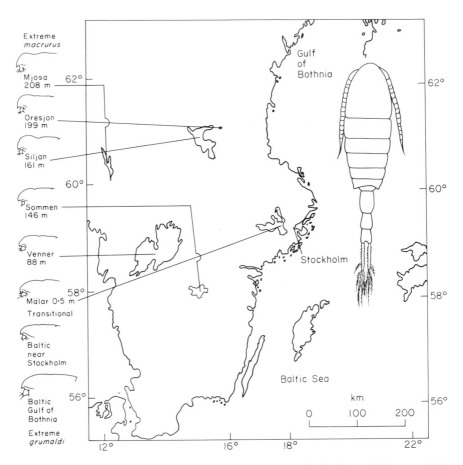

Fig. 2.5. Distribution of the copepod *Limnocalanus macrurus* in Sweden and Norway. This species shows a steady change from the brackish *L. macrurus* stranded in freshwater lakes formed after the last retreat of the polar ice. The longer the lake has been isolated, the greater the change from the brackish *L.m.grumaldi* forms to the *L.m.macrurus* forms. Altitudes of the lake surfaces (a measure of length of time since isolation from the Baltic as the land has risen in the post-glacial period) are shown under profiles of the head of the organism (seen from the side with the insertion of the antenna). An entire copepod, seen from the 'front' rather than 'side' view is shown. Lake Mälar has been cut off from the sea only since the early Middle Ages. (From Hutchinson [379].)

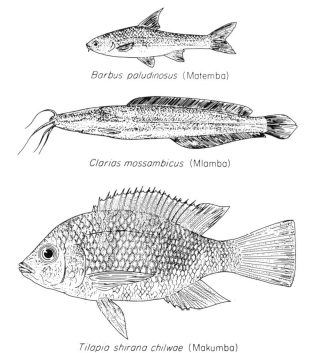

Barbus paludinosus (Matemba)

Clarias mossambicus (Mlamba)

Tilapia shirana chilwae (Makumba)

Stomach contents (% of total items)

	Higher plants	Green algae	Blue-green algae	Diatoms	Crustacea	Snails	Insects	Rotifers	Fish
Barbus	14	22	1	1	56	0	3	1	1
Clarias	12	4	1	1	47	1	15	0	20
Tilapia	28	23	13	10	14	0	0	7	5

Fig. 2.6. The three main species of fish in Lake Chilwa, Malawi, and the average stomach contents of samples of several hundred fish of each species. Lake Chilwa undergoes irregular phases of drying out; its fish species have very broad diets and unspecialized habitat requirements which allow them to cope with this variability. Each fish species, however, takes more of one particular item than the other two.

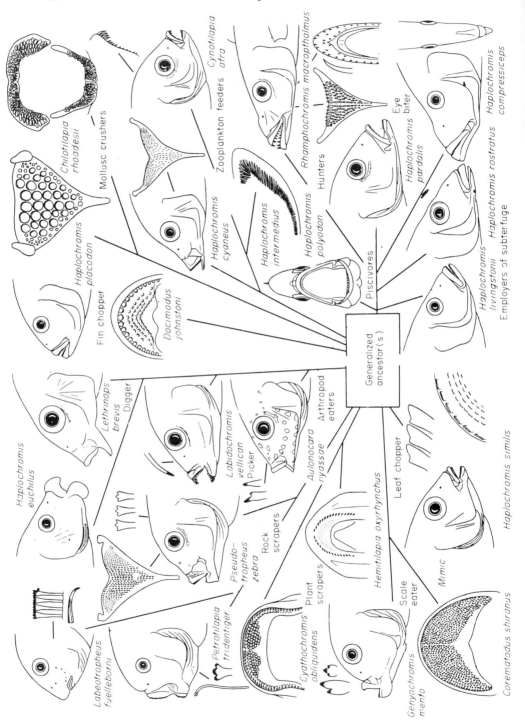

Chilotilapia rhoadesii — Mollusc crushers

Haplochromis placodon

Fin chopper

Docimodus johnstoni

Haplochromis cyaneus

Haplochromis intermedius

Haplochromis polyodon

Cynotilapia afra — Zooplankton feeders

Rhamphochromis macropthalmus — Hunters

Haplochromis macropthalmus

Haplochromis pardalis — Eye biter

Haplochromis rostratus

Haplochromis compressiceps

Haplochromis livingstoni — Employers of subterfuge

Piscivores

Generalized ancestor(s)

Haplochromis euchilus

Lethrinops brevis — Digger

Labidochromis vellican — Picker

Aulonocara nyassae

Arthropod eaters

Pseudotropheus zebra — Rock scrapers

Leaf chopper

Haplochromis similis

Hemitilapia oxyrhynchus

Mimic

Labeotropheus fuelleborni

Petrotilapia tridentiger

Cyathochromis obliquidens — Plant scrapers

Genyochromis mento — Scale eater

Corematodus shiranus

the resources of the habitat finely among themselves. The fish form an excellent example (Fig. 2.7). Many species in these old lakes are endemic—they have evolved there and occur nowhere else.

The old lakes will not be permanent in geological terms—eventually they will disappear in some phase of earth movement—but for the moment their diverse communities support the idea that the fundamental distinction behind fresh waters and the sea lies in the general impermanence of the former and the stability of the latter. Fresh waters have long been subjected to disturbance; the resilience they still have in their abilities to accommodate the impacts of man owes much to this feature of their collective history.

Fig. 2.7. Specialization that has occurred in diet for a number of closely related small species of the genus *Haplochromis* in Lake Malawi. In some cases whole heads, in others details of the pharyngeal bones (the triangular diagrams), upper part of the mouth (arch-shaped diagrams), or individual teeth are shown. Most descriptions are self-evident, but the mimic resembles a harmless (to other fish) plant eater, though in fact it scrapes scales from the bodies of other fish species, whilst the employers of subterfuge are fish-eaters (piscivores) which lie on the bottom, resembling rotting carcases. Curious small fish come to inspect them and are soon eaten as the carcase 'comes to life'. Pickers delicately pick small animals from rock surfaces, whilst diggers use their sensitive lips to feel for animals whilst probing in sand or sediment. The eye biter removes eyes from larger fish, but also eats whole small ones. (From Fryer and Iles [246].)

3

From Atmosphere to Stream—the Chemical Birth of Fresh Waters

Only for an instant is liquid water ever naturally pure. It distils to the atmosphere mostly from the ocean, is mixed by the winds, and eventually condenses into droplets or freezes into ice particles. Only at the moment of condensing or freezing is it ever likely to be pure water and probably not even then. Condensation probably requires nuclei of other substances—dust, or minute salt particles borne into the atmosphere from the breaking spray of the sea, or perhaps even single ions. Atmospheric gases immediately dissolve and as the droplet is moved in the air currents, salts and particles from dust and other gases are scavenged by it. The liquid droplet has become a very dilute fresh water. Snow crystals also pick up particles, ready to dissolve when the snow eventually melts.

The water which reaches the soils of the catchment areas thus already has a complex chemical nature. It is a dilute, weakly acidic, seawater solution modified by dust. Over towns and cities, or in rain falling from air which has passed over them, it may become more strongly acid and near deserts it may have much more dust. The solute composition of rain is not constant from place to place (Table 3.1) or time to time; it depends on the relative contributions of three main sources—the atmosphere, the sea and industrial pollution. These will be discussed in turn.

Table 3.1. Concentrations (mg l^{-1}) of several major ions in rain water, together with their percentages by moles (in parentheses) of the total sum of these ions compared with the percentage of the ions, on the same basis, in sea water

Ion	English Lake District, UK*	New Hampshire, USA†	Norwich, UK‡	Sea water
Sodium	3.1 (44.7)	0.12 (17.9)	1.2 (25.9)	(43.0)
Potassium	0.2 (1.66)	0.07 (6.43)	0.74 (9.5)	(0.9)
Calcium	0.2 (1.66)	0.16 (14.3)	3.7 (46.3)	(0.9)
Magnesium	0.3 (4.3)	0.04 (6.07)	0.21 (4.38)	(4.9)
Chloride	5.1 (47.7)	0.55 (55.4)	1.0 (13.9)	(50.2)

* Gorham [291]; † Likens *et al.* [469]; ‡ Edwards [193].

32

3.1 DISSOLVING OF ATMOSPHERIC GASES AND THE ACIDITY OF RAIN

Of the atmospheric gases which diffuse into the water droplets, carbon dioxide, with its high solubility and reactivity with the water, is the most important. Carbon dioxide is a major determinant of the acidity of the rain, as the following shows. It dissolves in water to form H_2CO_3, a weak acid which can dissociate as follows:

$$H_2CO_3 \text{ (aq)} \rightleftharpoons HCO_3^- \text{ (aq)} + H^+ \text{ (aq)} \qquad \text{(a)}$$

$$HCO_3^- \text{ (aq)} \rightleftharpoons CO_3^{2-} \text{ (aq)} + H^+ \text{ (aq)} \qquad \text{(b)}$$

Equilibrium constants for these reactions have been determined:

$$k_{(1)} = \frac{[H^+][HCO_3^-]}{[H_2CO_3]} = 4.5 \times 10^{-7} \qquad \text{(c)}$$

$$k_{(2)} = \frac{[CO_3^{2-}][H^+]}{[HCO_3^-]} = 4.7 \times 10^{-11} \qquad \text{(d)}$$

The square brackets indicate the concentration of each ion in moles l^{-1}. Equation (d) suggests that carbonate $[CO_3^{2-}]$ formation is only favoured when the concentration of hydrogen ion is very low (i.e. the pH is very high). Under most conditions the total amount of CO_2 bound up in these reactions, $[CO_{2\,tot}]$ is thus effectively:

$$[CO_{2\,tot}] = [H_2CO_3] + [HCO_3^-] \qquad \text{(e)}$$

The equilibrium between the CO_2 in the atmosphere and that which dissolves is defined by the solubility constant, K_H:

$$K_H = \frac{[H_2CO_3]}{[CO_2 \text{ (gas)}]} = 3.79 \times 10^{-2} \qquad \text{(f)}$$

Combining equations (c) and (f),

$$[HCO_3^-] = \frac{k_{(1)}[H_2CO_3]}{[H^+]} = \frac{k_{(1)} \cdot K_H[CO_2 \text{ (gas)}]}{[H^+]} \qquad \text{(g)}$$

and since $[HCO_3]$ arises in equivalent quantity to $[H^+]$ from the dissolution of H_2CO_3 (equation a):

$$[H^+] = \frac{k_{(1)} K_H[CO_2 \text{ (gas)}]}{[H^+]}$$

so, $[H^+]^2 = k_{(1)} K_H[CO_2 \text{ (gas)}]$

The concentration of CO_2 (gas), in the same molar terms in which the other concentrations are expressed, is also the partial pressure of the gas in the atmosphere (0.032%); therefore

$$[H^+]^2 = 4.5 \times 10^{-7} \times 3.79 \times 10^{-2} \times 0.00032$$

$$[H^+] = 2.3 \times 10^{-6} \text{ mol } 1^{-1}$$

and

$$\text{pH (which is } -\log [H^+]) = 5.64$$

This is the pH to be expected in naturally-forming rain, because CO_2 is the most abundant substance in the atmosphere which reacts with water to form hydrogen ions. The well known increase in CO_2 concentration in the atmosphere, which has resulted from the burning of coal and oil on a large scale in the last few decades, will have had some effect, but only a small one, on the equilibrium pH of rain water. The CO_2 content was about 0.029% in the nineteenth century, and 0.0332% in 1976. The equilibrium pH values for these concentrations are 5.65 and 5.62 respectively.

Other natural influences may serve to change the equilibrium pH of rain. Near active volcanoes, sulphur compounds, including SO_2 and H_2S are released into the atmosphere, and, in small concentration, but with a large total contribution, various sulphides of carbon, such as dimethyl sulphide [14, 819] are released by bacteria and algae in the ocean. These react with water to produce hydrogen ions, and, because some of these gases, particularly sulphur dioxide, are very soluble, the effects of even low atmospheric concentrations may be large.

For a SO_2 concentration (partial pressure) in the atmosphere of 10^{-7}, the equilibrium pH is 4.9. Charlson and Rodhe [109] believe that natural rain water (i.e. unaffected by human activities) is likely to have equilibrium pH values of 4.5–5.6 as a result of the combined action of carbon dioxide solution and that of sulphur compounds. Rain water in remote regions, however, usually has a pH close to the upper limit of this range.

3.2 CONTRIBUTION OF SEA SPRAY TO RAIN

The second major influence on the composition of rain water is that of sea spray. The evidence for this lies in often generally similar proportions of the most common ions in rain falling close to the coast to those of sea water (see Table 3.1). There is also a general decrease in the concentrations of these ions in rain with distance inland from the coast. Inland, however, the influence of local dust may mask the influence of sea salt.

The mechanism by which the salts reach the atmosphere is probably through the bursting of bubbles of air formed by wind action at the sea surface. On bursting the bubbles emit fine droplets which are readily carried away by the wind. Dust is, in general, likely to be much less important in determining rain composition than sea spray because dusty land surfaces are those over which rain forms least readily. Proportionately, they are also much smaller than the area of the ocean. Dust, in reflecting local geology, will also have a much more variable effect than sea spray. The high calcium concentration in Norwich rainfall (see Table 3.1) probably reflects the abundance of chalky soils in the area.

3.3 ATMOSPHERIC POLLUTION

The third major contributor to rain composition is human activity. This may take the form of smoke production, perhaps introducing heavy metals such as lead, chromium or zinc into the atmosphere from furnaces and smelters to be scavenged by the rain, but such effects are again likely to be local. The more general influence comes from gases released from fuel burning. Two are of particular importance—sulphur dioxide, which comes largely from industry, particularly power stations burning large quantities of sulphur-rich coal, and nitrogen oxides, released from vehicle engines and oil-burning power stations. These produce increased quantities of hydrogen ion when they react with water and thus make the rain more acid.

The product of sulphur dioxide solution, HSO_3^- may be oxidized by dissolved oxygen to form HSO_4^- which dissociates to form the very strong sulphuric acid. This must also be formed from natural SO_2 (see above) but calculations based on the amount of coal burned in the last few decades suggest that the latter is a greater source (perhaps 65×10^6 tonnes of sulphur per year—mostly as SO_2) than natural ones (30×10^6 tonnes per year, with about 7×10^6 as SO_2, the rest as carbon sulphides and other compounds). The burning is also concentrated into smaller areas. Sulphur dioxide has been linked with lung disease and much (and continuing) effort has been made to reduce the release of sulphur dioxide into the atmosphere. Sulphur dioxide emissions have been reduced by 15% in the European Economic Community since 1973 and are unlikely to increase in the next two decades. Whilst this has been happening, however, the burning of oil and the production of nitrogen oxides has generally increased. These oxides, NO_x, are also produced naturally in the atmosphere by reactions between nitrogen and oxygen energized by lightning sparks, ultimately forming NO_2. The reaction rate is very low, and although the supply of combined nitrogen from this source may be not insignificant for land ecosystems, the NO_2 has little overall effect on rain chemistry. Only half the global total of NO_x production comes from human activities (cf. 90% of SO_2 production) but again it is concentrated in the more industrial areas. In the atmosphere, NO_x is oxidized to the very strong nitric acid. There seems to be an increasing trend in nitrate concentrations in rain water particularly in the USA as a result [66].

The effects of solution and oxidation of SO_2 and NO_x have been to increase the H^+ concentration of rain and to reduce its pH to values below those naturally expected, in many parts of the world. Values down to 2.1–2.8 have been recorded in the USA, UK and Scandinavia, and averages are usually lower than 4.6 in these areas [36, 115, 467]. These values are usually compared with those obtained some decades ago (usually pre-1960) and found to be significantly lower. Some care is needed in making these comparisons, however, because techniques for determining pH, or bicarbonate concentration, which is often used as a surrogate for pH, have improved in recent decades. Older methods give slight over-estimates [438]. On balance, however, the changes in rain pH recorded from industrial areas have been greater than can be accounted for by problems of methods, and the rain that falls onto catchment areas downwind of intense human activity is more strongly acid than naturally expected. Snow may also be very acid, for droplets of quite concentrated sulphuric and nitric acids (often called 'dry' fallout) may be picked up by the snowflakes as they fall. If it does not melt almost immediately, snow ablates, that is vaporizes without liquid formation in very cold, dry air. The acid droplets then may become highly concentrated in long-lived snowbanks, so that run-off water as the snow melts in spring may be extremely acid.

3.4 THE COMPOSITION OF WATER DRAINING FROM THE CATCHMENTS

Rain and melted snow are thus complex, if dilute, chemical solutions. They move over and through the land surfaces of the catchment areas and become further modified as they pass eventually to the streams which drain the catchments. Every stream water will differ from the next and will vary within itself from time to time—certainly over days, sometimes over minutes. This is because the final composition depends on the interplay, unique in every sub-catchment, of several variables.

First there is the initial composition and amount of rain and snow, and the nature of the movement of water from the catchment (referred to as the hydrology). Secondly there is the availability of an unknown number, probably many thousands, of chemical substances which become dissolved in the water and the reactions between them. Thirdly there are the sources of these substances: local geology, soils and ecosystems. And lastly there are the many ways in which the catchment may be altered by human populations: forest removal, cultivation and fertilization of land, the building of settlements, industrial use and the disposal of wastes.

3.4.1 A chemical catalogue for run-off waters

It is useful to look first at the final product—the general chemical composition of the water—and then to derive this from the various sources in the groups listed above. A simple catalogue of groups of dissolved substances will include: major ions, atmospheric gases; key nutrient ions; trace nutrient ions; other trace ions; refractory (difficult-to-decompose) organic substances; and labile (very reactive) organic substances. The order of this list largely reflects the progressive difficulty of analysis of individual substances in the groups, and, consequent on that the historical order in which the groups have been investigated. No one has yet analysed completely a natural water sample. It does not reflect any particular order of importance to aquatic systems of the groups, or individual substances in them. Major ions, for example, are so-called only because they are present in greater concentrations than others; in some ecosystems the availability of one of the key nutrient ions may be paramount, in others a trace metal, if highly poisonous, may be of greatest significance.

The major ions include Na^+, K^+, Mg^{2+}, Ca^{2+}, SO_4^{2-}, Cl^- and HCO_3^-. They are generally dissolved in quantities of at least $mg\,l^{-1}$ (parts per million) and in general vary only a little in concentration during the year (though this may not be true for HCO_3^-, which may be absorbed by some plants and algae). For this reason they are called 'conservative', their concentrations being not greatly changed by the activities of living organisms, which may require them, but not in large quantities relative to those in which the substances are available.

In contrast the key nutrient ions, which include phosphates (PO_4^{3-}, HPO_4^{2-}, $H_2PO_4^-$), nitrate, ammonium, sometimes silicate, and occasionally iron, manganese, carbon dioxide plus bicarbonate and molybdenum are not conservative. Their concentrations are generally lower than those of the major ions ($\mu g\,l^{-1}$–$mg\,l^{-1}$) but more importantly the requirement for them by living organisms is high relative to the supply of them. Often they fluctuate greatly in concentration over the year.

The atmospheric gases have been discussed in Chapter 2. The inert gases are of least importance; nitrogen is of interest because of the contrast between its relative abundance, as N_2, and the relative natural scarcity of those of its compounds (nitrate, ammonium) which are available to living organisms. Oxygen has a key role, particularly when absent, in setting the chemical scene for many important reactions. In the context of drainage water, however, it is generally present and of less importance. Under temporary circumstances, in some waters, CO_2 may be a key nutrient, but generally supplies are more than adequate because of its atmospheric reserve and high solubility.

With the trace ions, analytical problems begin to become more important. Some, including Cu, V, Zn, B, F, Br, Co, Mo are known to be needed by organisms and are adequately present at ng l^{-1} or $\mu g\, l^{-1}$ concentrations. Others (e.g. Hg, Cd, Ag, As, Sb, Sn) are probably not required but are generally present at very low concentrations (ng l^{-1} or less). Probably almost all the natural elements not yet mentioned fall into this group. Both groups may be toxic if present at higher concentrations either through natural means (for example in springs draining mineral lodes or percolating new volcanic lava) or discharged by industry.

The chemistry of these elements is often complex, however, with several oxidation states being available, or with different ions formed at high and low pH or in various possible combinations with organic matter, for example cobalt in cyanocobalamin (vitamin B_{12}). Modest concentrations (say 100 μg l^{-1}) of iron, for example could be toxic if it is present as Fe^{2+} in conditions of low oxygen concentration or Fe^{3+} at low pH (3–4), but similar apparent concentrations at high pH (7–9) may be bound up in colloidal hydroxides largely inert to living organisms. Analytical methods for these trace substances often depend on conversion of the element to a highly soluble form for reaction and cannot distinguish the several individual states and substances in which the element is present in the natural water. These states are, however, distinguishable to living organisms and informative analysis may depend on a bio-assay in which the reaction of a test organism is used rather than a chemical analysis.

The analytical problems becomes even more difficult with the organic compounds. The chemistry of most elements is sufficiently constrained for some fairly comprehensive knowledge to have been attained of the range of their possible compounds, and certainly of those that dissolve in water. This is not true of carbon, however; its ability to form straight and branched chains, regular and irregular rings, and to combine with other elements not only in one form but in a variety of isomers, of similar elemental composition but different structures, means that an immense variety of organic compounds can exist.

Because these compounds are produced and modified by living organisms, small differences between them—even one isomer compared with another— may be crucially important, yet very difficult to detect in the laboratory. Some of the larger molecules may be relatively stable, e.g. the residual products of decomposition in soils of organic matter like wood, and hence more readily analysable. Their very stability, however, perhaps means that, despite concentrations of the order of mg l^{-1}, they are of least importance to the aquatic ecosystems. Other, smaller molecules—amino acids, alcohols, carboxylic acids, sugars, peptides—are more reactive, shorter-lived and present at low concentrations (ng l^{-1}–$\mu g\, l^{-1}$) but with a high turnover rate. They, and the more complex of the larger molecules which may be present, soluble proteins for example, pose the greatest analytical problems. Such labile substances may be important as nutrients to many microorganisms.

The organic compounds which reach stream waters represent the stages—

early and advanced—of decomposition of organic matter produced in the catchment (Table 3.2). The very earliest stages, in the form of leaf litter or smaller particles, may be present in suspension as well as the colloidal and soluble later derivatives. In parallel, the inorganic component of the water composition represents the early (particulate) and later (dissolved) stages of decomposition of the rocks of the earth's crust. From the mildly acid, very dilute and dust-changed sea water of rain, we need now to derive, through the catchment kitchen, the slightly saltier, more organic consommé of the stream waters.

Table 3.2. Origins and nature of organic substances washed into stream waters

Origin in living organisms in the catchment	Organic derivatives washed into draining waters
Proteins	Methane, peptides, amino acids, urea, phenols, indole, fatty acids, mercaptans. Melanin*, melanoidin*, yellow substances (gelbstoffe)*
Lipids (fats, waxes, oils, hydrocarbons)	Methane, aliphatic acids, acetic, lactic, citric, glycolic, malic, palmitic, stearic, oleic acids. Carbohydrates, hydrocarbons*
Carbohydrates (cellulose, starch, hemicellulose, lignin)	Methane, glucose, fructose, galactose, arabinose, ribose, xylose. Humic acids*, flavic acids*, tannins*
Porphyrins and plant pigments (chlorophylls, hemin, carotenoids)	Phytane*, pristane*, isoprenoid*, alcohols, ketones, acids, porphyrins

* Generally refractory in water.

3.4.2 Rock weathering

Rocks and rain give most of the inorganic substances which reach fresh waters. Rocks are made up of minerals, most of which are based on crystalline combinations of silicon and oxygen, the two most common elements in the Earth's crust, with small additions of other elements (see Table 3.4). The silica minerals are very varied for, like that of carbon, the silicon atom can combine with up to four other atoms in a tetrahedral formation. The tetrahedra may then form lattices—chains, sheets and three-dimensional structures. As the structure becomes more complex it becomes more stable and the number of available sites for the attachment of other elements is proportionately reduced. The Earth's crust is dominated by these minerals.

Following the formation of the Earth heavy molten iron and nickel sank to the core whilst the lighter silicon floated to cool and react within the surface layers to form the primary silica minerals in igneous rocks. Though stable under the pressure, and high temperature, of their birth and in the absence of oxygen and water deep in the crust, the minerals are not necessarily so at the Earth's surface. Their weathering to form products stable at the Earth's surface releases soluble ions to the drainage water. Weathering also produces the inorganic parts of soils, whose properties determine the nature of later reactions and releases.

When exposed at the surface, igneous rocks are often massive. Cooling and crystallization lead to tight, intimate mixtures of mineral crystals, not perforated by channels or pores. Released from overlying pressure, heated, cooled, frozen, thawed and penetrated by roots the rocks crack, exposing more surfaces to water and oxygen. The process is slow and the small surface areas initially presented for access to water mean that relatively few soluble ions are initially released. The ones that are, are those ionically (as opposed to covalently) held on the silicate lattices. These can be attracted by the polar water, and replaced by hydrogen in the lattice. The very dilute waters draining from these igneous rocks reflect in their composition the relative abundances of the more soluble of the more common metals bound in the minerals. Generally $Mg > Ca > Na > K$ but this is modified by the cations present already in the rain and derived from sea spray so that in general $Na > Mg > Ca > K$. Areas producing such waters include the Precambrian rocks of the Canadian Shield, and the central part of the English Lake District (Fig. 3.1).

Silicic acid, $HSiO_4$, is also released from hydrolysis of the lattice but often reprecipitates to form clay (see below) and appears in the drainage water in much lower quantities than those initially released. Lattice breakdown does not produce many soluble anions therefore, and the anions in the water draining igneous rocks are those provided by sea spray and the reaction of carbon dioxide. The low degree of attack on these resistant rocks usually leaves sufficient H^+, ultimately derived from rain, to give an acid drainage water.

The weathering rate of minerals is increased by increasing temperature, increasing acidity, a brisk percolation of water and the presence of oxygen. Increasing H^+ concentration not only increases the replacement of metal ions in the lattice but may also create conditions for dissolution of less soluble ions, like those of aluminium, which is a very common component of lattices. Al^{3+} is soluble below pH 4.5 and hence its concentration in drainage waters from igneous areas has been increased through the lower pH of rain acidified by the reactions of SO_2 and NO_x. Rapid percolation of water removes the soluble products of the lattice hydrolysis; it therefore promotes the reaction to replace these products according to Le Chatelier's principle. Oxygen promotes the removal of elements like iron and manganese from lattices; these elements can exist in several oxidation states, for example FeII, FeIII, and can be dislodged by oxidation.

Extensive breakdown of igneous rocks in the warm, humid tropics can leave residues (ultimately called laterites) containing only the more resistant

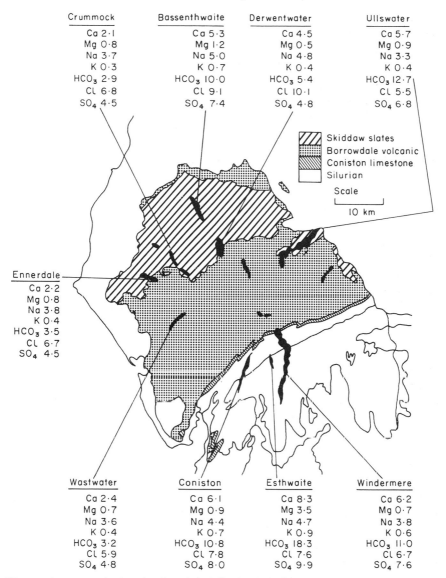

Crummock		Bassenthwaite		Derwentwater		Ullswater	
Ca	2·1	Ca	5·3	Ca	4·5	Ca	5·7
Mg	0·8	Mg	1·2	Mg	0·5	Mg	0·9
Na	3·7	Na	5·0	Na	4·8	Na	3·3
K	0·3	K	0·7	K	0·4	K	0·4
HCO_3	2·9	HCO_3	10·0	HCO_3	5·4	HCO_3	12·7
Cl	6·8	Cl	9·1	Cl	10·1	Cl	5·5
SO_4	4·5	SO_4	7·4	SO_4	4·8	SO_4	6·8

Skiddaw slates
Borrowdale volcanic
Coniston limestone
Silurian

Scale

10 km

Ennerdale

Ca	2·2
Mg	0·8
Na	3·8
K	0·4
HCO_3	3·5
Cl	6·7
SO_4	4·5

Wastwater		Coniston		Esthwaite		Windermere	
Ca	2·4	Ca	6·1	Ca	8·3	Ca	6·2
Mg	0·7	Mg	0·9	Mg	3·5	Mg	0·7
Na	3·6	Na	4·4	Na	4·7	Na	3·8
K	0·4	K	0·7	K	0·9	K	0·6
HCO_3	3·2	HCO_3	10·8	HCO_3	18·3	HCO_3	11·0
Cl	5·9	Cl	7·8	Cl	7·6	Cl	6·7
SO_4	4·8	SO_4	8·0	SO_4	9·9	SO_4	7·6

Fig. 3.1. Average major ion chemistry (mg l^{-1}) of waters of some of the lakes of the English Lake District. The paucity of cations and bicarbonate in those to the west, drawing water from the igneous Borrowdale volcanic rocks is well shown, whereas those to the south, served by sedimentary rocks have much higher concentrations. (Based on Macan [491].)

minerals like quartz $(SiO_2)_n$ and insoluble red or yellow oxides and hydroxides of iron III, whilst all the soluble ions are mobilized into streams. As the weathering goes on, the stock of such ions is progressively depleted, and the concentrations in the water fall. In cooler and drier temperate regions, the breakdown is less rapid and some of the soluble products like silicic acid are not immediately washed away but persist in the interstitial water where they may re-form secondary silicate minerals by precipitation. These include the clay minerals—flat plates of silica tetrahedra interleaved with plates of aluminium oxides.

Two most important properties of the clay minerals are that they generally have a net negative charge which attracts H^+ ions, and that they are small (generally less than 2 μm) and present a large surface area per unit weight for further chemical reaction. Cations in rain reaching the surfaces of clays may be adsorbed onto the clays, displacing H^+ into solution. The more acid the rain water, the less likely it is that H^+ will be displaced from the clay and the more likely it is that metal ions will be leached out to the streams. Conversely the higher the pH of the rain the more likely it is that metal cations will be adsorbed by the clay and retained in the soil. In this way, through cation exchange, the clay minerals can begin to regulate the composition of the drainage water.

From igneous areas thus are washed dilute solutions of major cations in which may be suspended fragments of unweathered minerals, resistant minerals and clays. There may also be dilute solutions of silicate and of other anions like phosphate derived from scarce phosphorus containing minerals present in the original rock. Phosphates, however, are only reasonably soluble at circum-neutral pH. At low pH they are precipitated as iron or manganese phosphate, so concentrations in the drainage water are generally very low (e.g. < 1–$5\ \mu$g PO_4–Pl^{-1}).

The weathered rock debris eventually comes to rest as an ocean sediment. To it may be added the remains of oceanic organisms which have concentrated, from the water, quantities of essential elements like sulphur and phosphorus and nitrogen, and of substances like calcium carbonate and polymerized silicate which they had used to form cell walls. As the mass of sediment accumulates it is compressed; water is squeezed out and calcium and other compounds may serve to bind it together into what will become a sedimentary rock. Eventually this rock may be raised above sea level by earth movements and exposed to a new cycle of weathering. It may also be buried, compressed, reheated and melted to emerge, cooled, as a metamorphic rock with silica minerals re-formed. A metamorphic rock will then weather again in much the same way as an igneous rock. The northern part of the English Lake District is of metamorphosed slates and produces waters similar to those of the Borrowdale volcanic rocks (see Fig. 3.1).

3.4.3 Weathering of sedimentary rocks

Weathering of an unmetamorphosed sedimentary rock will be very different from that of an igneous rock, and despite the 95% dominance of igneous rocks by volume in the Earth's crust, the surface of the continental crust, where most of the weathering takes place, is 70% covered by sedimentary rocks. Overall these have a much more widespread effect on freshwater chemistry. Sedimentary rocks are built from jumbled particles; they are often porous, presenting a large surface area for water to percolate and many lines of weakness for the rock to crack. The cements which hold them together are in general soluble and readily weather, releasing again the soluble ions, and further exposing any primary minerals to decomposition. Their manner of formation from inorganic and organic debris means also that anions such as sulphate, carbonate and phosphate are relatively abundant and this is reflected in the waters that drain from them. There is a particular abundance of calcium carbonate, derived from the cell walls and shells of many marine organisms in many sedimentary rocks. This means that calcium and bicarbonate, released from the rock by the acids in rain, are often dominant among the major ions in such waters. The abundance of carbonates and other anions leads to a ready neutralization of the H^+ in rain water, so that the drainage waters are neutral or even alkaline and ions such as aluminium are not so readily mobilized. The high pH of the soils which form from them means also that phosphates are again precipitated, though now as calcium phosphate, so that the run-off water is still poor in phoshate though enriched in other ions.

The ready weathering of sedimentary rocks leads to deep soils, easily cultivable and hence quickly modified by human activity. It is more difficult therefore to isolate the effect of geology in determining the contents of drainage water from sedimentary rocks than it is for the igneous rocks where landscapes, often mountainous and with thin soils, are least inhabited.

The example of the English Lake District (see Fig. 3.1), where the northern areas have igneous and metamorphic rocks and the southern parts sedimentary rocks, including bands of limestone, however, shows the distinction. The waters from the highest mountain lakes, the small tarns of the north and centre, are essentially collected rain waters little altered by rock weathering because the flow-through is very rapid. The larger lakes where the retention of water in the crystalline rocky catchments has had some influence, have greater ionic concentrations, but calcium and bicarbonate are scarce and H^+, Na, Mg, Cl and SO_4 dominate the solution. The southern waters on sedimentary Silurian rocks have a dominance by calcium and bicarbonate at higher overall concentrations.

Further south in England, the subdued sedimentary landscape of the east Norfolk rivers (Fig. 3.2), draining chalk to the west and glacial deposits derived from chalk and other sedimentary rocks to the east show the expected high major ionic concentrations. Closeness to the sea brings very high concentrations of sodium and chloride, through percolation near the coast and, to a lesser and decreasing extent, through spray to the rivers inland. The river concentrations of chloride are much higher than those in rain ($<$ 10 mg

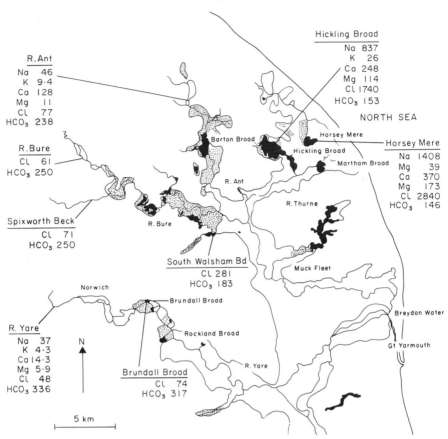

Fig. 3.2. Average major ion chemistry (mg l⁻¹) of the Norfolk Broadland. The area is floored by soft sedimentary rocks and alluvium, and the effect of the chalk to the west on bicarbonate concentrations is well shown. Chloride concentrations tend to decrease westwards as the influence of the sea, either through spray, tidal movement upstream (S. Walsham Broad), or percolation through nearby dunes (Horsey Mere, Hickling Broad) decreases. (Based on Moss [563].)

l⁻¹) because much of the rain is evaporated in the catchment, effectively concentrating the salt in the eventual smaller volume of run-off water.

3.5 EFFECTS OF SOIL DEVELOPMENT AND VEGETATION ON THE CHEMISTRY OF DRAINAGE WATERS

Living organisms (microorganisms, mosses and lichens) are usually present when rocks weather. As the rock flakes accumulate, more organisms colonize and contribute litter to form a maturing soil. This development changes the nature of the water that will drain from it. First it provides a source of combined nitrogen; secondly, the live biomass and undecomposed litter form a store into which soluble ions may be taken up, no longer to be so vulnerable to leaching. Thirdly, the organisms decomposing litter provide organic matter which may reach the streams.

3.5.1 Nitrogen fixation

Because nitrogen gas is relatively unreactive and volatile, little nitrogen is available in rocks. Atmospheric nitrogen gas, in contrast, is abundant but unavailable to most organisms. Most available nitrogen compounds ultimately reach living organisms through the mediation of nitrogen-fixers in the catchment soils. All of these are prokaryotic, including some blue-green algae, and other bacteria.

Fixation means the reduction of nitrogen by the addition of hydrogen; it is a complex process, requires energy, is inhibited by oxygen and eventually produces amino groups ($-NH_2$) bound into amino acids from which the fixers form their proteins. Nitrogen-fixers must protect their nitrogen-fixing enzymes (nitrogenases) from oxygen. Some operate in waterlogged, anaerobic places; others place the enzymes in specialized cells from which oxygen is excluded, or in cells deeply encased in jelly; yet others associate intimately with oxygen-consuming bacteria which create pockets of low oxygen tension in the soil. The amino-nitrogen is released on death of the fixers, and decomposed to ammonium ions, which may be adsorbed, as cations in clays, or taken up directly by other organisms. Ammonium is a reduced ion, from which energy may be obtained by oxidation; some bacteria, the nitrifiers, can oxidize it to nitrite and nitrate, in which latter form it may continue to be available for plant uptake.

Like most prokaryotes, the nitrogen fixers grow best at circum-neutral or alkaline pH. Nitrogen fixation is thus most prolific in soils derived from sedimentary rocks, rather than those derived from igneous ones. Both ammonium and nitrate are relatively soluble ions, and easily leached, so that the losses of them from soils to fresh waters to some extent reflect the amount of fixation in the soils.

3.5.2 Storage in the plant biomass

The effects of vegetation in modifying water chemistry can be studied by measuring the income of particular elements to a land ecosystem and then the subsequent loss to the drainage waters to obtain element budgets under vegetated conditions. If the vegetation is then removed and the exercise repeated, insight into the role of the vegetation can be gained. Such experiments must use self-contained catchments which are small enough to be uniform, yet large enough to have general value. All the water entering as rain or snow and leaving by evaporation or stream-flow must be accounted for; and the experiment must be carried out for several years to eliminate year-to-year effects of changing weather. This means that permeable sedimentary rocks, through which water may be lost to the deep ground waters cannot easily be used. Reliable results are most likely obtained from well-sealed basins of igneous or metamorphic rock.

Perhaps the fullest such study has been that on the sub-catchments of streams draining into the Hubbard Brook in New Hampshire, USA [469]. The underlying rock is igneous granite with some metamorphic shales; the hilly land gives well-defined catchments covered with natural deciduous (hardwood) forest of maple, beech and birch.

Elements enter in rain and snow, as dust and by direct uptake of gases, particularly SO_2, by fixation (of nitrogen) and weathering of the rock. All of the loss of elements, except some loss of nitrogen by denitrification, is by the draining streams. The atmospheric inputs (water and dry deposition (dust and gases)) can be collected and measured. The outputs can be determined by analysing the stream water and measuring its volume. The contributions from rock weathering of elements other than nitrogen are obtained by difference. Because the rocks contain almost no nitrogen, this difference for nitrogen is assumed to come from fixation. Table 3.3 shows the percentage contribution each year to the undisturbed forest by precipitation, dry deposition and weathering.

Table 3.3. Proportionate income (% of total) and fate (% of total) of various elements in the undisturbed forest ecosystem at Hubbard Brook, USA (Based on Likens *et al.* [469])

	Ca	K	Mg	Na	N	P	S
Income							
Rain and snow	9	11	15	22	31	1.4	65
Dry deposition (as gas)	—	—	—	—	—	—	31
Fixation*	—	—	—	—	69	—	—
Rock weathering*	91	89	85	78	—	98.6	4
Fate							
Incorporated into vegetation	35	68	17	1	43	82	6
Incorporated into litter and soil	6	4	5	<1	37	18	4
Lost, dissolved, to stream	59	22	74	95	19	0.25	90
Lost as suspended particles to stream	1	6	5	3	1	0.35	1

* Determined by the difference between atmospheric income and losses to the stream.

Weathering supplies most of the Ca, Mg, Na, K and P, whereas rain and snow contribute significant amounts of Na, N and S. Most of the nitrogen comes through fixation, however, and nearly a third of the sulphur from dry deposition of SO_2 or as sulphuric acid droplets derived from it. The fate of these elements is either to be incorporated into the ecosystem as live biomass, litter or soil or to be washed out dissolved or as suspended particles. Most of the Ca, Mg, Na and S is washed out, largely in dissolved form, which means that the vegetation has little effect on the concentrations of these elements in the drainage water. Potassium, however, and nitrogen are selectively retained by the ecosystem, though the high solubility of ions of these elements means that 20–30% is lost to the stream. Phosphorus is very strongly held, the stream losses being less than 1% of the annual supply.

These data can be looked at in another way—in terms of the net gain or loss of each element to the land ecosystem—by balancing the income from just the atmospheric sources (including nitrogen fixation) against the losses to the stream. This is a measure of the extent to which the rain chemistry is changed by the rock–soil–vegetation system. The elements fall into two groups—those for which there is a net loss from the soil–vegetation system (Na, K, Mg, Ca) and those for which there is a net gain (N, P, S). For nitrogen this gain is very substantial. This general picture has been obtained from other sites and seems to indicate a particularly important role, at least of forest vegetation, in affecting the concentrations of nitrogen and phosphorus compounds in the run-off water. These elements are emphasized because they are particularly important in determining the productivity of fresh waters (see later chapters).

This role of nutrient conservation is emphasized when the forest is removed. In a sub-catchment at the Hubbard Brook site, the trees were cut, and the wood left on the site. Re-development of vegetation was prevented with herbicides. Some of the effects of felling, over the 3 years following it, are shown in Fig, 3.3. They are quite dramatic, with increased gross losses in dissolved form of every element measured, except sulphur. The losses of nitrogen were particularly high. In the undisturbed ecosystem the fixed nitrogen released on decomposition in the soil is probably very quickly taken up as NH_4^+ by the grazing plants and not exposed to much risk of leaching. With the vegetation removed, the ammonium ion was probably oxidized by bacteria to nitrate, whose high solubility, and the lack of vegetation to take it up, led to the large losses. As well as the loss of dissolved substances, there was a fivefold increase in the particulate matter carried by the stream. This was much less than expected intuitively, but one feature of it was that the loss of phosphorus as soil particles was much greater than the dissolved loss of phosphorus and about fifteenfold greater than that in the undisturbed forest.

The Hubbard Brook study has been a particularly extensive one and similar studies have also concentrated on forest ecosystems. There may be dangers in extending conclusions from these too widely but the generalizations that can be made from the study are consistent with much other information. For example the natural retention of N, P and, to a greater extent than other cations, K, in natural systems suggests a general scarcity of these elements for plant growth. Of the elements required for plant and algal growth and

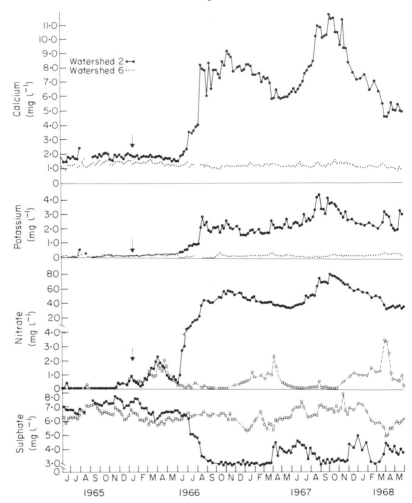

Fig. 3.3. Changes in the chemistry of stream water in two watersheds at Hubbard Brook in the White Mountains of New Hampshire. The forest in watershed 2 was removed in early 1966 (arrow) whilst watershed 6 remained intact. Note change in scale for nitrate concentration. (Based on Likens and Bormann [468].)

derived largely from the Earth's crust rather than its atmosphere, the supply of phosphorus is smallest in relation to need (Table 3.4). This is also reflected in the use of N, P and K as the main agricultural fertilizers. In relation to plant need, rather more nitrogen fertilizer than phosphorus is used and this reflects the high solubility of nitrate and greater loss of it by leaching (see below) than of phosphorus.

Much evidence now suggests that the felling of tropical forest for agriculture gives only a short period in which the soils are reasonably fertile

Table 3.4. The relative supply of various elements indicated by their abundance relative to phosphorus in the Earth's crust, and the relative requirements of these elements, again relative to phosphorus, by plants and algae. The ratio of supply to need for all elements is greater than that for phosphorus. (Modified from Hutchinson [381])

Element	Ratio of amount of element to that of P in Earth's crust (supply ratio)	Ratio of amount of element needed by plants and algae relative to that needed of P (need ratio)	Ratio of supply to need
Na	32.5	0.52	62.5
Mg	22.2	1.39	16.0
Si	268.1	0.65	413.0
P	1.0	1.0	1.0
K	19.9	6.1	3.3
Ca	39.5	7.8	5.1
Mn	0.9	0.27	3.3
Fe	53.6	0.06	893.0
Co	0.02	0.0002	100.0
Cu	0.05	0.006	8.3
Zn	0.07	0.04	1.8
Mo	0.0014	0.0004	3.5

[583]. Loss of the forest, and the mechanisms by which it holds elements, means that leaching losses, especially with very high rainfall, are greatly increased. Stream waters draining mature rain forest are very dilute indeed. For example, Amazonian forest streams average (in mg l^{-1}): Na, 0.22; K, 0.15; Mg, 0.04; Ca, 0.04; HCO_3, 5.5; Cl, 2.2 [247]. Most of the ions released by weathering are contained in the forest biomass. Destruction of the biomass, often with burning to produce a very soluble ash, leads to rapid loss of the nutrient stock, and little is left in the soil itself.

A consequence of the small leaching losses from the mature forest on its anciently weathered, now depauperate soils, is that a major source of elements for the aquatic communities of Amazonian rivers is in the form of plant biomass falling into the water rather than leached ions (see Chapter 5). Such tropical waters may on the other hand have quite high concentrations of silicate and relatively higher concentrations of phosphorus than expected because the rock weathering has reached a very advanced stage.

A third consistency, supporting generalization made from the Hubbard Brook results, is that where landscapes lacking vegetation are created—for example after glaciation—there appears to be a much greater run-off of ions and particles at first than subsequently when vegetation has developed. This change is a long-term one so evidence for it comes from the analysis of sediments in lakes. In the English Lake District, sediments laid down just after the ice retreated and before vegetation developed are rich in cations [499], and also the easily eroded clay minerals. Those deposited after the colonization, initially of tundra and then of birch forest, are poorer in both.

The net retention of elements by land ecosystems seems reasonable so long as the biomass of the vegetation is building up and soil is accumulating. Once the balance between build-up of biomass and annual consumption and decomposition of it, the 'climax' state, is reached, then as much of a given element that newly enters the biomass each year from the atmosphere or weathering should leave it, and wash-out of elements from the land ecosystem to streams might increase again. However, the ultimate source of most elements is rock weathering and this does not go on indefinitely. The minerals become progressively leached in the upper and middle layers of soil where they are accessible to water and air. The provision of new minerals becomes less easy as the weathering proceeds deeper and deeper, insulating the fresh rock from sun, water and oxygen. The theoretically very large supply of elements in the Earth's crust is thus not all available. An undisturbed land ecosystem should thus, in time, deliver smaller quantities of dissolved ions to the streams draining it. Lack of any disturbance by geological or climatic processes or by man is, however, never an indefinite state for any ecosystem.

3.5.3 Vegetation and the supply of suspended silt and dissolved and suspended organic matter to drainage waters

Dissolved inorganic substances are not the only contribution made by catchment areas to the drainage waters. Silt particles—eroded minerals and organic detritus from the vegetation and associated animals—are also important. So is dissolved organic matter.

Silt loss tends to increase with run-off of water up to a point (about 25 cm year^{-1}) where it declines because the increasing precipitation supports a fuller vegetation cover which reduces erosion. There is an inverse relationship between the amount of erosion and the vegetation cover, with forest being more effective (by about five times) at stopping erosion than grassland and other less bulky vegetation. The implications of this are that the drainage waters of arid areas will naturally bear quite high concentrations of suspended solids but that those of wetter areas will not. The Orange River in South Africa was probably not named because of the colour given to it by suspended silt long before any disturbance of its catchment area by farmers but is appropriately named nonetheless. Deeper in the African continent, in wetter climates, the rivers run red, not naturally, but because of removal of the natural vegetation for cultivation.

Organic matter may enter drainage waters as whole branches, leaf litter, bud scales, flowers, pollen and spores, the faeces and even carcasses of animals, and as finely divided soil humus. The amount of litter washed in will depend on the vegetation type; it will be high in dense, deciduous forest overhanging the stream valleys, and will enter mostly during autumn leaf-fall or the period of bud bursting and flower production in spring. It may be much less in very dense tropical forest where leaf-fall is constant year around.

The forest density prevents much litter blowing around and being deposited in the streams and decomposition is very rapid on the forest floor. There will be less where the vegetation is sparse or not overhanging the streams, for example in grasslands or open savannah. In these cases the wash-off of inorganic solids is likely to be the more important.

3.5.4 Dissolved organic matter

Problems of analysis have prevented much generalization about the dissolved organic matter. Almost all natural waters have a dissolved organic concentration much greater than the suspended organic concentration. The total dissolved organic concentration (perhaps $1-20$ mg l^{-1} expressed as carbon) may rival that of total dissolved inorganic matter. Much of it seems to be refractory. (The definition of this depends on whether or not it is decomposable by high intensity ultra-violet radiation in the laboratory). Refractory substances (often phenolic substances or carboxylic acids) are those which persist in soils because they are difficult to decompose further. The amounts dissolved may be sufficient to give an obvious brown stain to water from peaty areas, such as the upland moorlands of Britain or the heathy *fynbos* vegetation of South Africa. All drainage waters, however, will have some of these compounds. They are relatively inert, but may combine, for example with easily precipitated metals like iron, and maintain these metals in solution as organic complexes. In this role they are described as chelators.

Labile compounds will usually be decomposed rapidly in soil, but a proportion of them is leached to the streams where it may be very important despite usually very low concentrations in the order of $\mu g-ng$ l^{-1} (see Chapter 4). Such compounds (amino acids, sugars) may leach from insect-wounded leaves or carcasses, or from decomposing litter.

3.6 EFFECTS OF HUMAN ACTIVITIES ON THE COMPOSITION OF DRAINAGE WATERS

This will depend mostly on the numbers of people and the kind and scale of their activities. Rock weathering and sea spray still dominate the major ion composition of the world's fresh waters, but farming and settlement now probably have the greatest effect on key nutrient concentrations, whilst increasingly, atmospheric pollutants and industry influence those of trace elements.

3.6.1 Agriculture

Perhaps the greatest impact of human use of the land is the removal of the original vegetation cover and the upsetting of mechanisms which seem often to have conserved nitrogen and phosphorus in the land ecosystem. Thereafter, effects on the drainage water depend on whether there is continuous plant cover, in permanent grazing for example, or whether the soil is left bare for part of the year. They depend also on the extent to which fertilizers are used and the weather when they are applied; also on the number of stock kept and how the animals are managed.

For example, studies have been made of the effects of improvement of previously unfertilized upland grassland in the British Isles. Usually this means ploughing and re-seeding, often with a seed mixture including legumes, which are associated with nitrogen-fixing bacteria, and then liming and fertilization. The pasture is then left as permanent pasture, though more fertilizer may be added from time to time. In an upland catchment at the headwaters of the R. Wye in central Wales, three small sub-catchments have been compared. One had semi-natural vegetation with mat grass (*Nardus*), bent grass (*Agrostis*) and fescue (*Festuca*). A second had been ploughed 40 years previously. Magnesian limestone and basic slag (a phosphate fertilizer) had been added to it and it was re-seeded, but no further treatment had been given in the intervening years. The third was improved in the mid 1970s, in the same way that the second had been, and further doses of fertilizer had been given. The effects on the average chemistry of the streams draining the catchments in 1979–1981 were measured [360]. There was some effect of the treatment given to the most recently improved site, with increases in concentrations of all substances (major ions, nitrate, phosphate) over the control, but these were not large. The water draining from the previously improved, but then abandoned, catchment was intermediate between the other two, suggesting a steady return to the former condition. For all three catchments, soluble phosphate was always found to be at concentrations below $5 \, \mu g \, l^{-1}$ (as P). In general this semi-natural grassland regime was releasing a little more of most elements than undisturbed forest regions but not as much as the Hubbard Brook sub-catchment in the years after clear-felling of the forest. The original natural vegetation of the Welsh catchment studied would probably have been a mixed forest of birch and oak and other trees. The work suggests an eventual fall, with time and re-vegetation, in the catchment nutrient losses.

3.6.2 Lowland agriculture

Much more severe changes have occurred in lowland areas. Table 3.5 shows the effects of land use in four categories of catchment in the central states of the USA—forest ($>75\%$ natural vegetation); mostly forest ($>50\% <75\%$); mostly agriculture ($>50\% <75\%$); and agriculture ($>75\%$). The change from mostly forest to mostly agriculture has led to increases in stream concentrations of total phosphorus and soluble orthophosphate by about tenfold, of total nitrogen, fivefold and inorganic soluble nitrogen (nitrate plus ammonium) by nearly fourteenfold. 'Total' amounts of elements include all

Table 3.5. Effects of land use on the rates of loss of nitrogen and phosphorus from the catchments, and the concentrations of them in stream waters, for a range of catchments in the east and central states of the USA. Values given are means. (Based on Omernik [605])

Type of catchment	Number of catchments	Concentrations in stream waters			
		Total P (μg l^{-1})	Orthophos-phate-P (μg l^{-1})	Total N (mg l^{-1})	Inorganic N (mg l^{-1})
Forest	53	14	6	0.85	0.23
Mostly forest	170	35	14	0.89	0.35
Mostly agriculture	96	66	27	1.8	1.05
Agriculture	91	135	58	4.2	3.2

	Gross amounts lost per unit area of catchment (kg ha^{-1} year^{-1})			
	Total P	Orthophos-phate-P	Total N	Inorganic N
Forest	0.08	0.04	4.4	1.3
Mostly forest	0.17	0.07	4.5	1.8
Mostly agriculture	0.23	0.09	6.3	3.7
Agriculture	0.31	0.13	9.8	7.4

forms—dissolved inorganic substances, dissolved organic compounds of the element, and colloids and particles containing it.

These differences illustrate a general conclusion that disturbance of vegetation and soils by agriculture leads to rather greater losses of nitrogen than of phosphorus. Most of the increase in phosphorus in this example probably comes from the excreta of stock. Faeces and urine contain high concentrations both of phosphorus and nitrogen, whereas the drainage from a cultivated field may be enriched only a little in phosphorus (largely in particulate form). Phosphorus is fixed quite strongly by chemical reactions in soils, but nitrogen compounds are relatively soluble and are easily lost. A single cow will produce as much phosphorus (about 18 kg year^{-1}) as 212 ha of forest or more than 57 ha of crop land. It will also excrete as much nitrogen (58 kg year^{-1}) as 68 ha of forest or more than 6 ha of arable land.

In more traditional farming systems, where animals and crops were managed together, considerable use could be made, by manuring the fields, of the large amounts of phosphorus and nitrogen produced by the animals. There was consequently only a small loss of nutrients to the drainage water. Recent trends in agriculture in the developed world are leading to a separation of stock-keeping and arable farming. Stock is often kept in intensive units which produce a large quantity of 'slurry' which cannot often be disposed adequately to the land but penetrates eventually to streams and ground water. And large areas of land are devoted entirely to arable agriculture. Here the greatest profit is to be made from highly mechanized methods, high-yielding

cereal varieties which demand heavy fertilization, and the abandonment of
traditional crop rotations which left the land in pasture for at least one or two
years in a cycle of four. The use of pesticides now allows annual cultivation
for cereals. This leaves the ploughed soil bare for a part of each year, when
nutrients are more vulnerable to leaching and soil to erosion. Just as the
economic pressures for the intensive raising of stock (high production at low
labour cost) have led to increased losses, particularly of phosphorus, to the
drainage waters, so the intensive growing of cereals has increased the
nitrogen losses and increased the concentrations in streams and rivers.

In England and Wales, the amount of nitrogen fertilizer (usually
ammonium nitrate or ammonium phosphate) put on agricultural land has
increased from about 50 thousand tons in 1928 to nearly 13 hundred thousand
tons in the 1980s. Of the nitrogen reaching the fields, only about half is
incorporated into crops, nearly a quarter is lost by volatilization of ammonia
to the atmosphere and a seventh is lost by denitrification and incorporation
into refractory compounds in the soil. Another seventh is lost to the drainage
waters. This is equivalent, after denitrification, to about a third of the
chemical fertilizer added, though simply reducing the use of fertilizer would
not reduce the amount lost in exact proportion. Cultivation alone leads to a
significant loss, perhaps by disrupting the complex system of microorganisms
which process nitrogen originally derived from fixation and stored in the soil
as organic compounds. The effect of these changes in farming practice has
been to increase the nitrate concentrations in lowland drainage very much
(Fig. 3.4). This has followed particularly the 1947 Agriculture Act in Great
Britain, which gave much support to farming through limited subsidies, and
the European Community's Common Agricultural Policy which in 1977 gave
effectively unlimited subsidy, in proportion to production, for most crops.

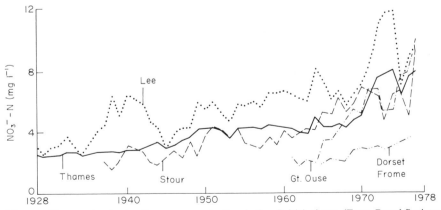

Fig. 3.4. Changes in nitrate-nitrogen concentration in five British rivers. (From Royal Society [696].)

Farmers now use a wide range of pesticides to control insect, fungal and other pests of crops. Most of these substances are not very soluble in water, being intended for uptake in the fatty membranes and tissues of their target organisms, but many are detectable in drainage waters at low concentrations. Pesticides are recognized as potentially very dangerous substances and most developed countries have regulation schemes which attempt to screen new products and make arrangements to minimize dispersal of the pesticide to the environment in general. Nonetheless the range of products available and the inevitability that rain falls on treated fields, from which water drains, means that pesticides will to some extent reach fresh waters.

3.6.3 Settlement

Expansion of villages, towns and cities has lead to increasing amounts of human sewage. Sewage is a mixture of wastes from laundry, bathing, cooking and the flushing of faeces and urine in lavatories. It is mostly water and to it may be added the rain water collecting on the streets and pouring down the drains. Sewage contains much labile organic matter and if released direct to streams will cause severe deoxygenation. In isolated houses or hamlets, a very simple system of allowing the sewage to decompose in an underground tank and the consequent effluent to seep away into the soil and ground water may be adequate. This is not practicable, however, where population densities are high. The populations of many countries are increasing and aggregating; the problem of disposal of sewage is thus also increasing.

Sewage treatment removes organic matter from the sewage but leaves an effluent, still containing some soluble and particulate organic matter, and greatly enriched in ammonium, nitrate and phosphate from the decomposition of the sewage. In western countries about half of the phosphate content of this effluent comes from domestic detergents in which sodium tripolyphosphate is used to remove calcium ions from solution and increase the efficiency of the surface-active cleaning agent, which would otherwise be precipitated by them.

The total amount of phosphorus released per person through the sewage works varies from country to country, depending on diet and devotion to laundry, but is in a range from about 0.5–1.5 kg year^{-1}. A typical effluent will contain up to 5 mg l^{-1} of ammonia-N, 40–50 mg l^{-1} of nitrate-N and 10–20 mg l^{-1} phosphate-P.

Sewage effluent is released, to a stream near the treatment works, usually at rates of only a few per cent of the total stream discharge. At such dilution the stream community should normally decompose any remaining organic matter without severe problems but the added phosphate in particular will greatly increase the stream phosphate concentration over that expected in water draining from natural vegetation or farmland. For example, an addition of 1% of effluent to a stream will increase the inorganic nitrogen concentration by about 0.5 mg l^{-1}—a significant increase over the concentration in natural drainage but not over that from farmland which may now bear up to 10, or in extreme cases more than 20 mg l^{-1}. On the other hand, the effluent will increase the phosphate-P concentration by 100–200 μg l^{-1}— double or treble that arising from farmland and vastly more than that usually coming from natural vegetation. Increases in phosphate concentration from sewage effluent are now very prominent in drainage waters and in many lowland areas the bulk of the phosphorus comes from this source, just as the bulk of the nitrogen comes from the use of the catchment for agriculture. Figure 3.5 shows the case of the River Bure, a small river in eastern England. Three points are prominent. First, almost every tributary of the river receives effluent from a sewage treatment works. Secondly, about three-quarters of the phosphorus came from sewage effluent and only a quarter from the land (of which more than half was in soil particles rather than solution). And thirdly, over 90% of the nitrogen came from the land, much of which was under arable agriculture.

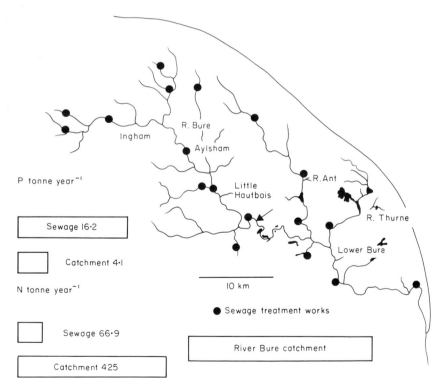

Fig. 3.5. The River Bure catchment. Norfolk with sewage treatment works discharging to the river shown and the total amounts of P and N arriving at Hoveton Bridge (arrowed). (Based on Moss *et al.* [567].)

3.6.4 **Industry**

A great many substances are released by industry to fresh waters. The nature and amount of these will vary with the particular local industries. It is estimated that some 30 000 compounds of commercial significance are manufactured within the catchments of the North American Great Lakes, and that 2–3000 new ones are added each year. About 500 have been determined to be potentially toxic in fresh waters.

A general list of discharged substances includes acids, alkalis, anions such as sulphide, cyanide and sulphite, detergents, labile organic waste (from food processing for example) chlorine, ammonia, heavy metals (particularly Pb, Zn, Cd, Cu, Hg, Ni, Cr), oil, phenols, formaldehyde, polychlorinated biphenyls and radionuclides. After the industrial revolution of the eighteenth century began in western Europe the most common discharge was of organic matter—largely untreated sewage as the town and city populations increased in response to the availability of work. Many rivers became completely deoxygenated for long stretches within and below towns as did the newly dug transport canals. Discharge of raw sewage brought with it problems also of disease, such as cholera, because drinking water supplies became contaminated, so that, from the middle nineteenth century, treatment works became common. However, the problems of organic pollution by sewage were not substantially solved until well into the twentieth century for some city rivers like the Thames and Rhine. Meanwhile, the variety of industries was increasing, and for ambitious industrialists the local stream was a cheap means of disposal. The rivers of the Greater Manchester area, for example, became, and remain, some of the worst polluted waters in the British Isles. I remember well, as a child in the 1950s, watching from a bridge over the River Mersey, near my home in Stockport, a thick, brown, astringent-smelling water, eddying, rather fascinatingly, around the bridge stanchions. It was taking the effluent of tanneries, food processing, a gas-works, electroplating, oil and grease processing, a slaughter-house, cotton factories, bleachworks, dyeworks, lead-acid battery manufacture, paint and rubber works and the manufacturers of paper and glue. As small boys we added our mite from the parapet with no guilt whatsoever.

Thirty-four per cent of the catchment of the Mersey is urban, with 828 000 people at a density of 1220 km^{-2}. It is still a severely polluted river [354], affected by heavy metals (zinc from paper treatment, chromium from tanning and lead from battery manufacture) and chlorine from the bleach works, among many other substances. In recent decades the policy has been to direct industrial discharges through a sewage treatment works where even highly toxic substances can be precipitated in a sludge and disposed on land. The discharge of these substances to a river should then be reduced. This assumes the works operates efficiently; some are old and do not. But this is not always possible because some wastes are so toxic they would upset the operation, which is essentially biological, of the sewage treatment works. In these cases in England and Wales, the Water Authorities issue 'consents to discharge' which limit the concentrations and amounts of particular substances that may be released. This may require a factory to treat its own wastes on the site but full treatment may be so costly that the factory might be put out of business if the standards were set too high. Dilution in the river thus remains a widespread solution on the one hand and problem on the other. In old industrial areas there are also many discharges which have gone undocumented and therefore uncontrolled.

Currently there is much concern about the discharge of heavy metals (e.g. Pb, Hg, Cd, Cu, Cr, Zn) and of chlorinated hydrocarbons (to some extent pesticides, and particularly a group of industrial substances, the polychlorinated biphenyls or PCBs). Both groups are very poisonous, though small quantities of heavy metals are weathered from rocks and enter waters and some, for example Cu, are essential trace nutrients for organisms.

There are some waters—hot springs in volcanic areas where water emerges at the surface after having reacted with minerals in the Earth's crust at high temperature and pressure—which may be extremely concentrated in metalloids like arsenic but which support a limited but tolerant community. In general, however, human activities now contribute most of the load of heavy metals to stream waters [433].

Like organochlorine and most other pesticides, the industrial polychlorinated biphenyls have no natural analogues. They comprise two linked benzene rings (biphenyl) in which chlorine replaces hydrogen in various combinations at up to ten sites on the molecule. About 200 such compounds are possible and mixtures of them have been sold under the trade name Arochlor. They have useful properties—high dielectric constants, stability, non-inflammability and cheapness and were used as insulating fluids in electrical transformers and capacitors, in cutting oils, for making brittle plastic pliable, and in paint, printing ink and carbonless copy paper. Their properties make them very difficult to decompose so they persist in the environment. Use of these substances has been restricted in the USA but a large reserve of them exists in waste dumps and derelict electrical machinery in the industrial world. PCBs are more soluble in fat than water, accumulate in organisms (see Chapter 4) and may sometimes be carcinogenic.

3.6.5 Industrial atmospheric sources

The acidification of rain and snow probably began with the industrial revolution when the burning of coal on a greatly increased scale released much sulphur dioxide to the atmosphere. This caused local problems of bronchitis among townspeople and environmental damage within radii of a few hundred kilometres. Major changes in the vegetation of the moorlands between Yorkshire and Lancashire in the UK may be due to local air pollution. Many interesting plant species have been lost and there has been increased erosion of the blanket bog peat which covers the flat moors. I remember a keen disappointment as a teenager at finding these moors so changed from descriptions made of them [572] earlier in the century.

The solution to the problem of high local sulphur dioxide concentrations was partly to encourage the burning of fuels of lower sulphur content by householders and partly to build high chimneys at power stations and other factories. The sulphur dioxide was thus released into higher and faster air streams and dispersed further away and more greatly diluted. It was not, of course, destroyed by this. The local air pollution problems of the 1950s in Britain had also been caused by smoke particles. Processes were introduced to remove these electrostatically before the effluent gas was released. The fly ash so removed, however, had contained calcium and other cations capable of neutralizing some of the acid produced when the sulphur dioxide reacted with water in the air. Effectively more acid was thus produced.

The measures taken to reduce national air pollution turned the problem into an international one, with movements of sulphur dioxide in the prevailing winds for distances of thousands of kilometres during which rain scavenged the acids. Meanwhile the local production of NO_x by vehicles had also been increasing as the number of cars increased. The combined effects of acidic rain and snow on the drainage waters were seen markedly from the 1960s onwards, particularly in areas of igneous and metamorphic rocks where there was little possibility of neutralization of the acid by carbonates and other weak acid salts. Such places have included the Laurentian Shield area of Canada, the New England States of North America, and hard-rock areas of Wales, Scotland, Norway and Sweden.

Effects are less clear in sedimentary rock areas where the buffering capacity of the soils is great. The drainage waters from the less well-buffered areas are not only decreased in pH but may also have high concentrations of aluminium ions which at concentrations as low as $100 \ \mu g \, l^{-1}$ may be toxic to fish. Other metal ions may also be mobilized by the acid but, in contrast to aluminium, the concentrations of those available for weathering in rocks is generally low (see Table 3.4). There is, however, now some suspicion that the bulk of the load of heavy metals reaching lakes such as the Laurentian Great Lakes may be derived from industrial smoke and may come via the atmosphere and rain [724]. The Great Lakes have proportionately small catchments and receive half of their water by direct rainfall.

4

Upland Streams and Rivers

4.1 INTRODUCTION

Most of the rain falling on a catchment does not immediately run off the land surface, but soaks into the soils giving a reservoir which supports the dry weather base-flow of the upland streams. To this small volume of water is then added a flood-flow, dependent on the amount of rain that has fallen, and the ability of the catchment soils to soak it up and store it. The chemistry of the base-flow may be different (usually it has greater concentrations of ions) from the flood-flow because the former has had much longer contact with the soil and rock. The flood-flow, on the other hand, may carry much more organic debris like surface leaf litter, and many more soil particles eroded from the land.

The flood-flow passes as a pulse downstream. It may move quite fast, carrying suspended matter from clays and fine organic debris to large boulders with it, the maximum size of grain or stone carried depending on the current speed. Only at infrequent times of very high flood do the largest stones move, eroding the valley sides and the stream channel. The deep, narrow valleys of many upland streams have been cut in this way. The relatively meagre base-flow moves fast enough only to carry very fine particles, so that in dry weather beds of sand and gravel are left among the larger rocks and boulders.

The 'ideal' stream lies in uniform geological terrain; it is most erosive in its head waters where the rocks and boulders left from a previous spate create an uneven or turbulent flow over a rough stream bottom. As the downward erosion of the stream bed continues, however, the slope of the bed is progressively reduced, and the stream's ability to erode its bed declines. The overall slope of the bed flattens until it reaches, over a sufficient distance, the horizontal when the stream (then usually called a river, but there is no absolute distinction) meets the sea.

In the ideal stream many changes steadily take place along this profile. The catchment area increases bringing a greater supply of water to the stream. The increasing area means that the effects of local storms are evened out so that spates characteristic of the headwaters are less frequent or eliminated. The water movement is less turbulent than on the steeper slopes upstream so that pockets of finer sediment can be deposited among the rocks and persist even during floods. Periods when current speeds are capable of moving larger stones and boulders become rare until eventually the bed becomes dominated by fine sediments leaving bare stones only in the centre of the channel.

The size and depth of the channel increase downstream as more water is discharged from an increasing catchment area. However, the channel does not usually increase in cross-sectional area at the same rate as the catchment area, for it is determined by the erosiveness of high floods, whose effects are muted downstream. The discharge (volume passing per unit time) and the cross-sectional area of the water passing in the channel are related by the average current speed:

$$\begin{array}{ccc} \text{Discharge} = & \text{cross sectional area} & \times \text{ speed} \\ (\text{m}^3\text{ s}^{-1}) & (\text{m}^2) & (\text{m s}^{-1}) \end{array}$$

Paradoxically, then, in the 'ideal' stream, the average current speed must increase downstream to accommodate the greater discharge. The less turbulent flow in the lower reaches, however, means that the range of velocities, over the cross-sectional area, can accommodate very low speeds in contact with the bed and very high speeds in the mid-water at the centre of the channel. The turbulent, foaming water in the headwaters may have higher speeds locally as it eddies in flood, but its average speed, largely a function of the base-flow, is lower than that downstream.

In the lowest reaches, the channel may increase its storage capacity by winding over a floodplain to accommodate the greater of its normal flows. The channel then moves from time to time as the river erodes soil from the outside of its bends (meanders), and deposits it on the inside. At very high flows the channel may be unable to accommodate all of the water and overtops its banks to cover its floodplain. The floodplain is not dry land damaged, in our perception, by an abnormal feature we call floods, but a natural part of the river bed which is used less frequently than the main channel to accommodate the highest river flows.

The 'ideal' river, then, falls in a smooth, graded profile from uplands to the sea but does not really exist. Changes in geology often prevent the establishment of a smoothly graded profile, because different rocks erode at different rates. Natural lake basins may also interrupt the sequence. Major uplifts of the Earth's crust may start lowland rivers cutting deeply into their beds as the land rises. And much management by man—the creation of dams, the alteration of the channel itself in the interests of flood control, has upset any smooth change downstream. The 'ideal' river, nonetheless, does have value as a simple framework on which to base studies of river ecology which fall into two chapters—one about erosive streams and small rivers (this one) the other about the larger rivers where sediment deposition is predominant (Chapter 5).

4.2 UPLAND STREAMS—THREE GENERAL QUESTIONS

Three groups of questions can be asked of any ecosystem. First, what organisms are found in it, and how are they adapted to the particular conditions found there—in this case the problems of coping with turbulent, erosive water? Secondly, by what pathways does energy flow in the ecosystem? And thirdly, what determines the distribution of different organisms in the ecosystem and the compositions of their communities?

4.2.1 The erosive stream habitat and survival in it

The stones and boulders, the pockets between them, and the interstices of gravel and sand provide a complex architecture for streams. It is not a stable architecture for high water flows may change it from time to time, though the boulders and larger rocks are likely to be moved very infrequently. Given this, and a very great talent in microorganisms for producing glues and gums often as the outer layers of their slimy cell walls, permanently wet rocks soon become covered by an organic layer of bacteria, fungi, protozoa and algae. Some organic matter may also be deposited as a film by purely chemical means. Some of these epilithic (Greek: *epi*, on, *lithos*, rock) organisms may be held flat to the rock, others may be attached by pads and protrude upwards (Fig. 4.1). Some may move through the mucilaginous matrix produced by the others. This film of microorganisms may be scoured away by suspended sand and silt during floods, but this does not often happen because of a peculiarity of the movement of water over surfaces. The surface creates a 'drag' or friction on the water movement which may slow its flow to near zero at and for a short distance above the surface. A smooth surface causes the formation of a 'boundary layer', a few millimetres thick, in which the flow is not turbulent and hence highly erosive, but smooth, or laminated, in lines parallel to the surface of the rock.

The composition of the organic layer depends on the chemistry of the stream water, and the amount of light reaching the stream. Overhanging vegetation may limit the growth of algae and favour predominance by bacteria dependent on dissolved organic matter washed in from the catchment. A more open stream may have fewer heterotrophic bacteria and more photosynthetic algae.

The surface films on stones are not the only microhabitats present. In crannies between the stones and rocks the flow may also be reduced and pockets of gravel, sand, leaf debris and twigs may collect. These support generally few higher plants or large animals, for these would not find a suitably permanent rooting medium or would protrude so far into the main current as to be torn free. Some small plants—mosses and liverworts (e.g. *Fontinalis*), a few genera of red algae (*Lemanea, Hildenbrandia, Bartrachospermum*) and in the subtropics and tropics one group of flowering plants, the

Fig. 4.1. Development of organic layers on stones in a New Zealand stream. (a)–(c) show development in darkness on a cleaned stone surface at the start (a) and after 1 month (b) and 3 months (c). Scattered slime can be seen in (b), with additional fungal hyphae and a diatom in (c). (d) shows development in the light after 2 months with the diatom *Cocconeis* (oval shape) and white structures, probably the cysts of protozoa. (From Rounick and Winterbourn [694].)

Podostemonaceae, may be found. These, like the microorganisms, form flat red or green crusts (*Hildenbrandia* and the Podostemonads) or small tufts, generally on the side of stones to the lee of the main current and sometimes enclosed within the boundary layer. Small invertebrate animals are generally abundant. They are attached to the rocks in the main flow (e.g. limpets) or exploit the gravel interstices, the underside of stones, or the leaves packed into crevices. In general the organisms of turbulent streams persist by avoiding the turbulence though they may be torn loose by it to form the 'drift'. The water itself carries no specifically-adapted, suspended organisms (plankton); there would be little time for their populations to grow before the water had moved far downstream.

4.2.2 **Adaptation to moving water**

The apparently obvious best way of an animal preventing itself from being washed downstream would be permanent attachment but most stream animals can move freely. Permanent attachment carries the risk of stranding during dry periods of low flow, and as a result permanently attached organisms, for example freshwater sponges, mostly survive on the undersides of submerged rocks. Motility gives flexibility, though with greater risk of displacement.

For some animals, such as the mayfly nymph *Rithrogena*, the risks have been reduced by evolution of a flattened body which does not project above the boundary layer on the tops of stones, or which, in the case of other mayfly nymphs like *Ecdyonurus*, allows the animal to crawl under stones. Streamlining, in which the greatest width of the body lies about 36% along its length, confers least resistance to current force and is found in *Baetis* spp. (mayfly nymphs). The long tails of mayfly and stonefly nymphs seem often to act as fins which turn the animal always head on into the current, just as a small boat fares best if headed into the waves of a rough sea.

Leeches have suckers with which to cling to the rocks, and the suction-pad feet of snails and freshwater limpets are similarly useful. Others, such as the water penny beetles, *Psephenus* spp., have friction pads of small movable spines which can be fitted into tiny irregularities of the surface of rocks. Hooks, grapples, and claw-like legs have all been evolved, and some caddis-fly larvae make cases with such heavy mineral particles that the case acts as ballast.

Some aquatic caterpillars (Lepidoptera) spin flat sheets of silk and attach them to rocks. Under the protective sheets they may form coccoons with low risk of displacement. Silk is also used by *Simulium* (blackfly) larvae. *Simulium* spins a pad of silk which adheres to the rock and to which it attaches with hooks at its hind end. If it is dislodged, it has a 'safety line' attached to the body and the pads on the rocks. By working it between the front proleg and rough spines on the head, the animal can 'climb' it to regain the security of its silk pad on the rock.

Despite these adaptations, displacement and drift downstream is common. Drift is complex [195, 386, 573, 814]. Not all species, nor even different size classes of the same species, drift to the same degree and the process is partly determined by external factors. Drift of some species is markedly seasonal, peaks of travel coinciding with high current flow, whereas in other cases it may coincide with low current speeds. For some species (e.g. *Gammarus, Baetis*) drift increases just after sunset, sometimes with peaks late in the night, while in others (e.g. Chironomids) there seems to be no distinct diurnal pattern.

Active movements upsteam along the bottom to some extent counteract the effects of drift, although only to the extent of perhaps a few per cent replacement when drift rates are high. Active downstream movement, however, has also been found for some insect larvae just prior to pupation and emergence of the adults. The adults of some aquatic insects (e.g. some mayflies and stoneflies) tend to fly upstream before depositing their eggs and this too must partly counteract the effects of drift.

Its complexity of patterns suggests that drift might not merely be a passive consequence of stream living but may have some adaptive advantages. It results in rapid re-colonization of newly wetted channels opened in a stream after drought and stretches denuded by violent spates. It is also relatively high when food supplies are scarce. In an experimental stream, Hildebrand [348] showed that drift of animals feeding on algae attached to stones was high when the algae were scarce; this allowed dispersal to sites possibly richer in food. Drift rates of a net-spinning (see below) caddis-fly (*Plectrocnemia conspersa*) and a leafpack-inhabiting stonefly (*Nemurella picteti*) were exceptionally high (20% and 43%) in a southern English stream when densities of the former were so high (100 m^{-2}) that net-spinning sites were very scarce, and when, in summer, leaf packs for the latter were few. In contrast, drift rates were low in the same period for another stonefly nymph, *Leuctra nigra*, where its food supply, iron bacteria, was very abundant [814].

4.3 SOURCES OF FOOD AND ENERGY FLOW IN UPLAND STREAMS

Many upland streams receive most of their energy from organic matter washed into the stream, largely as leaf litter, from the catchment [826]. The litter is then progressively processed to carbon dioxide by a succession of microorganisms and animals. They deal with successively smaller particles of it in a continuous sequence, like a factory production line, as the organic matter is moved downstream. This happens for areas where deciduous forest covers the catchment, for example in parts of North America [136], but not all streams depend so heavily on the processing of catchment-derived litter. In upland Britain it is more likely that the stream will be surrounded by grassland producing much less litter, for the land is grazed. Such a stream would not be overhung by trees, more light would reach its surface and a greater dependence on photosynthesis by epilithic algae might be expected. Here, several groups of streams will be compared.

4.3.1 Hot spring streams

In volcanic areas, ground water is superheated and forced through cracks back to the surface where it may emerge as steam, or water at boiling point or some lower temperature. Its contact with the underlying lava may have changed its composition so that as well as being hot it may be charged with hydrogen sulphide, sulphuric acid or high concentrations of silicate. Even at boiling point living organisms are present though they are absent in the superheated stream vents. This seems to be due only to a lack of liquid water, for bacteria have been isolated from such vents deep on the ocean floor [35], where pressure allows liquid water to exist at several hundred °C.

As it emerges, the water may have some dissolved organic matter, ultimately derived from the land surface, and this may support mats or tufts of filamentous bacteria at the edge of the spring boil. The water has cooled a little there, but sterilized cotton or microscrope slides placed at the boiling centre will also grow bacteria. If hydrogen sulphide is present, chemosynthetic bacteria, such as *Thiobacillus thiooxidans* and *Sulfolobus acidocaldarius* may grow attached to the rock at 85–90°C. They use the H_2S to reduce CO_2 to produce organic compounds. As the water flows from the boil to form a stream, it acquires an often V-shaped set of coloured patterns on the channel floor. These are of organisms successively able to colonize the cooling but still hot water [71, 72]. The water cools faster at the shallow edge than the deeper middle, and the V pattern, pointing downstream, represents earlier colonization at the edge than the centre.

The first obviously coloured V is usually of *Synechococcus* sp., a photosynthetic blue-green alga, capable of growing at up to 75°C. It is not merely surviving high temperatures, above 70°C, but if cultured in the laboratory grows best in a small range around 72°C. The mats of this alga grow actively, continually replacing cells which are dislodged by the current and washed downstream. The mat of *Synechococcus* often rests on an underlying one, sometimes several mm thick of filamentous photosynthetic or heterotrophic bacteria. One of these, *Chloroflexus*, is capable both of photosynthesis, based on its orange pigments and of feeding on organic matter produced by the *Synechococcus*. Further downstream more complex blue-green algae occur including *Mastigocladus laminosus* which is filamentous and has two sorts of cells, one of them, the heterocysts, probably capable of fixing nitrogen.

All of the organisms so far mentioned have been prokaryotes, with their relatively simple cell organization. Eukaryotes, with the insides of their cells divided by membranes, cannot colonize very hot water. This may be because the sorts of membranes which can allow movement through them of large molecules—a necessity for proper functioning within the cell—are too 'holey' to survive much heating without breakage. The first eukaryotes to colonize are fungi, mingled in the bacterial mats, at about 62°C; eukaryotic algae follow at about 60°C and protozoa (*Cercosulcifer* and *Vahlkampfia*), at 57–60°C.

Thus far only microorganisms are present, and one of the reasons for the prolific algal mats may be a lack of grazers, for no multicellular animals (or plants) can tolerate more than about 50°C. Vascular plants must wait until about 45°C and vertebrates 38°C. Only at 45–50°C are ostracods and the larvae of certain flies present and capable of chewing holes in the mats. The hottest springs therefore form very simple streams in their upper reaches, dominated by primary producers and heterotrophic microorganisms. Much of the energy fixed is washed downstream and not consumed *in situ*.

Stockner [776] studied a cooler hot spring (37°C), one of the Ohanapecosh hot springs near Mt. Rainier in Washington, USA. He fitted a wooden trough, 12 cm wide, 3 m long, to constrain the water so that precise measurements could be made. The trough soon became colonized by communities previously present in the more irregular natural channel. Ohanapecosh springs are alkaline, and supersaturated with carbonate as they emerge, so that on contact with air and through the action of algae in removing CO_2 (see Chapter 6) a deposit of calcium carbonate, in this case called travertine, is formed and has built up to depths of several metres in the area.

The stream was colonized by two filamentous blue-green algae, *Schizothrix calcicola* and *Phormidium* sp. and among mats of these the larvae of two flies, *Hedriodiscus trusquii* and *Caloparyphus* sp. were able to graze. Stockner measured the primary production of the stream by changes in its oxygen concentration as water flowed over the algal mats. He placed oxygen-detecting electrodes at each end of the channel and continually recorded the concentrations. From a baseline of the dawn concentration he could then calculate the total net increase in oxygen during the day, after a correction for that which would have diffused into the atmosphere. By repeating the measurements at night he could similarly calculate the oxygen used up in respiration. The sum of the net uptake by day and the net loss at night gave the gross photosynthesis. This is because the measured production by day is less than the total oxygen production because some of the oxygen is simultaneously respired. To calculate this respiratory 'loss', it was assumed that the respiration rate of the community was similar by day and by night. Extrapolation of the night respiration rate to the full 24 hours gave the total respiration of the community.

Some of the produced material must have been respired by the algae themselves, associated bacteria and the flies. These amounts cannot be separated, though the bulk of the biomass was algae and hence probably also the bulk of the respiration. Other of the produced material became dislodged by the current or the animals feeding. This was measured by sampling the water at the upper and lower ends of the stream, filtering it, and determining by weighing the material added over the stream stretch. Some material was grazed; the amount was estimated by finding the number of fly larvae present and measuring in the laboratory the average amount they ate each day. Finally, the formation of travertine incorporates organic matter as well, so some of the production is deposited in the stream bed. This was determined from the difference between the gross production and the sum of its various fates described above. Table 4.1 gives the results of the study made at

Table 4.1. Energy budget for the stream flowing from one of the Ohanapecosh hot springs in 1966. Values have been converted to common energy units of kilocalories m^{-2} $year^{-1}$. (After Stockner [776])

	kcal	Percentage of total
Input		
Gross primary production	4607	100
Output		
Community respiration	1158	25.1
Washed downstream	1461	31.7
Grazed	20	0.4
Deposited with travertine in stream bed	1968	42.8
Total	4607	100

Ohanapecosh hot spring. Two features of particular note at this stage are the dependence on *in situ* photosynthesis (external (called allochthonous) sources of organic matter were negligible) and that little of the material (0.4%) was consumed by grazers. The bulk was apparently deposited or washed downstream.

4.3.2 Bear Brook and other streams in wooded catchments

It is not an easy matter to account for all the sources of organic matter to a section of a stream, and to balance these inputs with equally good measures of what happens to them. Work of this detail has been carried out on a 1700 m section of Bear Brook, New Hampshire [218]. This stream section was ideal for the work since it lies on hard bedrock which prevents deep seepage and directs all the rain and snow which is not evaporated or transpired by the vegetation, to the stream. This allows full accounting of water movement. It is also a small stream uniformly surrounded by hardwood forest with a total area of about 0.6 ha. A maximum depth of about 60 cm eases sampling problems.

The forest edge overhangs the stream and direct fall of litter was measured by collection in suitable boxes placed along the stream bank. Litter includes not only leaves, but branches, bud scales, flowers, fruits and the exuviae and droppings (frass) of leaf-living animals. Although most litter fall is in autumn it is not negligible at other times of the year, for there is a continual turnover of leaves even in summer; bud scales fall in spring and frass is a constant source in summer. Wind and the weight of snow break off tree branches in winter. Not only is there direct litter fall, there is also a contribution blown sideways into the stream along the forest floor, and this was collected in traps placed at right angles to the stream. Rain dripping directly into the stream from the overhanging leaf canopy in summer picks up dissolved organic matter exuded from the leaves and from leaf insects. This throughfall was collected and measured by chemical means.

At the top of the steam section, the rate of entry of organic matter carried from upstream was measured in three categories—coarse particulate organic matter (CPOM), greater than 1 mm in size; fine particulate organic matter (FPOM), less than 1 mm in size but collectable by filtration through a glass fibre filter of pore size 1 μm, and dissolved organic matter (DOM) which passed through such a filter. DOM presents few problems if sampling is fairly frequent as its concentration is relatively steady and its total contribution can be calculated easily if the discharge of the stream is known.

Fine particulate organic matter, and to a much greater extent CPOM, present problems of measurement since they tend to come down the stream in pulses related to spates of water passing after thaws and rain storms. In Bear Brook, very frequent sampling was necessary to obtain a reliable measure of FPOM. Even this failed to give a reliable measure of CPOM when it comprised merely the spreading of a 1 mm mesh net across the stream for a short period each week or two. Fortunately, in a nearby similar stream, as part of another experiment, CPOM was continuously collected in a concrete ponding basin built into the stream. This collected all CPOM passing and gave a value some 20 times higher than that obtained by regular discrete sampling.

Dissolved organic matter entered from soil seepage along the length of Bear Brook. Samples of this water were collected from seeps for analysis and its volume determined from the difference between the total amount of water entering the stream at the top of the stretch and that leaving it at the bottom. Finally, estimates were made of the photosynthesis of the moss population of the stream (algae and higher plants were scarce) by a method based on the rate of production of oxygen by moss enclosed in glass bottles (see Chapter 5). The conversions of energy through animal consumption were estimated from biomass measurements and productivity data from other sites. Though these estimates were very crude (the methodology is discussed later) the picture of energy flow in the stream would be little altered if they were as much as ten times in error.

Table 4.2 shows the energy budget which has been constructed for the stretch of Bear Brook investigated. Most notable are the high contributions of litter (43.7%), particularly direct leaf fall, and of dissolved organic matter (46.3%). Autochthonous primary production was negligible in this stream.

The outputs of energy also show some startling features. First, if all the measured outputs (CPOM, FPOM, DOM, and ultilization by animals) are added, a total of 4013 kcal m^{-2} year^{-1} was accounted for, whereas 6039 kcal m^{-2} year^{-1} entered. The difference, 2025 kcal m^{-2} year^{-1}, must be attributed to the respiration of microorganisms feeding heterotrophically on the organic matter entering the stream, and thus processing about a third of it. The remaining two-thirds were mostly exported downstream. Such

Chapter 4

Table 4.2. Annual energy budget for Bear Brook, New Hampshire

Inputs (kcal m^{-2} year^{-1} (%))		Outputs (kcal m^{-2} year^{-1} (%))	
Direct litter fall:		Transport downstream:	
Leaves	1370 (22.7)	CPOM	930 (15)
Branches	520 (8.6)	FPOM	274 (5)
Miscellaneous	370 (6.1)	DOM	2800 (46)
Side blow litter	380 (6.3)	Respiration of micro-	2026 (34)
Throughfall (organic excretion, frass)	31 (0.5)	organisms*	
		Respiration of invertebrates	9 (0.2)
Transport from upstream		Total	6039 (100.2)
CPOM	430 (7.1)		
FPOM	128 (2.1)		
DOM	1300 (21.5)		
Ground water			
DOM	1500 (24.8)		
Moss photosynthesis	10 (0.2)		
Total	6039 (99.9)		

*Obtained by difference.

streams as this would appear therefore to be relatively efficient processing factories for the large amounts of organic matter they receive, and the next section considers the mechanics of these processes.

4.3.3 Mechanics of processing of organic matter in woodland streams

Leaf litter is very different, chemically, from the living vegetation it once was. Labile, reusable substances, both inorganic and organic, have been translocated back into the perennating organs and what is left is largely cellulose, lignin and resistant carbohydrates, plus substances like polyphenols which may be metabolic waste products. Microorganisms have already colonized the litter before it has fallen, and begun the decomposition which, for most leaves, will be completed in the soil of the land ecosystem.

Once the litter has entered most of any remaining soluble matter is leached out within a few days, or even hours [387a]. The speed of this depends on the leaf species, but deciduous leaves are leached more rapidly than those of conifers. This DOM, together with that washed out of the catchment soils, may be taken up by microorganisms associated with the stream bed, or may be washed downstream and used there. Some of it may be precipitated as part of the organic film on stones or aggregated into fine particles by apparently physico-chemical processes, and may thus join the FPOM fraction.

As soon as it enters the water there is a chance that the litter will be colonized by aquatic microorganisms, and over a few weeks there is an intricate succession of fungi present in and on the litter. Bacteria are present but do not appear to be important at this stage. The fungi concerned are largely from a group called the aquatic hyphomycetes. Their spores are often tetraradiate in shape (Fig. 4.2) which seems to favour their sticking to the litter like small grapnels when swept against it by the stream flow [394].

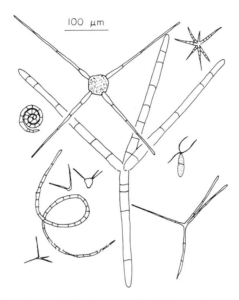

Fig. 4.2. Spores, mostly of aquatic Hyphomycete fungi, drawn from a sample taken below Sezibwa Falls, near Kampala, Uganda. (Modified from Ingold [394].)

Leaf litter provides only part of the growth requirements of the fungi which colonize it. Although it is rich in carbon and chemical energy, it contains little nitrogen and phosphorus and the fungi must obtain much of their supply, particularly of nitrogen, from the water. The nitrogen to carbon ratio of the colonized litter increases as the fungi build up their biomass [287, 424, 425]. This is important for the next stage in decomposition by invertebrate animals, for the fungal biomass, rich in nitrogen, is a 'better' food for invertebrate animals than uncolonized litter. The speed with which different sorts of litter are consumed largely depends on their food contents as expressed by the preferences of animals feeding on them.

4.3.4 The shredders

Mechanical abrasion breaks down some of the colonized (and uncolonized) leaf litter to FPOM, but much of it is chewed by coarse particle-feeding invertebrates, the 'shredders', which bite out the softer parts, between leaf veins for example, leaving the vascular skeleton for later abrasion or consumption [132, 134]. Shredders include insect larvae and nymphs and Crustacea. Much work has been carried out on the crustacean *Gammarus* for it is easily maintained in the laboratory.

When *Gammarus* were offered fungally colonized leaves of elm (*Ulmus americana*), alder (*Alnus rugosa*), white oak (*Quercus alba*), beech (*Fagus grandifolia*) and sugar maple (*Acer saccharum*), they showed a preference roughly in the same order as that in which the leaves support fungi—elm, maple, alder/oak, beech. This same order of preference appears to shared by stonefly and mayfly nymphs [425]. The intrinsic properties of the leaves themselves help determine these preferences, for the order is also maintained if uncolonized litter is offered. Preference is always for colonized over uncolonized leaves, however, and can be influenced by artificial inoculation of particular species of fungi [32, 33].

The role of fungi in the processing of leaf litter appears similar to that of the sandwich fillings and butter commonly used by us to increase the palatability and nutritional content of modern bread. There are usually several species of shredder present in a stream, and preference for different sorts of 'sandwiches' may explain why they all can coexist.

Microorganisms can degrade leaf litter alone, but the process is accelerated by a fifth or more through the action of shredder invertebrates [132], such that as much as 1.5% day^{-1} of the litter is converted to animal tissue, carbon dioxide or FPOM. Fine particulate organic matter results from abrasion by water movement and from waste during feeding. It also includes the egested faeces of the shredders. Fine particulate organic matter is additionally colonized by microorganisms as 'new' surfaces are exposed and forms a food source for a second series of invertebrates, the 'collectors', which collect the FPOM by filtration or deposit feeding. Sometimes half of the FPOM they ingest consists of the faeces of shredders.

4.3.5 Collectors, scrapers and carnivores

The finer debris that results from shredder and fungal activity, the fine material washed in directly from the catchment area, and particles eroded from the algal communities attached to rocks in the stream-bed, form a rich food source for those invertebrates that can efficiently use it. Most of these organisms are filter or deposit feeders though some scrape the particles from surfaces, together with the attached algae growing there, where the current is slow enough to allow some deposition. Typical scrapers include snails and freshwater limpets which rasp at the rock surfaces with toothed organs called radulas, and some caddis-fly larvae and a few mayfly nymphs whose mouth-parts have evolved stiff bristles with which they scour the rocks. Deposit-feeders, which include burrowing dipteran fly larvae (particularly chironomids) and some mayfly nymphs inhabit pockets of sediment in areas of slack flow.

Filter-feeders show remarkable ingenuity at gathering organic matter from flowing water. Sometimes they have fringes of fine hairs on the mouth-parts (blackfly larvae) or legs (some mayfly nymphs) in which particles collect before transfer to the mouth. Others (including some caddis-fly larvae) construct nets between stones and the bottom. A group of freshwater prawns in Dominican streams [242, 243] have long bristles on a pair of limbs, the chelipeds, near the front of the animal. These bristles are delicate but can be held together like the tip of an artist's wet paintbrush. They can then be used to sweep organic particles from deposits or surfaces, and ancillary appendages transfer these to the mouth. Alternatively, the bristles can be expanded into a fan which is held into the current and which, with the help of very fine setules on the bristles, acts as a filter. Apart from their intrinsic interest, these prawns provide a reminder that classification of stream (or any other) animals into feeding modes is but a convenience and that the categories are not always distinct. For example, shredders like *Gammarus* thrive better in laboratory experiments if allowed access to the FPOM of their own faeces as well as to fungally-colonized leaf material, than if given the latter alone.

The food web of upland streams (Fig. 4.3) is completed by carnivores, both invertebrate and vertebrate. The invertebrates include leeches, a variety of insect larvae and water mites (Hydracarina) among others, and just as shredders and collectors may ingest animal material with their predominantly detrital and fungal diet, the carnivores may also have a wider range of food than their name suggests. All the invertebrates are potential prey for fish in the stream.

4.3.6 New Zealand streams

The idea of streams as well-organized factories for the orderly processing of allochthonous organic matter is an attractive one. Based on their experience with mostly wooded North American streams, Vannote *et al.* have created the 'River continuum concept' (1980) to link and summarize information about this idea. They see an orderly sequence in which CPOM is first processed by shredders, which provide food for collectors downstream; the life histories of different animals in the two groups are seen to be synchronized to take advantage of the annual pulses of incoming organic material—the collectors timed slightly later than the shredders. In the upstream sections the ratio of *in situ* photosynthesis (P) to total community respiration (R) is found to be less than one, indicating a dependence on the allochthonous matter to feed the community. Further downstream, when the stream widens and is not so darkened by overhanging forest, so that the stone surfaces may be colonized by algae and mosses, the $P:R$ ratio may be greater than one. At the later, plains stage, when the river becomes large and is more likely to

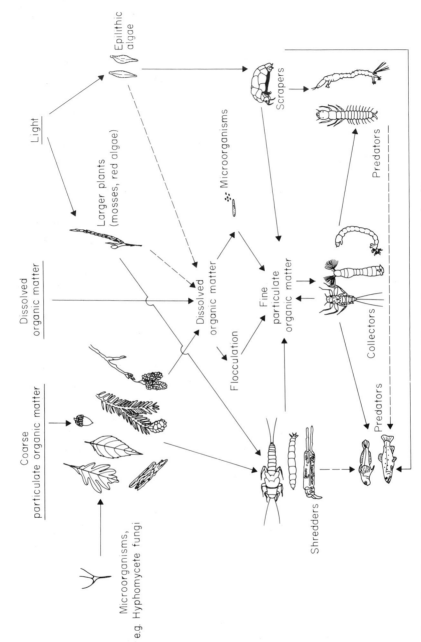

Fig. 4.3. Some food relationships in a turbulent stream. (Modified from Mahan and Cummins [513].)

collect inorganic silt, which decreases light penetration, the $P:R$ ratio may fall again, but this is a matter for Chapter 5. The river continuum concept has attracted much support.

The general idea that most stream systems depend, in one way or another, on allochthonous organic matter derived from the catchment probably is true, but whether the systems are quite so well ordered as the river continuum concept would suggest ought to be questioned. We are all prisoners of our own experience and the drawings on the cell wall, no matter how good a map of the prison, may tell little of the rest of the world.

Streams in New Zealand, for example, seem not to fit in with the continuum concept. Winterbourne *et al.* [868] in a challenge to the paper by Vannote *et al.* [826] point out that New Zealand has a low timber-line, with many streams having great lengths above it, that the slopes are steep, the rainfall heavy and irregular and that the stream bottoms hence do not retain coarse litter very readily. The forests of southern beech (*Nothofagus*) and podocarp-hardwoods are evergreen, and the litter they produce is not concentrated into the autumn. It is produced continually in small amounts and is very woody and difficult to decompose. The streams are thus deprived first of a major source of CPOM and secondly of the regular changes which would favour an orderly processing of it. They are short of shredders too so that the collectors must feed not on the products of shredder activity (which, in the river continuum concept, include quantities of shredder faeces) but on material eroded from the soils. In general this is inorganic rather than organic, so Winterbourne *et al.* emphasize the organic layer on the stones as a major food source [867], and point out that many New Zealand stream invertebrates are opportunists, taking different sorts of food as it becomes available. Their life histories are not synchronized either, so that there is a contrast with the North American streams. The latter seem to have a highly predictable regime which supports a complex, interdependent community; the New Zealand stream communities seem to have developed in a less predictable environment. Winterbourne *et al.* [868] do not deny the importance of allochthonous matter—the stone surface layers on which the New Zealand stream animals depend are derived partly from absorption of dissolved organic matter produced in the catchment soils, but they (and others [443]) do question any global generality of the North American scheme.

4.3.7 Fish in upland streams

To many people, fish are the most significant features of upland streams. Large amounts of money are spent in the maintenance and renting of upstream fisheries, largely for salmonid fish, like the Atlantic salmon (*Salmo salar*) and brown trout (*Salmo trutta*) in Britain, and the coho salmon (*Oncorhynchus kisutch*), chinook salmon (*O. tshawytscha*), rainbow trout

(*Salmo gairdneri*) and brook trout (*Salvelinus fontinalis*) among others in the USA. Some of these, particularly the *Salmo* species, are migrating fish which make most of their growth in the sea but move into upland waters to breed. They are narrow-bodied, streamlined fishes, able to swim sometimes for thousands of kilometres up-river, and, in their juvenile stages in the rivers capable of short bursts of very high speed to catch drifting prey or flies dipping at the surface to lay eggs.

The requirements of these fishes (Table 4.3) are for highly oxygenated, relatively cool water, especially for breeding, fast flow, which maintains the bottom clear of silt, and often for rising waters in autumn which stimulate the upstream movement of the adult fish. The eggs are usually laid in depressions excavated in the gravel on the stream bottom and then covered to prevent their being washed away or eaten. Sites are chosen where the stream flow infiltrates the gravel bringing well-oxygenated water continually to them. Silting is fatal for it blocks this flow and the relatively large eggs asphyxiate. Any major reduction in flow which leads to the gravel beds being uncovered and drying out is also deleterious.

Table 4.3. General habitat requirements of some salmonid fish species. (Based on Templeton [799])

Fish	Optimal temp. for growth (°C)	Max. temp. usually tolerated (°C)	Spawning temp. (°C)	Oxygen concentration, min. required (mg l^{-1})	pH range
Atlantic salmon (adults)	13–15	16–17	0–8	7.5	5–9
Brown trout	12	19	2–10	>5	5–9
Brook trout	12–14	19	2–10	4–4.5	4.5–9.5
Rainbow trout	14	20–21	4–10	4–4.5	5–9

Salmonid fish receive the most attention from fishery ecologists studying streams, and there is a very large literature upon them. They are not, however, the only, nor even the most numerous fish in the world's upland streams. A second group, of fish from many families, comprises small fishes living on the bottom and well fitted to do so. These include many cyprinid fish, catfish, sculpins, and gobies. In Europe the bullhead or miller's thumb (*Cottus gobio*), and loaches (*Nemacheilus* and *Cobitis* spp.) are found. In North America a very large number of species of the perch family, of several genera (e.g. *Percina, Ammocrypta* and *Etheostoma*) and often called darters are typical. In hill streams in the Tropics there is a very diverse collection [28, 386]. Bottom-living fish contrast with the salmonids with their generally rounded body, a flattened underside and an arched back in cross section. Their mouths are often turned downwards which eases bottom-feeding, and their pectoral fins may be muscular and spiny and used to wedge them across crevices in the stream bottom. Sometimes the fins move water from underneath the fish so as to keep it on the bottom or act as foils against which the current presses, to the same end. The fins may also be modified as suckers or friction pads. The swim-bladders, buoyancy devices in mid-water fish, are much reduced, and the skin is dark and mottled, perhaps to provide camouflage against larger predatory game fish. These small fishes also depend on swift currents for the successful production of young, for they lay eggs in small piles of gravel or stick them in groups under flat stones.

4.3.8 The Atlantic salmon

The Atlantic salmon is a well known species of great importance. As a representative of upstream fish, some details of its life history will illuminate the problems of its fishery, to be discussed later. Atlantic salmon spawn in rivers in the USA, Canada, Iceland, Norway, Ireland, Great Britain and France and later mingle in the North Atlantic off Greenland. The eggs hatch in April–May after 70–200 days in their gravel nest, or redd. On hatching, the larvae feed from their yolk sacs for about 6 weeks, then start to eat small invertebrates. They are then called parr and have a distinctive alternation of blue-grey 'finger marks' and red spots on their sides. The parr grow slowly and become silvery in colour, but after 1–5 years they move down to the river mouth and, now called smolts, feed on small fish and crustaceans. Eventually they move into the ocean and grow rapidly on sand-eels, herring, sprat and other fish and crustaceans. The Atlantic salmon spends 1–4 years in the sea then migrates back, often to the river in which it was spawned. Subtle chemical differences in organic compounds released by related fish to the river water may allow it to recognize this river system, but probably some mixing of fish from different systems occurs.

Most adult salmon move into the rivers in late winter. They are fat but do not feed, though they retain for a time a reflex to bite at suddenly appearing prey—this is the basis for the fly fishing on them. Movement up-river may mean their jumping waterfalls up to 3 metres high and ascending longer rapids in stages. During the journey the fat is used up, the sexual organs ripen and secondary sexual characters appear; the male develops a hook or kelp on the lower jaw.

Mating pairs choose suitable patches of gravel in 0.5–3 m of water and the female excavates the redd by flapping her tail vigorously to form a hollow 10–30 cm deep and up to 30 cm long. The animals lie side by side, with much violent trembling and jaw-gaping; the eggs are shed and fertilized, then covered with gravel as the female excavates another redd upstream. The eggs are slightly sticky and denser than water. Spawning and migration make great energy demands on the salmon; the eggs may account for 25% of the body weight. After spawning most of the fish are very weak, only half their original weight, and prone to fungal infection or standing in shallow water. Most soon die as they drift downstream, though a few (about 5%) reach the sea where they quickly recover and, called kelts, may return to spawn a second time in the next year.

Some information has now been given on the energy input into the stream ecosystem and on the animals consuming this energy. The production of those animals will next be considered.

4.3.9 Animal production in streams

Freshwater ecologists have spent much time measuring production. Because such measurements, done well, are very time-consuming, it is worth asking why they should be made. One reason is simply that of better understanding the ecosystem—the animal production is part of the processing of organic matter; a second is that of managing fisheries. Production measurement essentially involves measurement of the number of animals per unit area at successive times (perhaps the most difficult stage to do well) and the amount of growth the survivors have made in successive intervals of time.

Three general sampling methods are available for the invertebrates—nets, lift samplers, and placement of trays of cleaned substrata. Most interest has centred on the macro-invertebrates, those more than about 3 mm in size when adult, and the net mesh most commonly used, about 1 mm, reflects this. A net, mounted on a pole, is held in front of the feet of the observer who stands with his back to the current and rubs his feet vigorously on the bottom. The method is called kick sampling and animals so dislodged are swept into the net. A known area may be sampled, or, for relative measurements the sampling goes on for a set time as the observer works his way steadily upstream. The Surber sampler (Fig. 4.4) is a slightly more sophisticated version in which a frame is attached to the front of the net. Animals are dislodged by hand from the area of the frame and caught by the net as they are swept downstream by the current.

On sandy and gravelly bottoms one problem is that animals may penetrate to much greater depths (20–30 cm) than are sampled by nets [116]. Usually such deep-living animals are very small ones—microcrustacea and nematodes for example—and this has contributed to relative neglect of them. They can be sampled by first inserting a cylinder into the stream bottom to delineate the sampling area. Compressed air is then forced through a tube to dislodge a mixture of water, air, sediment and animals up through another tube and then through a series of filters (Fig. 4.4). Some animals may be damaged by this.

Artificial substrata, usually comprising cleaned rocks fixed to a tray, may be left on the stream bed for several weeks for them to become colonized. A large number of such trays, progressively removed at intervals for examination, allows a picture of seasonal fluctuations to be built up. It is

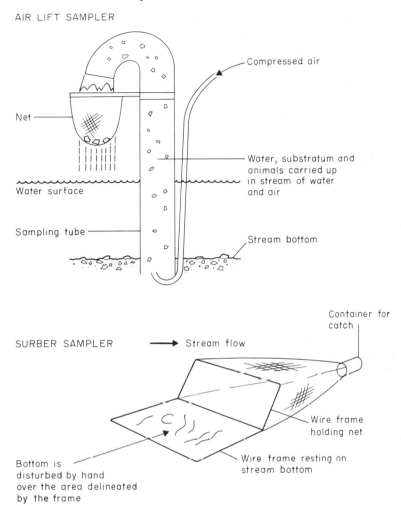

AIR LIFT SAMPLER

Compressed air

Net

Water, substratum and
animals carried up
in stream of water
and air

Water surface

Sampling tube

Stream bottom

Container for
catch

SURBER SAMPLER   ⟶   Stream flow

Wire frame
holding net

Bottom is
disturbed by hand
over the area delineated
by the frame

Wire frame resting on
stream bottom

Fig. 4.4. Principle of the operation of an air-lift sampler and of a Surber sampler for use in sampling the bottoms of gravelly streams.

important, of course, that the artificial substrata used closely reflect the natural substrata of the stream stretch under investigation, but for purely comparative purposes, the samplers may comprise simply piles of flat plates, spaced by washers and bolted together.

A relatively large number of separate samples must be randomly taken for numbers of an animal to be determined with reasonable precision on any occasion. The number depends on the population density of the animal. The complexities of the physical environment, of interactions between predator and prey, and the behaviour of individual species, may lead to markedly non-random distributions. Larvae of the caddis-fly *Potamophylax latipennis* (an algal scraper) were found, for example, only on the undersides of stones of diameter 11–22 cm in a Scottish river, and even then were grouped in particular parts of such stones [86].

Fish are difficult to sample. For stony rivers the best method uses electrofishing. In this, an electric field of 200–300 volts is created in the water between two electrodes, powered by a portable generator. With alternating current the fish are temporarily stunned and removed from the water by a net for measurement and recovery before being replaced. Many fish may, however, be overlooked. With direct current the fish swim towards the positive electrode and must be removed by net before they touch it. Streams are convenient in that sections can be isolated by nets placed across the width of the stream whilst fishing goes on. It is possible then to remove most of the fish population from the section, but smaller species and smaller individuals of larger species respond less to the electric field than larger ones. The method is also potentially dangerous to the operators, who can be stunned or killed if they are not fully insulated.

If there was no predation or drift, or death from miscellaneous causes, production could be measured simply by the change in weight of a population over a period. A very large proportion of the production may be eaten, however, and a more sophisticated approach is needed. The basic data are numbers of an animal per unit area at a given time and the mean individual weight of the animals. It is tedious to weigh all the animals from a large enough sample to be statistically acceptable, so a previously determined relationship between length of animal, or width of head, and weight is often used. The simplest case is for populations which develop from eggs all laid at the same time. These are known as cohorts.

If the cohort begins with N individuals of mean weight w_1 on hatching, the initial biomass will be $N w_1$. At a future time, t, some animals will not have survived, but the remainder N_2 will each be heavier with a mean weight w_2. The total production will thus be that of survivors, $N_2(w_2 - w_1)$, plus that of those which died:

$$[(N_1 - N_2)(w_2 - w_1)]/2$$

The assumption has to be made that all those that died were lost half-way through the period, and that weight per individual increased linearly during the period. Sampling at sufficiently frequent intervals reduces the error from this assumption. Expressed graphically, as it was first used (for fish) by Allen [8], the area under the Allen curve gives total production (Fig. 4.5). Number of organisms is plotted on the y axis and mean weight per individual on the x axis. The sum of production of each cohort present plotted separately in this way gives total production.

This method works adequately (given well designed sampling) for organisms with well understood life histories and synchronous reproduction, or for organisms which can be easily aged, for example by markings on the

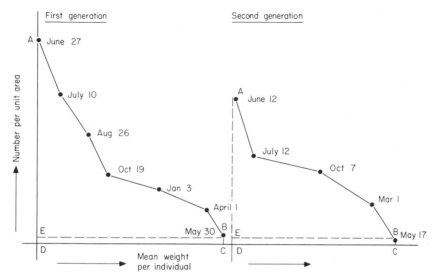

Fig. 4.5. Allen curve method for determining the production of an animal population whose members can be recognized as distinct cohorts, each starting life effectively simultaneously. Numbers per unit area are counted on each of a series of sampling dates and plotted against mean weight of animals on these dates. Area ABCD represents total production; ABE the production of animals which were eaten or died before reproduction; BCDE the production of adults surviving to breed, and ABCD:BCDE the turnover ratio.

shell in bivalve molluscs or scales or the annual rings on the ear bones (otoliths) in the case of fish. For animals which reproduce continuously, and which cannot be readily aged, this method cannot be used. For such creatures, determination of growth rates by serial weighing or measurement of captured individuals in laboratory conditions resembling as closely as possible those of the habitat may be used. Production is then given by multiplying the field biomass by the appropriate percentage increase in weight per day for the contemporary field conditions of temperature and other pertinent factors. An example, for sediment-living invertebrates, is given in Chapter 8. An alternative method using field data alone is given by Southwood [758], and a review of available methods in Rigler and Downing [683].

4.4 STREAM COMMUNITIES

Studies on animal production in streams (see, for example, [199, 442, 580, 602]) thus usually involve only one or a few species out of a total community which may number over a hundred in a small area; a complete study has probably not yet been done. It would be extremely tedious and might not justify the work involved in terms of the general understanding that might come from it. Individual production studies of selected species, on the other hand, may show how a variety of complementary life-histories and diets may allow coexistence of species in a given stream. This introduces the topic of the communities of organisms in streams, and the factors which determine which particular species grow where. Most interest has centred on the animal communities.

Some scientists believe that habitats are in general stable enough for their communities to be determined by competition between species, leading to a close-packing of the small realized niches of many species. Such communities would be resistant to invasion by new species, and would be similar in species composition from place to place under the same physico-chemical conditions. Such communities would also be rich in species, none of which would be extremely abundant because the competition of others would prevent their unbridled expansion.

Other workers believe that habitats vary a great deal, that disturbances of various kinds (e.g. unusually high temperature, low temperature, drought) frequently open up habitat by eliminating particular organisms which might or might not recolonize. Such communities, with organisms coping with wide ranges of conditions, would have fewer species, each occupying relatively large realized niches with less competition between them because of the more or less continual availability of temporarily unoccupied habitat. Such communities would have a random element in their composition and be less likely to be similar from place to place under the same average physico-chemical conditions. Differences between communities should also reflect differing physico-chemical conditions to a greater extent than if competition shapes their composition.

Despite spirited support for each of these hypotheses from their major protagonists, all real communities are probably determined to some extent by competition, physico-chemical tolerance and disturbance. The balance of these factors may differ however from time to time and from place to place in stream as in all ecosystems.

4.4.1 Do distinct stream communities exist?

To answer this question needs a listing of the communities in a very large number of streams, and a comparison between them. Wright *et al.* [876, 877] have analysed samples of macro-invertebrates from 340 places in Great Britain. All sites were of stony rivers and streams and the samples were standardized kick samples collected over 3 minutes. A total 587 different animal species was found (including 29 snails, 20 bivalve molluscs, 54 oligochaete worms, 16 crustaceans, 34 mayflies, 26 stoneflies, 24 water bugs, 88 water beetles, 87 caddis flies and 161 true flies (Diptera)). Data were also collected on a range of physical conditions, for example stone size, discharge channel width, depth, distance from headwater, and water chemistry. All sites were unpolluted in a conventional sense. They did not receive discharge of significant amounts of toxic waste or sewage, but of course many were surrounded by land used agriculturally in various ways, so that most had some influence of human activity.

The animal communities were analysed in two ways. The first, ordination, used a method called detrended correspondence analysis (DCA), which compares each community with each of the others and then calculates a position for it relative to three graph axes. The position is determined by the community's similarity to each of the other communities. This creates a three-dimensional graph with each community appearing like a star in a constellation. The constellation was a continuous one—groups of communities did not separate out distinctly from one another so that they could be segregated as separate and distinct entities. There was continuous change, and this continuum could be related to the physical conditions. The trend was best correlated with the size of the bottom stones, the slope of the bed, concentrations of phosphate and nitrate in the streams, and to a lesser extent with distance from the source of the stream. In general the features best correlated with the change in community were ones which steadily change as the river moves from upland to lowland. This analysis suggests that the compositions of the communities owe most to the steadily changing physical conditions in the streams.

However, a separate analysis was also made called two-way indicator species analysis (TWINSPAN), which attempts to classify the communities by progressively dividing the total into subgroups of two by means of the presence or absence of a suitable species in the two groups. This can be continued through a series which creates first two, then four, then eight, then sixteen groups until ultimately each original community rests in a separate classification of its own. That is the point at which the analysis started! The aim is to see if such classification is possible so the division into groups is stopped at a point where there is a considerable difference between the numbers of communities placed in each of the two groups following a division. This occurred after the fourth level of division, so the analysis was interrupted when sixteen groups had been determined. This, of course is arbitrary, and groups so formed cannot be used as evidence that discrete communities, determined through competition mechanisms, exist. Rather, the analysis confirms the relationships found by the ordination. The sixteen groups (Fig. 4.6) reflect, from left to right, a progression from upland to lowland sites. Stoneflies, many genera of mayflies, caddis-flies and some Dipterans dominated the uplands, and worms, molluscs, flatworms, leeches, crustaceans, water bugs and additional Diptera the more lowland stony streams. The differential distribution of some of the insect groups and the Crustacea and molluscs might reflect the different pathways by which these animals have invaded fresh waters—in the former case by land and air, in the latter from the sea via the lowland estuarine stretches of the rivers (see

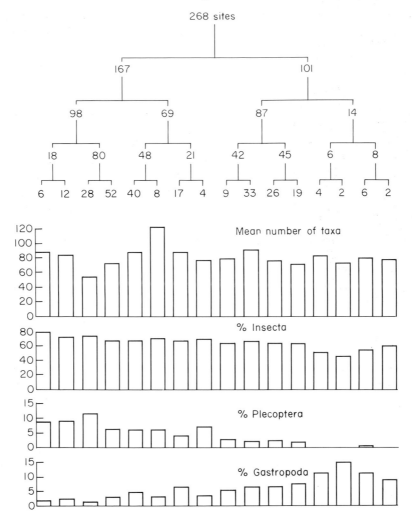

Fig. 4.6. Classification by TWINSPAN of invertebrate communities of rivers in Great Britain (Wright *et al.* [876]). Below the dendrogram, which shows the number of sites falling into each category, are the mean number of taxa and the percentage number of taxa (usually species) of certain invertebrate categories in each of the sixteen groups in the lowest level of the dendrogram.

Chapter 2). A further test of the relationship was made by using another technique, multiple discriminant analysis, which lists whether the community composition can be predicted from a knowledge of the environmental variables alone. This was used to predict which group of the 16 generated by TWINSPAN ought to occur at a particular site and to compare the prediction with what was actually found. Predictions were correct in 76.1% of cases, with the correct group being given as second most probable in a further 15.3%. Again this seems to confirm a strong relationship between the community and the physico-chemical environment.

The analyses made by Wright *et al.* did not set out to test hypotheses about what determines animal communities in streams although conclusions can be drawn from them about this. (Rather the work was intended to create a classification scheme against which changes resulting from alteration in river quality in future could be assessed by the Water Authorities.) A study on a smaller river system has set out to test these hypotheses more explicitly.

This study concerned the headwater streams of the Rivers Medway and Ouse in a heathy area, based on sandstone soils, in the Ashdown Forest of Sussex. The streams were all relatively acid (pH 4.8–6.5) though variable in calcium and nitrate concentrations because of differences in land use in the surrounding catchments. They were also iron-rich (0.2–2.8 mg l^{-1} of soluble Fe), and all had stony bottoms. Townsend, Hildrew and Francis [815] first carried out a similar analysis to that of Wright *et al.* [877] on 34 sites, using the same mathematical methods for ordination and classification of the communities. They also used a simpler technique of stepwise multiple regression which first relates the degree to which one factor—a feature of the community such as number of species in it—is correlated with another, say pH or calcium concentration. It then goes on to include information on additional factors and calculates how much (as a percentage) of the variability in the first feature is explained by an increasing list of additional variables.

Again a relationship was found with the physico-chemical environment. The multiple regression analysis found that the number of species was primarily related to pH with 48% of variation in it accountable by the pH alone.

For about 20 species (out of a total of 137), there were also significant correlations with physico-chemical variables. For numbers of *Gammarus pulex*, the freshwater shrimp, increases in pH explained 49.3% of the variation, pH and Ca together 56.2, and pH, Ca, and NO_3, 60.1. The most successful such correlation was for a stonefly, *Leuctra nigra*, in which successively pH, July temperature, iron concentration and maximum discharge explained 65.5%, 79.8%, 85.5% and 87.8% of the variation. Increasing numbers were positively related to iron concentration, negatively to the other three variables. Furthermore, fish were absent from the more acid sites but brown trout, bullhead and stone loach were found as pH increased.

The ordination and classification confirmed the importance of pH for the community, with July temperature and discharge being secondarily important. A dominance of stoneflies in the more acid waters and *Gammarus*, *Simulium* and others at higher pH, reflects in a general way the trends found by Wright *et al.* [877]. These trends, determined from large numbers of data and very comprehensive computer-based analyses, it might uncharitably be pointed out, were recognized some time ago by naturalists working intuitively [490]. As often, Ecclesiastes 1, 9–11 provides a succinct commentary.

The modern analyses, however, add detail though not necessarily increased comprehension. In both cases physico-chemical factors explained only about three-quarters of the variation and in the Ashdown Forest, most individual species' distributions could not be readily correlated with physico-chemical variables. Furthermore the existence of a correlation—with substratum type in the one case, with pH in the other—does not explain the mechanism. For example, pH was related to a number of other chemical variables (e.g. land use, Ca) only some of which were measured and included in the analyses and the nature of the bottom is just one of many interlinked factors concerning land use, erosion, water chemistry and stream flow, any of which might be more important in determining the causation mechanism than the mean size of the stones alone.

4.4.2 Competition in structuring stream communities

The implication of the first problem—the variation unexplained by physico-chemical factors—is that competition mechanisms might have some importance in explaining part of the community composition. Hildrew *et al.* [349] hypothesized that if competition is important, the number of individuals of a species present in both species-rich (high pH) and species-poor (low pH) communities should be greater in the species-poor community, and tested this for a series of three stream sites in the Ashdown Forest. There was an increase in the number of animals per species as the number of species decreased but this could equally reflect a lower diversity of food sources at the species-poor sites, leading to an exclusion of certain species on diet grounds alone.

Secondly, Hildrew *et al.* suggested that the sizes of niches should decrease with increasing species richness if competition was significant. The measurement of niche size is difficult since the niche is an abstract concept with many dimensions. However, size of organism is generally related to the size of food that can be taken, and a greater variation in size among the members of a species the greater the variety of food that can be taken. This might be used as a reflection of the overall 'niche width'. If competition is important, its effects should be to lead to greater specialization between potentially competing species and thus to a more restricted size range of each. For a group of stoneflies feeding on fine detritus, the variation in size was measured as the coefficient of variation (standard deviation as a percentage of the mean) of the width of the head at sites of different species richness and was found to decrease as species richness increased. A further test was made by calculating the degree of overlap in diet (a measure of niche overlap) as expressed in the overlap in head sizes between species. For any pair of species, x and y, the proportion p_i of each in a particular head size category can be found and expressed as a difference px_i-py_i. The niche overlap is then given by $1-0.5 (px_i-py_i)$. Apparently complete overlap (equal numbers of each) gives a value of one and suggests that the species are either competing so vigorously that one must soon be eliminated or that they are not competing at all—that food supplies are sufficient to support both—if they coexist indefinitely. However, if they are competing, natural selection should lead to specialization and a gradual decrease in overlap to a value of 0 if the two species, x and y become so different in size that there is no overlap at all. For a series of stream communities in the Ashdown Forest the niche overlap did significantly decrease as the numbers of potentially competing species of detritus-eating stoneflies increased.

Upland stream communities then seem to be structured primarily by physico-chemical factors but secondarily by competitive mechanisms. The problem still remains of explaining how the physico-chemical and the competition factors act. The answer is likely to differ for different species. *Gammarus pulex*, for example, seems to be excluded from some upland stream waters because there are insufficient sodium ions for it to be able to osmoregulate efficiently (see Chapter 2) [785, 786, 787].

4.5 UPLAND STREAMS AND HUMAN ACTIVITIES

Upland streams often lie far from the main centres of human population, and, as a group, are perhaps affected least of the main freshwater ecosystems by human activity. This does not mean that their communities are undisturbed, however. A simple classification into the effects the communities have on humans and the effects humans have on the communities includes in the first category the harbouring of a serious disease, river blindness, in the tropics, and the provision of recreational game fisheries. In the second are the effects of changing land management—pasture improvement, forestry, and timber felling—and of acid precipitation. Also included are the consequences of fishing and of damming of the river flow for downstream flood control, hydroelectric power generation and water storage.

4.5.1 River blindness

River blindness infects about 40 million people at any one time. For the large fractions of whole upland villages in tropical Africa made blind by it, it is a dominant part of their lives. It makes uninhabitable otherwise healthy, cool uplands. The disease is caused by a nematode worm, *Onchocerca volvulus*, whose larvae, or microfilariae, are carried by blackflies (buffalo flies), *Simulium* spp. This characteristic collector genus grows in swift-flowing stony rivers where the larvae are attached to stones, or sometimes freshwater crabs in East Africa, and collect fine particles from the flow.

The adult blackflies are blood-feeders and gouge a small hole in the human victim's skin from which to feed. In doing so they may also inject microfilariae incidentally sucked in from a previous victim. The microfilariae develop in the human host to form long adult worms, the female up to 50 cm, the male 2–4 cm, which rest in the surface layer of the body causing fibrous tissue to be formed around them and a large swelling to be made. Such swellings may block lymph nodes and cause further swelling as fluid collects around the groin, scrotum, or hips. The adult worms live for up to 20 years, continually producing more microfilariae which migrate to the skin, giving a dry, rough or papery symptom, and to the eyes. The skin population provides the source for infection of others via the blackflies, whilst blindness may result after a time as filariae move into the cornea and sometimes die there. This causes a reaction in which opaque protein is produced.

Attempts at control of *Simulium* with pesticides generally have most effect when the pesticide is adsorbed onto the fine particles which the *Simulium* larvae collect, and inevitably have only a short-term value because they are washed downstream so quickly. The upper reaches of the White Nile below the Owen Falls Dam in Uganda were treated with DDT in the early 1960s [78, 834] and *Simulium* was killed, but the present situation is not known. The disease itself can be treated by surgical removal of the adult worms and drug treatment to kill the microfilariae. Such medical sophistication is often unavailable, however, in the remote hilly regions of poor countries, and it is unlikely that the disease will be controlled soon, if ever.

4.5.2 Game fisheries

The provision of angling is a second way in which upland rivers affect human societies. Because of the size and behaviour of the fish, it is the salmonids which attract freshwater anglers. Fisheries for them support a large ancillary industry of tackle dealers, bailiffs and river managers, and hotel keepers and others concerned with the welfare and accommodation of the anglers. There are also commercial fisheries for the migratory fish. In the open sea and coastal waters these use drift nets into which the fish become held about the head by their gill covers as they blunder into them at night. In estuaries beach nets or fixed traps catch the fish migrating upstream.

The simplest fishery to manage is one that is discrete—the fish are confined to a known area, all parts of which are accessible for the collection of data. A river fishery for brown trout is an example. A first essential for management is that the habitat should be maintained suitable, with good water flows and quality, and gravel beds for spawning. Secondly, the fishing mortality must be regulated. It must never be greater than that which the natural mortality would have been in the absence of fishing. This is often not known, nor is it fully understood the extent to which an increased mortality of adults is compensated for by greater survival rates of the young. A very conservative view is usually desirable.

Fishing mortality can be controlled by use of a licence system which limits the number of fishermen, the size of fish they are allowed to remove if they catch them, and the season in which they may fish. The latter two ways help protect the spawners by preventing removal of fish which have not yet spawned, and minimizing disturbance during the spawning season when they are vulnerable due to the energy demands of spawning itself.

Natural mortality poses greater problems. The fish are part of an ecosystem which may have other values than as a fishery. It may support otters, or fish-eating birds like mergansers. A wise fisheries manager will accept that some of the fish will be taken by these predators. One of narrower mind will want to remove natural predators and will meet considerable problems from other people in doing so. Fishermen in general believe that natural predators make huge inroads on the desired game-fish stock. In general this is not true and the major threat comes from habitat deterioration.

The pressures for predator removal can nonetheless be enormous and may extend to other fish species. An interesting example is that of the dolly varden charr (*Salvelinus malma*) in Alaskan rivers [556]. The dolly varden was believed to be a major predator on the eggs, alevins (larvae) and juveniles of the Pacific salmon, *Onchorhyncus nerka*, particularly when its smolts moved downstream to the sea. Between 1920 and 1941 $300 000 were spent in Western Alaska alone as bounty on the killing of the dolly varden. A second species, the Arctic charr (*Salvelinus alpinus*) was also incriminated. Examination of about 5000 charr stomachs, however, showed only 42 with any salmon remains in them. It would seem that an elaborate and expensive control programme based on belief rather than proper data had been a complete waste. A little thought would have suggested that it was most unlikely that fish at the upper end of a food chain would depend much on a similar predator, supplies of which, by the very nature of energy loss along food chains, would be relatively scarce.

A fisheries manager may also be unsatisfied with the natural growth rates of the fish. There is little that can sensibly be done about this for the growth rates depend on a complex of environmental conditions over which the manager can have little control. Food supply and temperature are particularly important, and the maximum attainable growth rates of brown trout are particularly well understood. Natural rates were found [194] to be usually 60–90% of the maximum determined under ideal conditions in a fish nursery [196, 198] and to depend largely on temperature.

The management of fish that migrate is more difficult, for very little control can be had over the marine stages. When the major feeding area of Atlantic salmon was discovered off Greenland, there was an immediate concentration of a salmon fishery by many countries in the area, and the numbers of fish returning to spawn in the rivers fell in most of the home countries. The fishery has now been regulated by international agreement—largely by the extension of a restricted zone off the Greenland coast. Attention has now passed to sea-fisheries for the salmon in coastal waters. There are also suggestions that seals remove many salmon off the Scottish coast, and that too many fish are taken by inshore drift-net fishermen in England and Ireland before the fish enter the estuaries.

The Canadian Government has stopped drift net fishing off its Atlantic coast so as to allow stocks to increase [684], but has been faced with problems in doing so. The fishermen came from small communities which had a traditional history of such fishing on which they largely depended, and although financial compensation could be given, there was some social discord. A drift net fishery, though banned in Scottish waters since the 1960s, is maintained in the British Isles off the Northumberland coast—an area of high unemployment so that removal of it would have strong political overtones, and a similar situation exists in Ireland. The extent to which the drift net fishermen off the English and Irish coasts affect the runs of salmon in Scottish rivers is a controversial problem not helped in its solution by an increase in illegal poaching both at sea and in the rivers.

The angling associations and fly-fishermen are a powerful and articulate group, so firmly convinced that drift-net fishing is responsible for reduced numbers of rod-caught fish that they recently suggested a complete ban on Atlantic salmon fishing within 12 miles of the coast [847]. The sea-fishermen are much less influential politically. They are not helped by the fact that a single salmon is worth far more, in economic terms, as a river-caught fish, on which a whole economic infra-structure of management, equipment and accommodation depends, than as a sea-caught food fish [539]. The situation is now further complicated by the increasing development of salmon farms in which the fish can be reared relatively cheaply in floating pens in suitable sheltered estuaries or fiords. In the long-run, just as the salmon have been prevented from successful spawning in many English rivers by changes in the habitat, so also may this become the key issue for survival of salmon fisheries in the upland rivers of Scotland, Wales, Scandinavia and North America, and it is to this topic which I now turn.

4.6 ALTERATIONS TO UPLAND STREAMS BY HUMAN ACTIVITIES

Streams have been altered in two main ways—through changes in their water chemistry, and through changes in the physical structure of the habitat. In the former category are effects of acidification and eutrophication—the addition of acid and nutrients respectively. In the latter are the impacts of dam building and alteration of the river flow for the generation of hydroelectric power, the transport of water for industrial use and the lessening of flooding downstream. Linking the two general categories are effects, such as siltation, of changes in land use which influence both water quality and the physical habitat.

4.6.1 Acidification

There seems little doubt that many rivers with catchments on poorly weathered rocks, with thin, base-deficient soils, have become more acid over the past 30 to 40 years or possibly longer. A degree of neutralization of rain more acid than normal takes place in the catchment soils, but nonetheless there is evidence of rivers in New England, Nova Scotia, Norway, Sweden, southern Scotland and central Wales with pH values below 5.0. For example, a group of southern Norwegian rivers had pH values of 5.0–6.5 in 1940, but had pH 4.6–5.0 in 1976–1978. pH values may fall to extremes of 3.0 for short periods after dry spells, or after snow melts in spring.

The causes of the acidification may be several. Prime are those dependent on acidification of rain and snow by release of SO_2 and NO_x into the atmosphere (Chapter 3), but there are other possibilities. Catchments may become more acid as the bases in their soils are leached out (Chapters 3 and 10) so that in the absence of fertilizer and lime additions to soil for agriculture, there is an inevitable acidification over periods of thousands of years [628]. This, however, seems unlikely to explain the recent marked increase in acidification. Afforestation with conifers in the uplands may also lead to acidification. The trees take up and store, either in their foliage and wood or in the refractory litter which accumulates under them, much of the small stock of cations in upland soils. In doing so H^+ must be released to maintain electrical neutrality. The trees also, however, take up anions and release OH^- so the net effect is not clear. The large surface area of evergreen tree foliage may act as a collector of 'dry deposition' particles (Chapter 3) so that as rain washes these to the ground, the water may become more acid than it would otherwise have been [318, 319]. Nonetheless, examination of changes in the diatom floras of lakes in the Galloway area of southern Scotland has shown that acidification has occurred whether or not the catchment was forested [224, 225].

Most concern for the consequences of acidification has been with the loss of fish stocks, but there are effects on the stream communities as a whole. One weakness in the available information [234] is that it is mainly about physiological effects on individual organisms studied in the laboratory rather than on whole ecosystems studies as such. The latter are the more useful studies. Hall *et al.* [311] acidified a section of the Norris Brook in the White Mountains of New Hampshire from April to September 1977 by steadily dripping in sulphuric acid. The pH fell from greater than 5.4 to 4.0. There were immediate effects. First, Al, Ca, Mg, and K were mobilized from the sediments of the channel. Secondly, there was an increase in invertebrate drift within 30 minutes of acidification, with the pattern and composition of the drift differing from an upstream control section. The drift rate declined after a few days, but largely because the populations of invertebrates had also declined—by as much as 75% for chironomids, tipulids, ceratopogonids and mayflies. Fewer of the latter, and also of stoneflies and dipteran flies emerged as adults during the summer. In the leaf litter, hyphomycete fungal populations fell, and leaf decomposition rates were probably reduced, despite the spreading of patches of an alien basidiomycete fungus. Algal growth on the stones of the stream increased, probably because of the loss of the grazer-scrapers, and there was an overall decrease in diversity of the stream community. The major fish species present, brook trout, migrated from the stream section, though a few, trapped in pools during the summer, survived the low pH with no apparent damage.

Such fish were unusual, however, for a major symptom of acidification has been loss of fish stocks, particularly of salmonid fish. Partly these losses come from a failure of reproduction, partly from death of the adult fish. The mechanisms involve both a direct one of hydrogen ions, and a secondary one of the high aluminium concentrations which accompany pH values around 5.0.

The direct effects of pH are felt particularly during episodes of very acid stream water following the melting of snow banks in spring. This period coincides with the hatching and juvenile stages of many salmonid fish. Although hatchery fish can be conditioned to low pH, wild fish eggs of Atlantic salmon fail to hatch if exposed to pH 4.0–5.5 at the stage where the embryo has developed eye pigment. This appears to be because the enzyme, chorionase, secreted by the larva's snout, needs a pH of 8.5 to carry out its work of dissolving the egg membrane to allow the larva to escape. At pH 5.2 its efficiency is reduced to 10% [633]. A second influence of low pH is the inadequate storage of proteins in the yolk of the egg and in the number of eggs produced by the adult fish. Compared with controls at pH 6.7, flagfish (*Jordanella floridae*) produced only 2.1–8.2% of eggs at pH 4.5–5.0 [697]. The critical pH above which there was no effect on reproduction was 6.5 in this species.

Low pH can also upset oxygen uptake through the gills, and ion regulation. Values below about 5.0 cause an alteration in the permeability of the gills which allows H^+ to move in and Na^+ to leak out; high blood pH also reduces the efficiency of haemoglobin to combine with oxygen, and mucus deposited on the gills at low pH also increases the length of the pathway over which oxygen must diffuse to the gills. The fish may thus die from oxygen starvation [234]. In invertebrates, low pH may interfere with calcium uptake, so that the moulting of Crustacea, which require much calcium for their exoskeletons, may be prevented [516]. Crayfish disappear from streams at about pH 5.3, and *Asellus* at 5.2. *Gammarus* spp. need a pH of greater than 6.0. There is also evidence that birds such as pied flycatchers feeding on calcium-deficient and aluminium-rich aquatic insects may form thin-shelled eggs which survive less well than normal eggs [595, 596]. There is also evidence that populations of another bird, the dipper, have declined in acidified streams [606, 607, 608].

Aluminium ions appear to raise the pH thresholds below which the physiological effects discussed above occur. Toxic aluminium concentrations vary with pH and with species but $200 \, \mu g \, l^{-1}$ appears lethal to brown trout at pH 5.0. Harriman and Morrison [319] found that forested streams in Scotland with aluminium concentrations up to $350 \, \mu g \, l^{-1}$ would not support fish nor allow hatching of trout eggs placed in them, whilst in the R. Tywi in Wales, Stoner *et al.* [779] found a correlation between the mortality of fish placed in cages in the river, and the dissolved aluminium concentration. The relationship was complex, with 50% dying in 5–6 days at pH 4.9 and Al^{3+}, $450 \, \mu g \, l^{-1}$, and 50% dying in 14 days at pH 4.8 and $215 \, \mu g \, l^{-1}$. Aluminium salts were found deposited within the abundant mucus coating the gills.

The reversal of acidification effects in streams is difficult. The addition of lime (calcium carbonate) has been effective in lakes, but additions to streams are short-lived and widespread liming of snowbanks to prevent the highly damaging effects of their melting would be impracticable and highly expensive. Restocking of fish populations is pointless unless the acidity is removed, even if the restocked fish have been conditioned or bred to live at low pH. The stream community as a whole and hence the fishes' food supply is affected. In any case such measures—a response to symptoms not to causes—almost certainly will lead to further problems. The only sensible way to solve the problem is to tackle it at source by reducing SO_x concentrations from industry and NO_x emissions from vehicles.

4.6.2 Changes in land use

Many changes are now happening to upland landscapes. Where natural forest remains it is increasingly used for intensive timber production, with much ground disturbance by heavy machinery; where forest has been removed it is sometimes being replaced by forest plantations, with accompanying drainage of peaty soils by the cutting of channels. Elsewhere, upland pastures are being fertilized to increase grass production. In the past 30 years or so about 150 000 ha or 8% of the total upland moorland in England and Wales has been variously reclaimed [587].

The consequences of forest removal have been related in Chapter 3 and a bibliography of papers on this subject is Blackie *et al.* [54a]. The effects on the receiving rivers are a major increase in nitrogen concentrations and perhaps a lesser increase in phosphorus concentrations, also an increase in suspended matter which may later settle out on the bottom. There is also increase in overall discharge, because evapotranspiration of the forest releases more water vapour to the air than open moorland or grassland. The distribution of flow during the year, however, may be greatly altered with reduced flows in summer in some cases [259]. Deforestation may also cause problems through the higher temperatures that follow removal of shading. The Atlantic salmon seems to have been excluded from streams flowing into Lake Ontario by coverage of the bottom by waterlogged sawdust in the nineteenth century; attempts to reintroduce it have failed because the waters are now too warm, and the summer flows are too low [749, 750].

One unexpected consequence of modern forestry may be to reduce the nutrient inputs to upland streams and the lakes they feed through a mechanism concerning migrating salmon. The spawning fish of many species die in the streams and release phosphorus and nitrogen compounds in an environment which is generally devoid of them. The carcasses, however, must remain in the stream for this to happen and not be washed downstream. Woody debris in North American streams acts to retain the bodies. If it is removed to ease the flow and minimize flooding or if it never reaches the stream because of intensive forestry practice which converts even the brushwood eventually to chipboard, the carcasses are not retained.

Cederholm and Peterson [107] found a significant correlation between retention rate of coho salmon carcasses in streams of the Olympic Peninsula, Washington and the amount of large organic debris present. Large organic debris included logs at least 3 m long and 10 cm in diameter lying in the water and within a metre of its edge. The carcasses were often removed to the edge by scavengers like raccoons and black bear, and release of nutrients via mammal excretion could be as important as direct release by microorganisms. The streams most productive of coho salmon were those which had the most debris and readily retained carcasses and their nutrient contents. A number of separate observations [80, 680] now seem to indicate an important role for retention and suggest that interference with the natural debris-strewn stream system is unwise. In one case [439] overfishing of salmon at sea resulted in ultimate impoverishment of the spawning headwater lakes.

Re-establishment of forest plantations, paradoxically, may also cause problems [538]. In Scotland, where more than 10% of the country is now covered in plantation, the dense shade cast by overhanging introduced conifers such as Norway Spruce and the refractory litter they produce has reduced invertebrate populations [536]. It has also diminished river flows, and has led to clogging of gravel bottoms and the filling in of deep pools favoured by adult fish. After heavy rain, however, the drainage channels cut through the plantations allow rapid run-off, scouring both the silt and the gravel and causing severe erosion. The spates, however, do not last long enough to serve the upstream migration of salmon which requires steadily rising waters.

4.6.3 **Physical alteration of the streams**

Upland areas are sources of two valuable commodities—water and potential energy. The supply of both is irregular for it depends on the weather and season; the need for electrical power by human societies also varies throughout the day. One solution to provision of a reliable supply of both water and power, recoverable at will, is to dam upland rivers to provide reservoirs of water for drinking, industry and the generation of hydroelectric power. At the same time such reservoirs can be used to regulate the amount of water flowing downstream so as to avoid flooding in the lower reaches (see Chapter 5). The river below the dam forms a ready-made conduit to deliver the water from the reservoir to where it will be used. Dammed and regulated rivers are now the norm (see Chapter 10); free-flowing ones are the exception.

Damming can cause many changes to the river. It causes a blockage against fish movements, upstream or downstream; it may change the flow downstream by making it more regular, in response to the need for power generation, but more extreme. It may make the flow more even if the reservoir is used for water storage or flood control. The nature of the river bottom will change in response to these changes in flow.

The water quality may change too. If water is released to the river from the upper layers of the reservoir where it has equilibrated with the atmosphere over several days, it may be better oxygenated than if it has seeped only a short time before from the catchment peats and soils; conversely if it has come from the deeper layers (see Chapter 6) it may be poorly oxygenated and contain suspended iron and manganese hydroxides, or dissolved hydrogen sulphide. The nature of the suspended organic matter will have changed also. Leaf litter will have sedimented out in the reservoir but planktonic (suspended) organisms may be washed into the river in large numbers.

The first problem, of fish migration, is not difficult to solve although it was not always recognized in the past. Many fish moving upstream can negotiate barriers of 1–1.5 m, and salmon species can often jump up to 3 m. Most dams are far higher, but fish ladders can break the ascent into manageable steps. They comprise a series of stepped pools in an incline beginning some distance below the dam and rising up the valley side or climbing, perhaps as a spiral, within the dam itself. Some dams have electrically powered lifts in which the fish are transported from bottom to top. If no provision has been made when the dam was constructed, fish can be netted at the foot of the dam and carried by road around it at the peak of the spawning season. Hatcheries may replace the natural spawning areas, perhaps now covered by lake sediment.

The main problem for the design of fish ladders is to encourage the fish to enter, because the flow of water down them is often much less than that coming through the main dam sluices, or out of the turbines in a dam used for power generation. Series of baffles may help. Scottish law has demanded provision of fish passes on all dams since 1860, but even as late as 1933, the Grand Coulee Dam on the Columbia river was built without a fish pass, and blocked 1600 km of river spawning sites to migrating salmon.

Passage downstream also may be difficult. Fish moving into turbine intakes may be killed by the blades, and those taking the fall of perhaps tens of metres over the dam wall may be smashed on the rocks below. Turbines can be designed, and operating procedures arranged to minimize damage, but death rates may reach 10–20% even so. A series of dams on a river may thus kill almost all the migrating smolts. The fish may not even reach the dam and ideally the head of the fish ladder because the reservoir lacks the strong currents which guide them down-river. They can again be netted and moved by road, but increasingly recourse can easily be made to a completely artificial system of stocking of the river with adults of non-migrating salmonid fish reared in hatcheries.

The fishes' problem may not be over even when they are in the river below the dam [73]. Though in general a minimum flow to be released from the dam in stipulated by law, the contrasts between low and high flows may be very great and change rapidly, especially with hydroelectric dams. The Kennebec River in Maine experiences flows of about $8.5 \text{ m}^3 \text{ s}^{-1}$ at night when little power is being generated, but up to $170 \text{ m}^3 \text{ s}^{-1}$ by day when the turbines are working fully. Twenty-five per cent of the river bed may be uncovered at night, stranding small fish, whilst the sudden high flows may damage them by abrasion against rocks.

4.6.4 River regulating reservoirs

The problems of fish passage are very obvious; the more subtle effects that dams may have on the fish through their food supply and growth are less so. In some cases there will be advantages, in others, disadvantages.

For example, the Cow Green reservoir was built across the R. Tees in the north-east of England to store water for industrial use further downstream. The estuary of the river is too polluted to encourage migration of salmon, but the upstream reaches, lying in open peat moorland, support populations of non-migrating brown trout and some smaller fish, including bullheads. The Cow Green dam was built just above a waterfall, Cauldron Snout, over which the Tees flows then passes for several hundred metres before it is joined by a main tributary, the Maize Beck, which has not been dammed. This provided an opportunity [17, 129], through a study of the regulated Tees and the unregulated Maize Beck to assess the effects of the dam.

Storage of water in the Cow Green reservoir caused a number of changes in it. Most prominent were changes in temperature and flow. The water discharged to the river was 1–2°C warmer in winter and cooler in summer, as a result of the large volume of water in the reservoir. Very high flows (>8 times the mean) were eliminated as were very low ones (<0.1 of the mean), so that overall the river water varied much less over the year in the R. Tees than it had previously done and continued to do in the Maize Beck.

The first major effect of the changed regime was a great increase in the biomass of algae and mosses among the stones of the river. Previously these had been scoured away by the higher flows. With the mosses came an increased number of bottom-living animals and a change in community. The number of animals per standard 1-minute kick sample increased from 56 to 420 after regulation of the Tees, but remained about the same (99 to 77) in the Maize Beck. There were no changes in the diversity of the communities, but whilst mayflies remained dominant in the Maize Beck, and persisted in the Tees, the latter also supported increased numbers of *Gammarus pulex*, the coelenterate *Hydra*, a group of oligochaete worms, the Naididae, and one of flies, the Orthocladinae. The waterfall may have been important in maintaining turbulent flow and helping to keep the stones clear of silt deposition—a frequent problem a little way downstream of dams which create reduced flows. The nature of the invertebrate drift changed also. It became dominated by zooplankton organisms from the reservoir in the R. Tees, and water fleas (Cladocera) became prominent in the diet of brown trout.

Regulation of the Tees was reflected not so much in increased growth of the fish but in an increased population density (Table 4.4). Growth rate of trout was about 80% of the theoretical maximum and was probably determined by the continuing low temperature, even though temperature fluctuations had been reduced. Population numbers of both trout and bullhead increased in the Tees and the bullhead bred earlier and more than tripled their net production.

No very serious effects on the Tees can thus be attributed to the Cow Green Dam. A diverse river community was maintained. Other schemes may not be so fortunate. One such is the Caban Coch (Craig Goch) scheme on the R. Elan, a tributary of the R. Wye in central Wales. The reservoir is used for water supply and flood regulation. A comparison of the bottom communities of the Elan and Wye [730] just above their confluence suggests some deleterious effects of regulation on the Elan. The problem, reflected in reduced diversity of the invertebrate communities, rests in clogging of the bed with inorganic deposits of iron and manganese released from the reservoir. These are formed in the bottom waters and also discharged from a water treatment works associated with the reservoir, where they accumulate in the filter beds. Probably they are formed by iron bacteria, converting Fe^{2+} to Fe^{3+} in the deoxygenated deeper waters of the reservoir. Two solutions are possible. First, the deoxygenation can be prevented by artificially mixing the reservoir (see Chapter 7) and secondly, water could be released to the river from the surface rather than the bottom waters. However, surface water falls farther and may more easily damage concrete structures at the foot of the dam, so release from the bottom is preferred.

Table 4.4. Changes in the fish population of the River Tees, below the Cow Green Reservoir, and in the unregulated Maize Beck, before and after building of the dam

	Brown trout		Bullhead	
	Before	After	Before	After
Maize Beck Growth rate				
Mean length of fish < 1 year old in May (cm)	No change		4.4	4.3
Population density (fish per 100 m², mean in parenthesis)	0.4–5.0 (2.0)	0.4–4.0 (1.8)	2.3–7.8 (4.2)	0.7–13.8 (5.8)
Age at sexual maturity	No change		No change	
Net production (g m⁻² year⁻¹)	0.15	0.15	0.4	0.55
R. Tees Growth rate		Small increase		
Mean length of fish < 1 year in May (cm)			3.7	3.5
Population density (fish per 100 m², mean in parenthesis)	1.4–5.1 (3.5)	1.9–8.1 (4.9)	5.1–19.5 (9.9)	11.4–93.8 (33.4)
Age at sexual maturity	No change		50% mature after second birthday (230 eggs m⁻² year⁻¹)	84% mature after second birthday (540 eggs m⁻² year⁻¹)
Net production (g m⁻² year⁻¹)	0.3	0.4 (mainly by increase in number of young (first and second year) fish)	1.06	3.64

4.6.5 Alterations by man of the fish community

The upland rivers, and the lowland reaches (Chapter 5) through which migrating fish must pass are for many reasons becoming less favourable for fish populations. At the same time the demand for salmonid fish by anglers, and for the table, has been growing. The native salmon and brown trout stocks in Britain are dwindling because of changes in their habitat and this trend is perhaps reinforced by the release of exotic, hatchery bred species which may compete with them [540].

The American brook trout (*Salvelinus fontinalis*) and the rainbow trout (*Salmo gairdneri*) were introduced to Britain in the late nineteenth century. Initially they were confined to hatcheries, for breeding was not successful in the wild. The fish as adults have been used to stock fisheries where the voraciousness of the rainbow trout, and the ability of both to tolerate greater extremes of temperature, dissolved oxygen and, for brook trout, pH, than the native brown trout, make them especially favoured by anglers and fishery managers.

Some populations have begun to breed in the wild—at about 50 localities in 1940, 550 by 1971—and it is not clear whether stable coexistence of these more resilient fish with the brown trout and salmon is possible. There has also been a great increase in fish-farming of rainbow trout for the food trade. From about 40 farms in 1970, producing less than 1000 tonnes per year, the industry has grown to over 400 farms producing nearly 7500 tonnes in the early 1980s. Fish farms take water from the rivers to supply the farm and return an effluent which may have a greater biological oxygen demand (from fish faeces). It may also contain antibiotics, and disinfectants, used to protect eggs against microbial disease. Fish also escape from the farms, and may introduce diseases, always liable to be rife under hatchery conditions, to the wild stocks of other species. Viral haemorrhagic septicaemia (VHS) and infectious pancreatic necrosis (IPN) are two such examples.

Furthermore, the hatcheries now favour genetically-tailored strains, selected for high growth rates, and disease resistance under hatchery conditions. Such fish may also be treated with steroids to produce only females, which grow fast, and to delay breeding so that spawning does not weaken fish grown fat for angling before they have been caught. There are fears that such fish may interbreed, even hybridize, with established stocks of their own species or with the native species respectively. Progeny might survive very well in most years, but a particularly severe year might result in widespread kills under conditions which naturally selected wild fish would survive.

The tendency to stock rivers with exotic fish for anglers is more widespread in the USA where game-fishing commands large sums of money, and where natural communities may have suffered considerably. Training of fishery managers in the past has emphasized the production of large fish for anglers, though with some change in philosophy, a more sympathetic attitude to conservation of the smaller native species may prevail. An article by E. P. Pister [639], a fisheries manager in the California Department of Fisheries and Game, describing his training and subsequent maturation to become more interested in conserving endemic small fish of the desert springs and streams is enlightening reading. Interestingly, one of the introductions he managed in the Owens River valley area of California to the detriment and near extinction of four native fishes, was the European brown trout.

5

Lowland Rivers and their Floodplains

There is a world of difference between the upland bubbling brook, with its rocky bottom and foaming-white water and the treacly floodplain river with its fringing swamps, bearing millions of tons of sediment to the sea each year. But this distinction is developed gradually. The river bed may at first develop pockets of finer sediment which may support submerged plants. The widening channel in the middle stages, even if overhung at the edges by trees, will become well lit over an increasing fraction of the bed, which will also allow more plants to grow in water still quite shallow.

Thereafter, in the lower reaches four gradual changes may take place. First the water depth will increase, and the water may have accumulated enough silt from erosion of the catchment soils to prevent much light penetration to the bottom, so the submerged plants may disappear. Secondly, at the edges, rooted plants which cope with the turbid water by emerging into the air above it may start to form permanent swamps. In the bed itself the animal community, still dependent on fine organic debris, will largely be deposit feeding.

Thirdly, the river will meander more and more. Its channel size is determined by the average annual flow so that at very high flows it may overtop its banks and spread over its natural floodplain, taking fish and other animals with it to exploit the seasonal swamps of the floodplain. Slightly raised banks or levees are created along its edges as coarser material borne by the flood settles out first.

And lastly, as the river becomes very large and water is retained in the channels for longer periods, there may be time for plankton to develop in the water before it can be washed downriver to the sea. This community may at first be of animals, the zooplankton, feeding on suspended fine organic debris. But in parts, enough of this silt may settle out to allow photosynthetic plankton (phytoplankton) to grow as light penetrates. Submerged plants may also recolonize the bottom. At this stage the lowland river will be lake-like.

This chapter looks at the colonization of middle stage rivers by submerged plants, then the development of swamps and floodplain ecosystems and the impact of man on them.

5.1 SUBMERGED PLANTS

Survival under water poses greatly different problems from survival on land yet the appearance of submerged plants is much like that of their land relatives. Most are angiosperms (with a few pteridophytes, like *Isoetes*) and the basic pattern of roots, stems, leaves and flowers adapted for wind or insect pollination above the water surface, is preserved. The picture of their evolution is one of sporadic and recent colonization of the water by a few lines more or less indiscriminately drawn from among the available land families.

Particular features of submerged plants (Fig. 5.1) lie in often reduced root systems, decreased proportions of woody tissues in stems and in the production of large air spaces (lacunae) among the tissues. Cuticles are spare, covering thin and sometimes dissected leaves and there is an emphasis on vegetative spread by stolons, rhizomes or contracted shoots called turions or

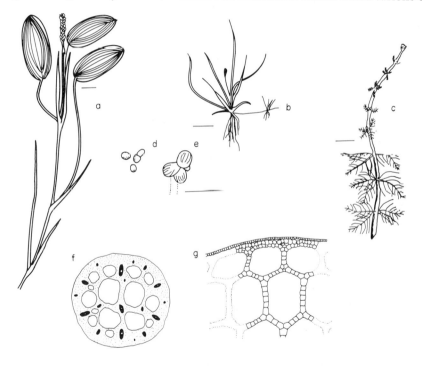

Fig. 5.1. Some features of aquatic plants. (a) *Potamogeton natans*, a submerged species with thin leaves which also has thicker, oval floating leaves and emergent inflorescences; (b) a small submerged species, *Littorella uniflora*; (c) *Myriophyllum spicatum*, with much dissected leaves; (d) *Wolffia columbiana*; and (e) *Spirodela polyrhiza*, two species of duckweed, with much reduced thalli, which float on the surfaces in quiet waters; (f) and (g) cross sections of the petiole of white water lily, *Nymphaea alba* and part of the stem of mare's tail (*Hippuris vulgaris*), showing the abundant air spaces (lacunae). Scale bars represent 1 cm.

winter-buds. The latter feature is often ascribed to supposed difficulties of pollination and sexual reproduction under water, yet a few species have evolved thread-like pollen which is transported by water currents. Perhaps the main reason is the great ease of spread by water currents of vegetative fragments. Other characteristics perhaps reflect the advantages for support of the plant of its living in a dense, wet medium. Further features reflect the disadvantages of deoxygenation in sediment, low diffusion rates of gases in water, and the shaded underwater environment (because of light absorption by the water itself).

Reduced root systems are often related to the relatively high concentration of nutrients at least in fine sediments and the abilities of water plants to take up nutrients such as nitrogen and phosphorus compounds through their leaves. Despite the shortage and lack of branching in roots of many aquatic plants, experiments with radioactive phosphorus tracers and isotopes of nitrogen have often shown ample uptake from the sediment by roots and translocation to the shoots [163].

Root systems depend on active uptake of most nutrients against a concentration gradient from the sediment to the interior tissues. This needs energy and maintenance of vigorous root respiration but may be difficult in sediments which are waterlogged and deoxygenated by bacterial activity. The production of lacunae and air-tubes through which oxygen may be moved to the roots is one solution to the problem (see later).

The general lack of woody tissues in submerged plants is understandable because xylem vessels and fibrous tissues, predominant in land plants, are not so much needed to cope with water movement and support as they are in a dry aerial environment. The thinness and dissectedness of submerged plant leaves is not so straightforwardly explained, however. Turbulent water tends to pull plants from their anchorage, and to move and batter them against adjacent rocks. Small species, like the shore weed, *Littorella uniflora*, might avoid this problem by growing tight mats of leaves on small plants close to the bottom, but stand the risk of being covered by shifting sand during spates. Taller plants can avoid burial but at the risk of mechanical damage. Flexible, narrow leaves would seem to be well fitted to cope with turbulence, but are no less common in still lake-waters than they are in swift streams. Perhaps the main reason for selection of such leaves lies in the problems of photosynthesis. Diffusion of carbon dioxide into the leaves is slow and light is scarce.

Submerged plant leaves show many features like those of the 'shade' leaves deep in the canopy of woodland trees. Their thinness increases the ratio of outer photosynthetic tissue to inner tissue where internal shading reduces or prevents photosynthesis but where respiration still continues. Chloroplasts are often present in the epidermes of shade and submerged leaves, but not in those of plants of well-lit places. This places photosynthetic tissue where it can best benefit from restricted light, and where the pathway for diffusion of CO_2 from the water is smallest. The problem of obtaining CO_2 is particularly acute in waters of low pH where there is little store of bicarbonate ions, either to act as a direct source of carbon for photosynthesis, or as an indirect source through their dissociation to CO_2. Some water plants, for example *Lobelia dortmanna* [869], draw carbon dioxide through their roots from the sediments, where decomposition of organic matter provides an extra supply. Others additionally use crassulacean acid metabolism (CAM) to store carbon dioxide temporarily as carboxylic acids at night when they can absorb it but not use it in photosynthesis [61, 62, 428, 679, 706].

The physiology of submerged plants is particularly interesting in itself; in turn, the colonizing plants provide a much more complex habitat for other organisms than is available in the upstream section dependent largely on rock and gravel for its architecture. However, details of these will be discussed in Chapter 8, when plant beds in lakes are also considered.

5.2 GROWTH OF SUBMERGED PLANTS

Several factors may determine the growth of submerged plants in rivers. Plants need adequate light, an inorganic carbon source and other mineral nutrients to grow in addition to particular features of the environment which may favour a particular species.

Light is readily absorbed by water, more particularly by the dissolved organic substances in it, and in most rivers by suspended particles. The rate of absorption in well-mixed water is exponential and follows the equation:

$$I = Io\,e^{-kz}$$

where I is the light intensity (or more correctly the photon flux density) at a given depth, Io the light intensity in the water column z metres above it, e the base of natural logarithms and k, the absorption coefficient, which defines the rate of absorption, in log units m^{-1}. Values of k express the proportion of the incoming light absorbed as it passes through a metre of water and vary with the wavelength (see Chapter 6) as well as the nature of the water. They may have values of around 0.01 or less in very clear waters, but of as much as 20 in very turbid ones. Middle-stage rivers, with beds of aquatic plants might have values of k of between 1 and 2 for the water, whilst the absorption by the plants themselves will increase these values within a plant bed.

Values of light intensity at which net photosynthesis (and hence growth) is just possible (gross photosynthesis $\geqslant$ respiration) vary with plant species but have been determined in two English rivers [851] to be between about 10 and 40 J $m^{-2}s^{-1}$. Maximum summer irradiance was about 320 J $m^{-2}s^{-1}$ so a fraction of at least 3–13% of the available light was required.

By use of an expansion of the equation above:

$$z = \frac{1}{k}(\log_e Io - \log_e I)$$

and substitution of values of 320 for *Io*, 10 or 40 for *I*, and 1 or 2 for *k*, some idea can be obtained about the maximum depth at which net growth will still be possible in the rivers quoted above. The calculated depths range from 1 to 3.5 m, with three of the four values below 2.1 m. Many middle-stage rivers have depths much less than 3 m and although the calculations must be made separately for each case, this example suggests that there will usually be enough light for plant growth to be possible on their beds. It also suggests that in deeper rivers the underwater light availability will be approaching the critical limit at the bottom and that much reduction in overhead light, for example by tree shading, could prevent growth even in shallow water.

The low rate of CO_2 diffusion into the bulky tissues of aquatic plants as well as the relatively low concentrations of CO_2, found at high pH, could possibly lead to limitation of photosynthesis by carbon shortage. This does not often seem to happen, however, either because light is in even shorter supply or because the plants can use the often more abundant HCO_3 directly as a carbon source. This is difficult to demonstrate unequivocally. At pH 4.5 essentially all the carbon is present as CO_2 or as H_2CO_3, whereas at pH 9 the chemical equilibria markedly favour HCO_3^-. Experiments are set up in sealed containers, with no gas phase, where equal amounts of total inorganic carbon at pH 5 or pH 9 are supplied to replicate plants. If photosynthesis (measured as ^{14}C uptake, or sometimes oxygen production) is significantly greater at the higher pH than at the lower one, then direct bicarbonate use is believed to have occurred [661]. This has been demonstrated in a number of species characteristically occurring in hard, bicarbonate-rich waters, for example *Ceratophyllum demersum*, *Myriophyllum spicatum*, *Elodea canadensis*, *Potamogeton crispus*, *Lemna trisulca* and *Chara* spp. [382], while aquatic mosses and plants of more acid, soft waters, *Lobelia dortmanna* and *Isoetes lacustris* seem to be confined to use of free CO_2. Because of their bulkiness and hence long diffusion pathways, aquatic plants may nonetheless be at severe disadvantages compared with algae when CO_2 is scarce [5, 488, 742].

The rate of supply of either CO_2 or HCO_3 to the leaf surfaces will be influenced by flow rate and photosynthesis might be expected to increase with current speed up to some value where factors other than the rate of supply of carbon become important. Westlake [849] found increasing net photosynthesis of aquatic plants with flow rate at velocities up to 0.5 cm s^{-1}. This is a very low flow rate for rivers, and although the flow within a plant bed will be much reduced the flow outside the bed would need to fall well below 10 cm s^{-1} to have much effect. This might happen in dry periods, but the self-shading within a plant bed makes it more likely that light would have become limiting long before carbon supply.

In studies of the tissue contents of nitrogen and phosphorus in aquatic plants, concentrations have been found which are high enough to make it unlikely that their productivity is limited by nutrient supply [266]. The plants were grown in glasshouses and their growth rates related to the nitrogen and phosphorus contents of their tissues. Growth of a variety of species increased with contents up to about 0.13% (as dry weight) of phosphorus and to 1.3% of nitrogen, when increasing tissue concentrations gave no increase in growth rate. A survey of wild-growing plants showed tissue concentrations greater than these critical values in most cases, except for plants rooted in sands [97, 382]. Many river species, however, will be rooted in fine gravel or sand and nutrient availability may be low. Work on algae in upland streams [63] suggests a shortage of phosphorus, and, despite the constant flow of water past the leaves of water plants, nutrients may be scarce. These conclusions have also been challenged by Schmitt and Adams [726] who found much higher phosphorus threshold concentrations for maximum photosynthesis of *Myriophyllum spicatum* and by Christiansen *et al.* [111] for leaf production in *Littorella uniflora*. The apparent increase in weed and algal biomass in lowland British rivers in recent years, where most rivers have been fertilized by agricultural run-off or sewage effluent also suggests natural nutrient limitation.

The conditions which allow a particular plant to grow will vary greatly from species to species. Plants which are generally upstream or downstream species, or favoured in rivers of chalky as opposed to clay areas or of greater rather than lesser flows can be recognized [328] though the particular traits which favour them, with the exception of CO_2 or HCO_3^- usage usually can not. The plant community may also change from time to time because of the effects of one species on another.

For example, in chalk streams in Dorset, UK, good early growth of the water crowfoot, *Ranunculus penicillatus* var *calcareus*, accumulates silt in which the watercress, *Rorippa nasturtium-aquaticum* can easily root. The *Rorippa* shades the *Ranunculus*, so that the biomass of it left to survive the winter is low. In turn the early growth the following year is low and conditions for good *Rorippa* growth are not produced and its establishment is low. The *Ranunculus* growth in late summer is thus high, as is the overwintering biomass; the cycle then starts again the following spring, if it is not interrupted by floods washing either species away at critical times [156].

5.3 METHODS OF MEASURING THE PRIMARY PRODUCTIVITY OF SUBMERGED PLANTS

The method to be used for determining the productivity of aquatic plants in a river will depend on why the measurement is needed. A view of the overall production of the mixed plant beds, and their associated algae may be wanted or specific information may be sought, plant by plant, with or without the complications of the algal epiphytes which grow on them. Methods for either of these can involve the enclosure of plants in experimental containers (with consequent partial control of conditions) or can cope with the plants in their natural, but varying environment.

5.3.1 Whole community methods

The earliest 'whole-community' method was simple and involved the cropping, drying and weighing of the plant biomass at the time of its peak growth. The assumption was that losses (to grazing and mechanical damage) during the growing season were negligible (say 5–10%) and thus that the difference between biomasses at the start and the end represented the net growth or net photosynthesis. A practical problem was that, if the natural variability of the habitat was to be allowed for, large numbers of bulky samples had to be taken. More important, however, were the problems that underground parts (rhizomes, roots) were usually ignored and that losses of biomass during the season were often not negligible. For six available estimates for submerged plants, a mean value of 0.65 ± standard deviation 0.53 has been obtained for the ratio of below-ground to above-ground biomass [851]. A similar survey of losses of biomass during the growth season gave values for the turnover rate—the ratio of total annual production (derived from various methods discussed below) to maximum biomass—generally greater than 1, with the mean of twelve values 1.9 ± 0.85. One plant *Lobelia dortmanna*, had values of 0.7–0.8. Values less than 1, of course, are just as much drawbacks to the biomass harvesting method of determining production as values greater than one. They imply over-wintering of considerable biomass, usually as roots and rhizomes.

A better method of determining overall community production is the upstream–downstream oxygen change method [308, 600] previously mentioned in Chapter 4 for thermal streams. In the upstream–downstream method a uniform stretch of river perhaps 100 m long is chosen and over 24 hours the concentrations of oxygen are very frequently (sometimes continuously) measured at its upstream and downstream limits. From the difference between the two oxygen curves so obtained, after rephasing to allow for the time it took for water to travel the length of the stretch, the net change in oxygen concentration over the 24 hours can be calculated. It is taken to equal the gross photosynthesis minus the sum of (a) respiration, (b) the net effects of diffusion between water and atmosphere, and (c) accrual, the addition of oxygen in seepage water along the banks. Accrual, which cannot easily be measured, is usually assumed to be negligible, diffusion is estimated from the oxygen saturation levels and temperatures during the period; respiration is calculated from the oxygen changes during the night extrapolated to the whole period.

5.3.2 Enclosure methods

The advantage of the above method is that it studies an unenclosed community, the disadvantage is that diffusion must be estimated rather than measured. It gives, as net oxygen change by day plus net change by night plus net diffusion change, the gross primary productivity of the community. The net primary production cannot be calculated because the respiration estimate includes the activities of organisms (bacteria, animals) associated with the photosynthesizers.

This objection, however, also applies to methods of determining productivity of individual plant species by their oxygen production in closed containers. The method was first developed for phytoplankton by Gaarder and Gran in 1927 [249]. Replicate samples of the plant material are placed in clear containers (usually bottles) and in containers made opaque with paint or black tape, and the bottles are incubated in the natural habitat for several hours.

Oxygen is released in the clear container (light bottle, LB) by photosynthesis, but is simultaneously absorbed for respiration by both the plants and associated microorganisms and any animals present. The change in oxygen concentration in the light bottle is thus:

$$\Delta O_2{}^{LB} = [O_2{}^{LB}] - [O_2{}^{I}]$$

where $[O_2{}^{I}]$ is the initial oxygen concentration and $[O_2{}^{LB}]$ the oxygen concentration measured in the light bottle after an exposure of t hours.

In the dark bottle, only respiration has taken place so the oxygen concentration will have decreased, with a change:

$$\Delta O_2{}^{DB} = [O_2{}^{I}] - [O_2]^{DB}$$

where $O_2{}^{DB}$ is the oxygen concentration after t hours. It is assumed that the respiration rates in the dark and in the light bottles are similar, though this may not always be so, and that oxygen produced or taken up by the plants is reflected in changes in the concentrations in the water. The air spaces (see below) in the plants may cause a lag in this [325] though not always [850]. On these assumptions the gross oxygen production per hour and per unit dry weight of plant is calculated as the sum of the increase in oxygen concentration in the light bottle plus the decrease in the dark bottle, divided by the weight of plant used (w) and the time of incubation (t).

$$\text{Gross photosynthesis} = \frac{[O_2{}^{LB}] - [O_2{}^{I}] + ([O_2]^{I} - [O_2]^{DB})}{t.w}$$

The initial oxygen concentration cancels out and thus need not be known. If the plant material is bulky compared with the amounts of microorganisms and animals present, the change in the light bottle approximates to net plant production and that in the dark bottle to plant respiration, and a measure of $[O_2{}^I]$ will allow these to be separately calculated.

Enclosure in bottles greatly changes the environment of plants normally growing in flowing waters and uptake rates of inorganic carbon at the plant surface could be altered [860]; there are also difficulties in extrapolating the results of what must, for practical reasons, be infrequent estimates made over only part of a day, to estimates of whole-day and entire season production. In general the amount of photosynthetically useful radiation is continuously monitored and the ratio of whole-day radiation to that during the experiment is used to scale-up the values. This assumes that the amount of photosynthesis is governed entirely by the supply of light.

A second method using plants enclosed in bottles measures the rate of uptake of carbon as its radioactive isotope, C^{14} [769]. The isotope is supplied as $NaH^{14}CO_3$ and as the method is very sensitive, incubation periods can be much shorter than in the oxygen method. The plants are thus more likely to show rates close to their unconfined natural ones. The bicarbonate rapidly equilibrates with the CO_2-bicarbonate system in the water and the addition is usually very small in relation to the amount of total inorganic carbon present. The isotope is taken up probably a little slower, because it is heavier than the more abundant C_{12} and corrections may be made for this. After a time the plant is removed, and exposed to fumes of concentrated hydrochloric acid to remove any adherent inorganic ^{14}C. Epiphytes can be scraped or shaken off and the amounts of ^{14}C incorporated into the epiphytes and plant separately measured, either after dissolution in a scintillation mixture or conversion by burning to CO_2. Calculation of primary production assumes that the ratio of ^{14}C uptake to total C uptake (which is what it is intended to measure) is in the same ratio as the supply of ^{14}C to the total amount of inorganic carbon present:

$$\frac{\text{Total carbon uptake}}{^{14}C \text{ taken up}} = \frac{\text{Total inorganic carbon}}{^{14}C \text{ supplied}}$$

The total inorganic carbon present can be determined by routine analytical methods. The advantages of this method (sensitivity, and more reliable separate estimates of epiphyte and plant uptake because the epiphytes can be left *in situ* during the experiment) are offset by the problem that what is actually measured—gross or net production—is not known accurately.

If the ^{14}C taken up stayed in the cells and was not respired at all, gross production would be measured; if it was taken up in ratio $^{14}C:^{12}C = x$, and respired in the same ratio, then net production would be measured. In practice, it is probably taken up in ratio x but respired in some different ratio y. It is assumed that respiration does occur but the difference between x and y will determine how close or how far from the true net rate of production the estimate is.

Paradoxically it is the approaches which use little equipment, but more ingenuity which may give the most reliable estimates of production for aquatic plants. Frequently they depend on a knowledge of plant structure and a willingness to carry out tedious but simple measurements.

An example is that of Dawson's [153, 155] work on *Ranunculus penicillatus* var *calcareus*. This is a common plant in lowland chalk rivers in southern England and grows stems of up to 4 m bearing leaves, each with many fine segments every 20 cm or so along the stems. Determining the biomass is not difficult with a sampler which cuts the weed from a known area, and the underground biomass is relatively small. A problem is that leaves continually break off or are cut to provide the cases of caddis-fly larvae such as *Limnephilus*. Stems are broken by water voles and moorhens to line their burrows and nests respectively.

Dawson estimated production from the total losses of leaves and stems from an area plus the biomass left intact. He measured the rate of leaf loss by counting the number of nodes which had lost their leaves along stems sampled at frequent intervals. He found that leaf loss was about 8.5% of the maximum biomass obtained. Whole-stem loss was determined by marking a sample of stems with coloured plastic rings. From the rate of loss of marked stems, an estimate of 77% of the maximum biomass was found for the stem loss. However, by placing nets downstream of the areas under study, Dawson found remarkably little 'export' of plant material. The detached stems caught on other stems in the beds and often re-rooted, so that the total net loss was small—about 30 g dry wt m^{-2} compared with a total production of 300–400 g m^{-2} during the year.

Apparently most of the produced material even at the end of summer was caught up and decomposed in the area in which it was produced, suggesting that downstream export was relatively unimportant. This brings us to a consideration of the roles the plants play in the middle-stage river ecosystem. Their rate of production varies from place to place, but a comparative view of the results that have been obtained is shown in Table 5.1

5.4 SUBMERGED PLANTS AND THE RIVER ECOSYSTEM

Submerged plants have many roles in middle-stage river ecosystems. Apart from the structures they provide for invertebrate microhabitats, for fish and invertebrates to lay eggs on, fish fry to find cover from predators, and fish predators to lurk in, they also supply labile organic matter and entangle organic matter such as leaf litter washed from upstream.

The plant beds may also affect the chemistry of the water, particularly by the removal of nitrate ions. This may be particularly useful in agricultural areas where increasing nitrate run-off from the land (Chapter 3) may cause

Table 5.1. Annual productivity of various aquatic plant communities. Values are given in grams of organic matter (ash-free dry wt) m^{-2} year^{-1}. (Modified from Teal [798]. Westlake [851], Bradbury and Grace [64] and Woodwell [872])

Communities	Average	Range	Maximum
Freshwater phytoplankton		Negl.–3000	
Submerged plants			
temperate	650		1300
tropical			1700
Floating plants			
duckweed	150		1500
water hyacinth		4000–6000	
papyrus		6000–9000	15000
Reedswamps			
Typha (reedmace, cattail)	2700		3700
Carex (sedge)		340–1700	1700
Phragmites	2100		3000
Tree swamps			
alder/ash		570–640	
spruce bog	500		
cypress		692–4000	
hardwood	1600	692–4000	
Comparisons			
tropical rain forest	2250		
boreal forest	900		
savanna	790		
temperate grassland	560		
open ocean phytoplankton	140		

problems in downstream lakes and for water supply (see Chapter 7). Nitrate removal may come from the growth of the plants themselves, in which case it is only temporary as the nitrogen will be released on decomposition of the plants, or it may be by bacterial denitrification. In this process bacteria use nitrate as an oxidizing agent to release energy from organic matter, but commonly do so at low oxygen concentrations (hypoxia) or in the absence of oxygen. This need not mean that the entire water body is hypoxic, but only parts of it. Usually these are sediment surfaces. The plant beds, in encouraging the accumulation of mud and in providing organic matter for bacterial decomposition may maintain extensive habitats where denitrification can take place.

The proportion of nitrate removed will vary greatly with the size and discharge of the streams. In the Whangamata stream in New Zealand with discharges of about 0.1 m^{3}s^{-1}, and with watercress beds, nitrate falls from about 0.6 mg NO$_3$–N l^{-1} to about 0.1 mg l^{-1} over a 2 km length. Plants could take up about 560 mg N m^{-2} of bed per day, whilst denitrification could account for about 56 mg N m^{-2} day^{-1} [369]. In another study [367], plant uptake was 1.14 g N m^{-2} day^{-1} and accounted for all of the nitrate loss from the stream.

In larger rivers, with both discharges and nitrate concentrations an order of magnitude higher than that of these New Zealand streams, the proportion of nitrogen taken up by the plants may be negligible in comparison [153], and any significant nitrate loss may be due mostly to denitrification. Owens *et al.* [616] found losses of 274 g m^{-2}year^{-1}, attributable 34% to plant uptake and 66% to denitrification in the R. Great Ouse.

There is lastly an aesthetic role of the river vegetation. Swards of water buttercups and the fringing brooklime (*Veronica beccabunga*), forget-me-not (*Myosotis*) and other species which characterize the shallow edges of small rivers are very attractive. In parts of Europe, favoured rivers for trout fishing may be carefully managed by hand-cutting of parts of the plant beds, and the leaving of others, to maintain ideal fishing habitats. Elsewhere, plant beds in rivers are regarded with less favour and are extensively cut or removed with herbicides in the interest of flood management.

5.4.1 Plant bed management in rivers

Extensive beds of submerged plants retard the flow and increase the depth of small rivers in spring and summer. In doing so they may cause local flooding of riverside land now used for the growth of crops. Summer flooding leads to deoxygenation of soil and deaths of the crop roots. In the past, in Great Britain, such flooding was acceptable for the flood-covered 'water meadows' were used for grazing with a flood-tolerant native flora, annually fertilized by the river silt. Many of these ancient habitats, very rich in plant species, have now been destroyed by ploughing [587, 740]. The fields have been seeded with crop grasses or cereals and flooding is unacceptable to the farmers.

Summer flooding may indeed have increased in frequency also as a result of these changed farming operations. First the removal of woodland and the drainage of riverside swampy land may have increased the summer flows (through reduced water storage and evapotranspiration). Secondly, the increased run-off of nitrogen and discharge of phosphorus compounds to the rivers may have increased the growth of *Cladophora glomerata* (blanket weed) which may form an important part of the plant beds. Thirdly, in the interests of cultivation to the river edge and of preventing brushwood falling into the channel and accentuating the flood risk by blocking it, previously overhanging trees and bushes have usually been removed. Loss of this shade lifts any previous light-limitation on the rooted plants and encourages their greater growth. The plant beds are managed in about a third of the main rivers and about 32 300 km of ditches and dykes in Britain.

Traditionally the plant beds (weed-beds if you regard them as a nuisance to your activities) are cut by hand or in deeper waters with a mechanical cutter from a boat. Cutting is done two or three times a year and can be regulated to preserve some of the plants for fisheries and aesthetic purposes. Removal of the cut material is needed to prevent its decomposition in the water, causing deoxygenation.

There is some evidence that cutting increases the growth and maximum biomass of the plants. Many of the aquatic plants are perennials with a yearly rhythm of growth in spring and early summer, then flowering in mid-summer and senescence in late summer. Cutting just before flowering removes the suppressive effect on growth of the flowering hormones and encourages extension of the side shoots. It also relieves the plant of the effects of its own self-shading which also encourages further growth. The effect is much like that of cutting a lawn. It determines the need for its own continuation. Beds of *Ranunculus* in Dorset rivers [153] were reduced to half their previous biomass when cutting ceased. The biomass is closely related to the efficiency of the beds in ponding back the water and creating floods.

Cutting is relatively labour-intensive and hence costly, and so there has been a movement towards the use of herbicides in recent years, especially in slower-flowing waters. These are believed to be less costly but because they are often over-used, the apparent savings may be small or not made at all. About eight compounds (Table 5.2) are officially cleared under the Pesticide Safety Precaution Scheme in the UK though rather more compounds are actually used. One in particular, diquat alginate, is favoured for treating submerged vegetation. The safety scheme in clearing a compound attends largely to the risks to stock and people and has little regard to the ultimate effects on natural communities. Indeed it cannot for data on such long-term effects are not available.

Herbicides are in general unselective and kill most aquatic plants. In turn this destroys the habitat for invertebrates and may cause deoxygenation as the plants rot. Herbicides may also be directly toxic to some animals. Repeated application can permanently destroy the habitat; recovery is often complete from light applications, and the disturbance caused by the creation of habitat 'space' during the effective period may allow additional species to colonize [578].

The effects of chemicals are, however, unpredictable in any precise way despite a plethora of information on their toxicities to selected animals in the laboratory. Usually the concentration (LC_{50}) necessary to kill 50% of a batch of test organisms in a given time (24–48 hours) is measured and from such experiments decisions are made as to safe dosages. However, the reactions of an organism may be quite different in its natural habitat where it is in competition with others for food or living space and where sub-lethal effects on reproduction or behaviour may result at concentrations far below the lethal dose in the laboratory. Enormous efforts have been made to standardize laboratory toxicity tests and the precision obtainable lends authority to this approach among those who would wish to use such chemicals in environmental management. I believe, however, that the accuracy of almost all of these data in predicting the consequences of chemical use for natural environments is very low and that this approach is useful only in screening out chemicals so overtly toxic even to people that they should never be used.

Table 5.2. Characteristics of some herbicides used for control of submerged plants. (Based on Newbold [590], Brooker and Edwards [74] and Eaton, Murphy and Hyde [184]

Herbicide	Notes
Diquat (6,7-dihydropyrido [1–2, a:2^1, −C] pyrozidiinium ion)	The cladocerans *Simocephalus serrulatus* and *Daphnia pulex* have LC$_{50}$ (48 hour) values of 4.0 and 3.7 mg l^{-1}. Variable effects on other invertebrates (LC$_{50}$ ranges from 0.12–>100 mg l^{-1}). Usually applied at 1 mg l^{-1}. *Simocephalus retulus* readily killed at 1 mg l^{-1}. *Polygonum amphibium*, and filamentous algae are restricted. Persists several days in water, months to years in mud
Dichlobenil (2,6 dichlorobenzonitrile) (breaks down to Chlorthiamid)	Most plants susceptible. Low toxicity to invertebrates (LC$_{50}$ values 3.7–24.2 mg l^{-1}) but higher for fish. Applied at 1 mg l^{-1} as granules with Fuller's Earth which are persistent. Its use favours replacement of plants by filamentous algae
Chlorthiamid	LC$_{50}$ (48 hour) values to *Gammarus lacustris* 1.5 mg l^{-1}, *Daphnia pulex* 3.7 mg l^{-1}, roach, 1.6 mg l^{-1}. More persistent than Dichlobenil
Terbutyrene	Controls plants at 0.05 mg l^{-1} and filamentous algae at 0.1 mg l^{-1}. LC$_{50}$ (48 hour) for *Daphnia magna* is 1.4 mg l^{-1}. Emergent plants are not controlled, only submerged. Very persistent: half life of 25 days in water, months in mud; many invertebrates not affected but herbicide drifts readily downstream
Dalapon (2,2 dichlorpropionic acid)	Used for emergent plants, is readily decomposed (in 2–3 days) and has low toxicity to fish. LC$_{50}$ (48 hour) for rainbow trout is 490 mg l^{-1}; LC$_{50}$ (48 hour) values for invertebrates 11–16 mg l^{-1}
Diuron (3-(3,4 dichlorophenyl)-1, 1-dimethylurea) (Monuron, fenuron and neburon are similar)	Used at 0.25–0.5 mg l^{-1} kills algae and floating plants. LC$_{50}$ (4 hour) values for *Daphnia pulex* and *Simocephalus serrulatus* are 1.4–2 mg l^{-1}. High toxicity to rainbow trout (growth retarded at 0.5 mg l^{-1}). Dragon- and damselfly nymphs killed at 1.5 mg l^{-1}
Copper sulphate	Mainly an algicide used at 0.5–1 mg l^{-1}. Very toxic to zooplankton and to insect larvae. LC$_{50}$ (48 hour) for rainbow trout is 0.14 mg l^{-1}. Carp survive to 0.33 mg l^{-1} and perch to 0.67 mg l^{-1}. Survival is greater in hard water because the Cu^{++} ion is precipitated as carbonate
Maleic hydrazide (2–4D; chlorpropham) (2,4 dichlorophenoxy acetic acid)	Used on floating leaved plants (e.g. *Nuphar*) and bank emergents, particularly broad-leaved species. Toxic at 1.0–4 mg l^{-1} to insects, oligochaetes, leeches and molluscs. LC$_{50}$ (48 hour) values for invertebrates 1.8–4.9 mg l^{-1}

A wiser approach to plant management has been proposed by Dawson [154, 155] and his collaborators [157, 158], who suggest the judicious planting of bankside trees to reduce the water plant biomass by partial shading. Trees on a northerly bank may cut down the incident light by about 20% in southern England and reduce the plant biomass by half. Trees and bushes planted on southerly banks will decrease light and biomass even more but may reduce the organic matter available to the animal community very significantly. The leaving of gaps in the plantings of say 20 m in 100 m allows some compromise. The replacement of trees along parts of the banks also has other advantages for wildlife such as riverine birds and otters, who make their holts or dens among the roots of large trees, and as an attractive landscape feature. Furthermore, the costs of tree planting per unit length of river are about the same as the costs of 1 year's cutting or herbicide treatment and additional maintenance over 10–20 years only doubles the initial costs.

5.5 FURTHER DOWNSTREAM—SWAMPS AND FLOODPLAINS

As the river widens and begins to deposit large areas of silt where its flow is reduced, emergent aquatic plants with the bulk of their photosynthetic biomass above water may come to occupy large areas known as swamps or marshes. There is no precise definition of either term but in general we can imagine swamps as permanently wet with peaty soils, often dominated by stands of a single tall herb species, usually a sedge or grass, or by trees and growing closest to the main channel of a river. Marshes, on the other hand, form the more distant short grasslands on more mineral soils flooded perhaps for only part of the year.

The combination of swamp and marsh in undisturbed river valleys may occupy all of a huge valley floor, the basis of a rich ecosystem involving indigenous peoples totally dependent on it. Alternatively it may have been long drained to form rich agricultural soils protected by embankments to prevent the river flood. The swamps act as huge filters for silt, change the chemical composition of water passing through them and are highly productive ecosystems.

5.5.1 Productivity of swamps and floodplain grasslands

Swamps are among the greatest producers of organic matter per unit area of the world's ecosystems (see Table 5.1). Values for the drier floodplain grasslands are scarce but these too are probably highly productive. The high swamp production is not surprising because of several factors. First the swamps tend to be dominated for large tracts by single, vigorous species (Fig. 5.2) so that little energy is used in competition with other species. In

a

b

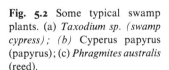

c

Fig. 5.2 Some typical swamp plants. (a) *Taxodium sp. (swamp cypress)*; (b) Cyperus papyrus (papyrus); (c) *Phragmites australis* (reed).

temperate regions of the northern hemisphere, the reed, *Phragmites australis*, reedmaces or cat tails, *Typha* spp., bullrushes, *Scirpus* spp. and other monocotyledons often predominate, with tree-swamp of alder (*Alnus*) and willow (*Salix*). In warmer temperate climates, members of the Restiaceae (South Africa, New Zealand), the saw-sedge (*Cladium jamaicensis*) or swamp trees—the swamp cypresses, *Taxodium* and tupelo, *Nyssa* (southern USA and Caribbean) are found. In the tropics, the igapo is a more diverse swamp forest in parts of the valleys of the Amazon and its tributaries, whilst tropical species of *Phragmites* and *Typha* and papyrus (*Cyperus papyrus*) characterize many African swamps. A full list would be much more extensive and the floodplain grasslands may be very diverse. A mixture of submerged plants and floating plants like duckweeds will be present in channels in the swamp, though the emergent vegetation will provide most of the biomass.

The second reason for high productivity is the favourable environment for growth. Only in extreme drought is water short; a continuous supply of river-borne silt brings in abundant nutrients; carbon dioxide is readily available from the atmosphere to the emergent parts and through the water to the submerged; and light probably becomes limiting during the growing season only when the stands become dense enough for self-shading to occur.

Measurement of the productivity of swamps has usually used methods which depend on the harvesting of biomass, partly because of a belief that little of the production is directly grazed—most entering the food webs as detritus. The few estimates made of turnover of the biomass, however, suggest that losses during the growth season may be quite high. Mathews and Westlake [526] applied the Allen curve technique (Chapter 4) to cohorts of emerging shoots of a grass *Glyceria maxima* and found an annual ratio of above-ground production to maximum above-ground biomass of greater than 1.5. Values of this ratio for some other temperate emergent plants (*Phragmites, Typha, Scirpus*) [851] do approximate to 1 but values for papyrus in African swamps were 1.8–3.6 [803]. A comparison [851] using the data of Schierup [716] among different methods of calculating the productivity of a stand of *Phragmites australis* showed that the lowest estimate (measurement of maximum above ground standing crop) gave a value of 1143 g dry wt m^{-2} year^{-1} which was only 55% of the best estimate (2085 g dry wt m^{-2} year^{-1})) obtained when turnover of shoots and roots and rhizomes was measured. Many estimates of swamp production are probably thus underestimates and this strengthens the case for particularly high production in swamps relative to other vegetation types.

5.5.2 Swamp soils and the fate of the high primary production

A continuous supply of dead organic material falls to the swamp floor. The plants themselves impede water movement so relatively little is likely to be washed downstream. Much of the production must thus be oxidized by secondary producers or stored in the swamp soil.

Oxidation through the food webs in the oxygen-rich environment of the emergent shoots appears negligible. The budgets that have been produced (see, for example, Table 5.3) suggest that a negligible fraction of green tissue is eaten by insects, such as grasshoppers and leaf miners, and that although birds such as geese may eat both green tissue and seeds and rodents like muskrat and coypu will excavate rhizomes, the impact of these activities is small. Perhaps this is because the aquatic plant biomass is bulky, because of the air spaces in it (see below) and has a high water content, hence a low energy content per unit volume.

If direct grazing is a negligible pathway by which the organic matter is oxidized, an alternative is fire. Particularly tropical and sub-tropical swamps, like those of the Florida Everglades, may lose much organic matter through occasional lightning-induced fires in the dry season [687]. The relative importance of fire in swamps is not known—it may be the reason why swamps in seasonally dry and hot regions do not accumulate much organic matter on the swamp floor. In cool and wet regions the storage in soil of this partly decomposed organic matter or peat is a major fate of the production. The other is oxidation through detritus pathways by microorganisms and

Table 5.3. Energy budget of a freshwater swamp. Values are in kcal m^{-2} year^{-1}. (Compiled by Howard-Williams [363] from Dokulil [171], Imhof [390] and Imhof and Burian [391]

Primary producers	Net primary production
Phragmites communis (australis)	15000
Utricularia vulgaris	150
Planktonic algae	500
Periphyton	1000
Total	16650

Herbivores	Energy uptake
Plant-eating insects	40–60
Muskrats	<10
Total	<70

Algal feeders	Net production
Chironomidae	10
Snails	30
Asellus aquaticus	10
Total	50

animals living in the water, and the upper layers of peaty soil, and the balance between the two will depend on the oxygen supply.

5.5.3 Oxygen supply and soil chemistry in swamps

In swamps oxygen may not diffuse into the system as rapidly as microbial and animal respiration uses it up. This is especially true of the soil which is often anaerobic from the surface downwards. In the most stagnant, warmer waters it may be true of the overlying water also. The deoxygenation limits the rate of decomposition and favours rapid peat build up so that if the flow of water is very low the soil may have little inorganic content (Fig. 5.3). The waterlogged soil will, however, have certain chemical characteristics which pose problems for the plants growing in it and change the chemistry of the overlying water [207, 741].

Figure 5.3 illustrates effects on soils dependent on flow rates, in swamps. In both, the soil itself—a semi-solid medium in which oxygen diffusion is low, will be reducing. The degree of reduction can be measured as the redox potential. If a calomel ($Hg_2 Cl_2$) platinum electrode is immersed in water, wet soil or sediment whilst connected into a circuit which also includes a standard hydrogen electrode, electrons will tend to accumulate on the calomel-Pt electrode to an extent reflecting the number of free electrons available. Oxidations are chemical processes in which electrons are removed from a compound whilst in reductions electrons are added. A large number

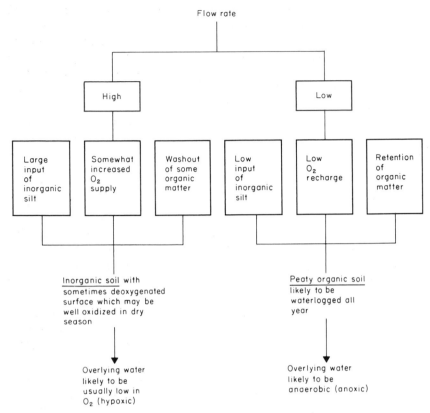

Fig. 5.3. Consequences of differences of flow rate in swamps for water chemistry and peat accumulation. High flows are usually seasonal; low flow implies year around stagnation.

of free electrons available for donation indicates a reducing environment. In the electrode system, the electrons accumulating on the calomel-Pt electrode relative to the hydrogen electrode can be measured in millivolts (mv) on a galvanometer placed in the circuit. In an oxidizing medium, electrons will tend to move away from the electrode giving a relatively positive electrode potential difference; in a reducing medium their accumulation will give an increasing negative potential. Whatever the value it is conveniently known as the redox potential. Well-oxygenated waters may have potentials greater than + 500 mv.

As heterotrophic bacteria in sediments decompose organic matter and the oxygen concentration falls, so does the redox potential to values of about + 200 mv when anoxic conditions are reached. At this point decomposition will have produced inorganic ions such as phosphate and ammonium to add to the ions already present in the interstitial soil water. Bacterial respiration, becoming anaerobic, may produce acids such as butyric and acetic acids as end products. The carbon dioxide content will have risen and the pH will consequently have fallen. As oxygen is depleted either generally, or locally in the middles of crumbs of soil, other oxidized ions may be used by microorganisms to oxidize organic matter. Some protozoa [217] and many bacteria (e.g. *Achromatium, Bacillus, Pseudomonas*) reduce nitrate, if any is present, to N_2O or N_2 through denitrification. At the same time, deoxygenation favours nitrogen fixation by free-living bacteria (*Clostridium pasteurianum*) and bacteria in nodules on the roots of swamp trees like alder (*Alnus*), so that after decomposition of these organisms there is a supply of ammonium ions in the soil.

Denitrification is generally complete as the redox potential falls to $+$ 100 mv, at which point anaerobic bacteria, unable to oxidize carbon compounds further, may release ethylene, and bacteria or perhaps inorganic chemical processes begin the reduction of Mn^{3+} ions to the more soluble (and toxic) Mn^{2+}. At slightly lower redox potentials *Clostridium* and other bacteria reduce the orange-red ferric ion (Fe^{3+}) to green-grey, also more soluble, ferrous (Fe^{2+}) ions. If water is moving through the soil these soluble ions may be washed out leaving a pale green or grey appearance in what are known as gley soils.

Just below a redox potential of o mv, bacteria produce and release methane, (*Methanobacterium, Methanomonas*) hydrogen (*Clostridium* spp.) or phosphine (PH_3) (marsh gases) which may spontaneously ignite as 'wills o' the wisp'. *Desulphovibrio delsulphuricans* and *Desulphomaculatum* oxidize H_2 or organic matter by reducing sulphate in the interstitial water to sulphide. In turn the sulphide may precipitate Fe^{++} to form black iron-sulphide in the most intensely reducing soils. The results of these processes thus produce a soil with a number of potentially toxic ions (Fe^{2+}, Mn^{2+}, HS^-) and organic acids, no oxygen, and ammonium rather than nitrate as a nitrogen source for roots. An example—a mild one for there is some water flow under the mats of papyrus—is shown in Fig. 5.4 of how these processes may affect in turn the composition of the overlying water.

Some of the chemical changes which occur on waterlogging may be reversed if the flooding is seasonal and the soil dries out from time to time. For example, CO_2 concentration may decrease, pH may rise and oxygen may diffuse in to greater depths. The reduced iron may be oxidized to Fe^{3+} to release energy by bacteria of the genera *Gallionella* and *Sphaerotilus* (iron bacteria) giving characteristic deposits of rust or ochre. The oxidation of the sulphides back to sulphate by sulphur bacteria (e.g. *Beggiatoa*) may produce insoluble sulphates (gypsum) if calcium is abundant, or sulphuric acid if it is scarce, as in many peats. This reduces the pH to 2 or 3 and forms acid sulphate soils. Ammonium ions may be oxidized by nitrifying bacteria (*Nitrosomonas*) to nitrite and then (*Nitrobacter*) to nitrate. On reflooding the nitrate will be denitrified so a swamp with a seasonally fluctuating water level may ultimately remove much combined nitrogen from the water.

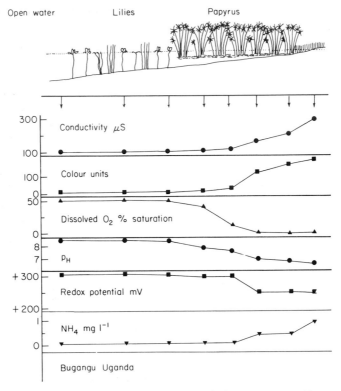

Fig. 5.4. Environmental conditions at the fringe of a tropical papyrus swamp. Decomposition in the stagnant swamp water leads to increased dissolved ion levels, high coloured humic acid levels, and ammonia levels. Oxygen is used up rapidly, redox potentials are decreased and CO_2 production decreases pH. (Redrawn from Carter [94].)

This interlinked set of inorganic chemical and bacterially mediated reactions thus produces a swamp soil, and an overlying water, which are 'difficult', for the higher organisms which colonize them, in more ways than mere oxygen lack. Some of the ways in which plant roots may cope with conditions in the sediment, and animals with those in the water will now be reviewed.

5.5.4 Emergent plants and flooded soils

Flooding kills many plants because their roots are unable to respire aerobically, and the toxic products of anaerobic metabolism—acetaldehyde or alcohol—usually build up to lethal concentrations in the tissues. Swamp plants often grow better in drained soils than in waterlogged ones, if freed of competition with other plants, but can persist in swamps because they are able to provide a supply of oxygen to the roots from the atmosphere via the emergent leaves and stems. They may be able to limit the formation of toxic anaerobic products, excrete them to the water or form non-toxic ones instead of ethanol. They may also be able to detoxify ions like Fe^{2+} and Mn^{2+} which diffuse into the roots from the soil.

That aquatic plants have systems of internal air spaces or lacunae has long been known and it is often assumed that these must act as channels for the diffusion of oxygen to the roots and rhizomes. In some cases this is true, though diffusion alone may be too slow a process to give a ready supply. In one species of water lily there is evidence of a mass flow of air from the emergent leaves to the rhizome [140, 141]. If polyethylene bags containing air are sealed over the young leaves by day, the bags collapse as air is withdrawn from them. An internal pressure, slightly greater than atmospheric, is created in the air spaces of the leaf by absorption of heat radiation. The inner parts of the leaf are in contact with the atmosphere through stomata on the upper surfaces, but the rate of diffusion of air through these is insufficient to prevent build up of pressure in the young leaves. As the leaves age the stomata enlarge and diffusion through them is much faster and prevents the pressure increasing. The pressure in the young leaves forces air down through the petioles and into the rhizome from which air is exhaled via the older leaves. In another case, rice, there seems to be a mass flow of air through the continuous bubbles which coat the outside surfaces of the submerged parts of the leaves and stems [660].

Not all emergent plants, however, have such bubbles or unoccluded internal passages. Sometimes plates of cells interrupt the passages — diffusion through these is very slow and mass flow is prevented [382, 731]. Some marsh plants, for example *Filipendula ulmaria* and *Phalalaris arundinacea*, do not have any air spaces. The role of lacunae in oxygenating the roots of emergent plants may thus often be minor, and biochemical mechanisms may sometimes better explain flood tolerance. However, one possibility is that the lacunae provide internal surfaces in contact with air on which toxic reduced ions (e.g. Mn^{2+}, Fe^{2+}, HS^-) diffusing into the root from the soil, can be oxidized. Diffusion of oxygen from some roots may have this effect at the outer surface in contact with the soil where iron compounds may be precipitated [870].

There are three groups of ways in which the biochemistry of roots may be modified to allow plants to grow in waterlogged soils [126]. First, many finely divided adventitious roots (those arising along a buried or submerged stem) may be produced providing a large surface area through which alcohol and acetaldehyde can diffuse into the soil water. This happens in several flood-tolerant trees, especially where there is moving water to remove the excreted compound. Secondly, the rate of anaerobic metabolism can be controlled to minimize the rate of production of alcohol [128a]. And thirdly, the end products of anaerobic metabolism can be diverted to less toxic compounds— pyruvic acid and glycolic acid (willow), malic acid, shikimic acid (iris and water-lily roots) [128] and several amino acids—alanine, aspartic acid, glutamic acid, serine and proline. Eventually all of these compounds must be oxidized if root function is to continue and this must either be by translocating them to better aerated parts of the plant or during parts of the day when oxygen movement to the roots is greatest. Not all flood-tolerant plants produce such compounds [747] and when they are produced, they may not entirely replace ethanol [427] and may have other roles, such as in combining with toxic reduced ions to form less toxic compounds [208].

5.6 SWAMP AND MARSH ANIMALS

Floodplains often have spectacularly large populations of birds and mammals; fish may be equally abundant. The reasons are the great diversity of habitat to be found on the floodplain (Fig. 5.5) and the high productivity of the floodplain vegetation.

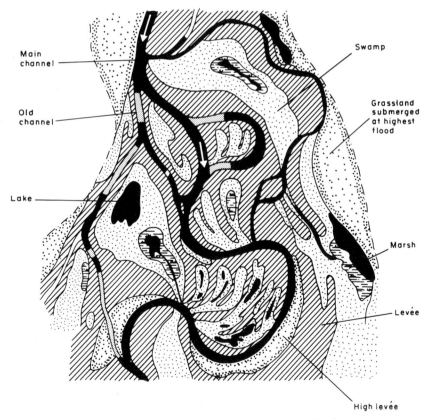

Fig. 5.5. Features of a floodplain environment. (Based on Welcomme [845].)

The truly aquatic invertebrates of the swamps are remarkable. Groups of air-breathers (Chapter 2)—beetles, some Diptera, pulmonate snails—are predominant, together with specialized members of some otherwise water-breathing groups. For example, an oligochaete worm, *Alma*, has developed a deeply grooved tail which is richly supplied with blood vessels. The tail can be extended to the surface of a waterlogged, deoxygenated floating mat of vegetation and the groove flattened out in contact with the air. Some snails, for example *Biomphalaria sudanica*, have both gills and lungs.

Among the fish, airbreathing is crucial for those which stay in the swamp all year. In the low-water season they may become confined to the deeper channels and pools where the stagnation of the water, and the confinement of large numbers to a small volume make deoxygenation an even more severe problem than in the flood season. It may also make, for the observer, a spectacular collection of fish, turtles, water snakes and alligators in, for example, the sloughs (pools) of the Florida Everglades.

Modifications for breathing in swamp fish (Fig. 5.6) range from a flattening of the head, allowing the fish to remain in the thin layer of water just at the surface, where oxygen concentrations are greatest, to a development of lungs (the bichir, *Polypterus*, and the lungfish, *Protopterus*, *Lepidosira* and *Neoceratodus*) used for no other purpose than the breathing of atmospheric air. In the latter cases the fish will usually drown if denied access to the atmosphere. Between these are many modifications of existing organs for air breathing. Some catfish (*Clarias*, *Anabas*) and the electric eel (*Electrophorus*) have bony supports which minimize the collapse of the gill filaments (a collapse would reduce their surface area for absorption) and allow the gills to be used in both air and water. The swim-bladders, normally buoyancy regulators, and one or other parts of the gut have in some species become richly provided with blood vessels to allow swallowed air to be absorbed through them. One of the world's largest freshwater fish species, the Amazonian *Arapaima* and the North American bowfin (*Amia*) and garpike (*Lepisosteus*) have modified swim-bladders; and the catfish *Plecostomus* and *Hoplosternum* have vascularized stomachs and intestines respectively.

These air breathing fish normally remain within the swamp for the whole year. Despite its chemical extremes, the swamp is a permanent and predictable habitat, compared with the temporary aquatic conditions of the floodplain grasslands, in that its seasonal conditions are repeated on a regular cycle. This favours an economy in reproduction, with relatively few, but large eggs each with a high chance of survival being produced and often guarded by the parents. Some make floating nests of vegetation and mucus froth. Where the swamp does dry out seasonally, some annual species may survive as eggs buried in damp mud (e.g. some Cyprinodont fishes [475, 476]) or even, in the case of some lungfish, as adults cocooned in a muddy chamber lined with body slime. The fish breathe through a tube emerging at the mud surface.

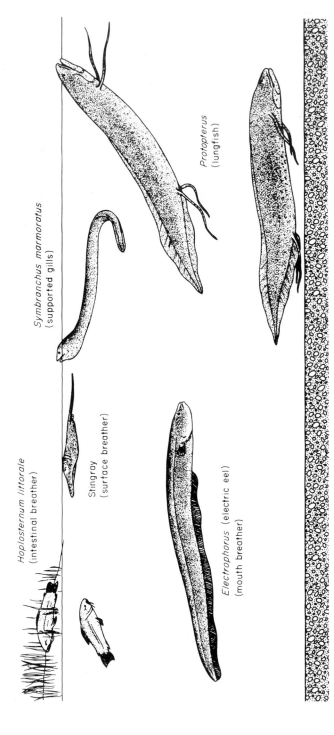

Fig. 5.6. Some air breathing fish from S. America and Africa (*Protopterus*).

5.6.1 Whitefish and blackfish

In the Mekong river floodplain the swamp-tolerant fish are referred to as 'blackfish'. Often they are dark in colour with small or few scales and come from a group of families which include the siluroids and anabantids (catfish), channids (ophiocephalids), osteoglossids, bichir and lungfish. In contrast there are the silvery, scaled 'whitefish' of other families, which have very different characteristics. The terms 'blackfish' and 'whitefish' conveniently describe similar functional groups in most floodplains.

The whitefish are migratory. They may move upstream from the dwindling channel of the river in the dry season and breed in the headwaters, or an upstream floodplain in high flood. Here they move out to the fringes where oxygenation is highest. The adults and young of the headstream spawners move back to the floodplain to feed and, with those that bred on the fringes, retreat to the main river as the flood goes down. The whitefish tend to produce a large number of small eggs which are scattered and each of which stands only a small chance of survival. This seems to be a strategy which is successful where the conditions for breeding—well-oxygenated water—may be short-lived and where rapid reproduction is needed in an environment likely to change rapidly and unpredictably.

The contrast between the tolerators of the permanent swamp, like the blackfish, and the migratory whitefish who take advantage of it seasonally, is reflected to some extent in the mammals and birds. Among the former, the sitatunga, a central African antelope is a more or less permanent swamp-dweller with feet splayed to allow it to walk over soft, often floating, beds of vegetation, whereas another antelope, the red lechwe, is a migrator onto the floodplain grasslands as the water retreats and fresh growth begins. Something of the same pattern can be seen in the behaviour of indigenous human societies associated with the floodplain wetlands.

5.7 HUMAN SOCIETIES OF FLOODPLAINS

5.7.1 The Marsh Arabs

For the Madan, the Marsh Arabs of Iraq (Fig. 5.7), Wilfred Thesiger [801, 802] Gavin Maxwell [527] and Gavin Young (Young and Wheeler [882]) paint a picture of a people more or less permanently dwelling in tall, permanent reed swamps. These span the 6000 square miles around the confluence of the Tigris and Euphrates at Qurna inland of Basra near the Arabian Gulf. *Phragmites australis* grows to 4 m in height and dominates the swamp; it is flanked by seasonal swamps of *Typha angustata* and a sedge *Scirpus brachyceras*. The Madan are descended, in part, from the Sumerians who founded the great Mesopotamian civilization at the edge of the marshes

Fig. 5-7. Part of a Madan village, Iraq. (From Thesiger [802].)

6000 years ago; a Sumerian legend has it that Marduk the Great God built a platform of reeds on the surface of the waters and created the world.

Perhaps this was the way that the ancient Marsh Arabs started to build their houses as it is the way today. On the platform simple houses of reed bundles are built and also huge and elaborate structures, *mudhifs*, or guest houses, which may be 60 feet long and 20 wide. The bases of their construction are pillared arches of reed 2 metres in girth and tightly bound. Inside these structures an elaborate social life is possible in comfortable carpeted surroundings.

Outside the buildings, the platform houses the three to eight water buffalo which are the mainstays of a marsh family. These placid animals are taken to graze, where the water is shallow enough for them to touch bottom, on sedge or water plants (*Polygonum senegalense, Jussiaea diffusa, Potamogeton lucens, Cyperus rotundus*) and on return at nightfall are fed the young shoots of sedge or reed (*hashish*) cut during the day by the Arabs. Buffalo provide milk, meat and dung for fuel, whilst birds are shot for meat, or, in the case of pelicans for the soft pouch skin which, once cured, forms excellent drum and tambourine skins.

Fish, particularly *Barbus sharpeyi*, are speared with forks from lines of boats which block channels in the swamp, or may be poisoned with flour and dung bait laced with extracts of the toxic plants, *Digitalis* or *Datura*. Increasingly, gill nets are laid, and the fish sold to the cities. Boats are usually of wood and brought in from the coast but, as elsewhere, simple canoes are built from *Typha* bundles bound together.

On the edges of the swamp, in shallow water, rice is cultivated and provides a staple diet; the ricefields also attract the large wild boar which are frequent in the marshes. At a metre or more at the shoulder and aggressive, the boar account for many deaths and injuries to Madan who surprise them whilst punting through the channels or cutting reed for the buffalo.

The Madan were originally largely self-contained; all their European visitors comment on the hospitality they received and the complexity of clan and tribal relationships. They suffered also from poor health—a high infant mortality, yaws, schistosomiasis, dysentry, tuberculosis. Increasingly, modern health care is being made available to them and there is more trade in woven reed mats and fish with areas outside the marsh. The Madan represent an instance of human beings who have been a part of a self-contained and perpetuating swamp ecosystem. So also do the Nuer, who, in contrast, are migrators, rather than permanent dwellers of the swamps.

5.7.2 The Nuer of Southern Sudan

The Nuer are a nilotic people living in the valleys of the White Nile (Bahr el Jebel), Bahr el Ghazal and R. Sobat around their confluences in the Sudd

swamp between Bor and Malakal in the Southern Sudan (Fig. 5.8). Like the
Dinka who occupy the eastern part of this region, to the Nuer's western, they
have a mixed economy and a social system which rests heavily on a climatic
and food cycle [110, 210, 592].

The area receives much rain (over 800 mm per year) which falls, between
April and September, onto a flat plain floored by clay, lacking trees and

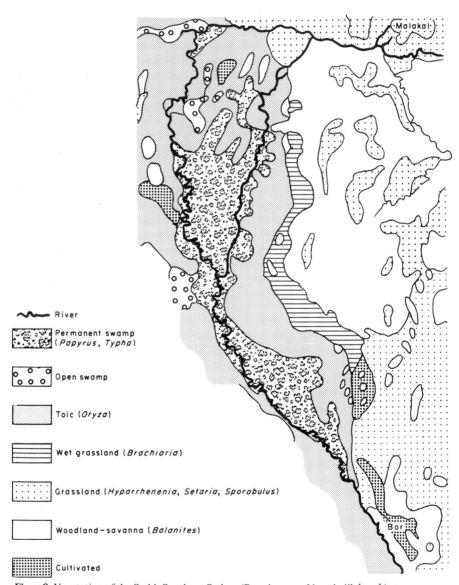

River

Permanent swamp
(*Papyrus, Typha*)

Open swamp

Toic (*Oryza*)

Wet grassland (*Brachiaria*)

Grassland (*Hyparrhenenia, Setaria, Sporobulus*)

Woodland–savanna (*Balanites*)

Cultivated

Fig. 5.8. Vegetation of the Sudd, Southern Sudan. (Based on van Noordwijk [592].)

crossed by large rivers. Close to the rivers is a permanent swamp of papyrus, reed, *Phragmites mauritianus* and floating or part-floating beds of water hyacinth (*Eichhornia crassipes*) and hippo grass (*Vossia cuspidata*). To the edge of the permanent swamp are regularly flooded swamps with *Typha australis*, which lack surface water in part of the dry season, and regularly flooded grasslands called toic. The toic has grasses such as *Echinocloa pyramidalis*, *E. stagnina* and wild rice, *Oryza barthis*, and receives silt-laden river water which fertilizes it annually. Further distant from the rivers the toic is replaced by grasslands which flood—to depths of up to a metre and largely with rain water—in the wet season and are less nutrient-rich. The upland, which is not flooded, is reached often some tens of kilometres from the rivers. The area is rich in both large aquatic animals and many bird species which move onto the grasslands as the water recedes. These include most of the world population of the Nile lechwe and several hundred thousand tiang, an antelope sub-species of the tsessabe (*Damaliscus korrigonus*).

The Nuer have their permanent villages at the edge of the upland, close to the flat, flooded plain. They are a cattle-keeping people and movement further into the upland is discouraged by the risk to their cattle of catching nagana, or cattle trypanosomiasis, a protozoan blood disease carried by the tsetse fly *Glossina morsitans*. The fly requires shady woodland for reproduction. The flooded clay-plain is clearly unsuitable for permanent villages, but the Nuer must stay close to it for it provides grazing for the cattle. The swamps beyond it are a source of fish when the milk yields are low in the dry season.

After the rains have begun, in May, the village gardens are sown with the staple cereal, millet (*Sorghum*) and perhaps some maize, beans, gourds and tobacco, which will be harvested in July and August. This work is done mostly by older people, the younger being still in temporary camps near the swamp where, at the end of the dry season, there is still some grazing and where the pools and lagoons and diminished stream channels are a concentrated source of fish. Everyone returns to the village by June and the wet season provides an ample food supply based on milk and its soured products from the cattle, and meat from some sheep and goats which are kept. Blood is drawn from the necks of the cattle, coagulated and roasted.

Meanwhile fishing is difficult for the fish are dispersed to feed and spawn in the flooded grasslands, which are difficult to cross. The lack of trees means that boats are scarce.

Especially when the millet is harvested, wedding, initiation and other ceremonies take place and the social organization is at its most complex, though the eventual need to move out in small groups has kept this to a low degree, with very little tribal hierarchy. A second sowing of millet in August is harvestable by December, when the rains have stopped and when, particularly if the rains have been poor, food may be very scarce.

The cattle are taken by the younger people onto the plain in August whilst the older remain to harvest the millet. The grass produced on the rain-flooded grassland is not of high quality for cattle and soon has dried. Some areas may be burnt to bring new shoots, but the main grazing is in the toic. Temporary camps are established on mounds at the swamp edge of this grassland, to which the older people also move after the millet has been harvested. These camps are primitive, comprising largely wind-screens of swamp grass, and the social life of the groups is family based and geared to the business of survival. Milk yields fall as the toic grasses shrivel and the millet store may be very meagre. This brings into emphasis fishing, the third focus of Nuer economy, with perhaps some collecting of wild dates (*Balanites aegyptica*), seeds of wild rice and rhizomes of a water lily (*Nymphaea lotus*) to eat. The swamp fish include catfish (*Bagrus, Clarias*) and lungfish (*Protopterus*) among others and are caught by primitive methods. These involve making dams across streams or lagoons and spearing the fish as they attempt to move back to the river. A line of withies—thin branches, probably from the ambatch bush (*Aeschynomene*) which grows in the swamps, may be pushed into the mud in front of the dam. The quivering of these as the fish move into them signals the target for the spearsman. Lungfish aestivating in the mud may be sounded out by tapping the surface with a pole or by recognizing sounds the buried fish make when a finger is scraped over a gourd. They can then be dug for. There may be a little hunting—largely for products such as skins of waterbuck for bedding or of hippo to make shields. Cattle stealing, especially by the Nuer from the Dinka, is not unknown.

As the rains begin again the older people move back to the villages to prepare the gardens and the cycle is completed. Its three parts are each essential—diseases like rinderpest prevent any great expansion of the cattle population, though the Nuer regard cattle keeping as the most honourable of their activities; the climate precludes total dependence on grain, the crops of which often fail; and the hydrological cycle of dispersion of the fish in the wet season means that fishing cannot be the sole source of food. Since 1961, however, a rise in the discharge of the Nile, linked with a 1–2 m rise in the levels of L. Victoria at the head of the river, has led to an extension of the permanent swamp and a marked reduction in the grassland. The Nuer have turned more to fishing though they regard themselves primarily still as cattle keepers.

5.8 FLOODPLAIN FISHERIES

For many peoples of the world inland fisheries are their major source of protein and, among fresh waters, the floodplains provide the most productive and diverse fisheries. Part of the Amazon floodplain illustrates this well.

The Amazon basin is rained upon all year round but the rains are seasonal and water levels change remarkably between the wetter season (December to June) and the drier (July to November). Differences in level may be as much as 16–20 m between high and low water. This means that the forests are flooded, sometimes to the very tops of the trees, and the lagoons and lakes alongside the rivers may vary in depth from 16 m to less than 1 m, or may even be dry in the drier season.

The fish fauna is very diverse and productive with strong tendencies to migrate which allow the fish to move seasonally between habitats. Indeed one of its particular characteristics is its dependence on the food supplied by the forests at high water. Among the characins, which, with the catfish, constitute about 80% of the species list of at least 1300 species, are many species feeding on seeds, fruits, flowers, insects and monkey dung falling from the forest canopy (Table 5.4). Fish returning to the main channels as the water recedes are fat from this source. Fruit and seed eating appears to benefit the trees also, because often only the fleshy fruit is eaten and the seed is not digested but is defaecated to germinate elsewhere [294, 295]. The epiphytic community of microorganisms (periphyton) which grows on the trees and on the tree debris is another important food, especially for some large catfish species. In an area around Porto Velho, on the Rio Madeira, 87% of the commercial catch comes from nine genera of which a third of the species depend directly on forest seeds, and a quarter on forest detritus and periphyton; the rest are predators on the seed and detritus eaters.

Other vertebrates—dolphins, manatees, caimans, turtles, anaconda, tapirs and capybaras are also integral members of the system. Where their numbers have been reduced by hunting, it has sometimes been noticed that fish production has fallen [219, 220]. They may mobilize nutrients from the sediments and soils to the water as they feed on plants (manatee, capybara, tapir) or larger fish (dolphins, reptiles). They move into the swamps or flooded forest at high water. This is where most of the fish spawn, producing young which require small planktonic organisms as their first food. The pulse of nutrients excreted by the pursuing predators may support planktonic production at the appropriate time.

A varied fish fauna demands a varied fishery and in the Porto Velho area of the Amazonian Rio Madeira [294, 295] methods vary from those that are simple, but cunning and dependent on a deep knowledge of the habits of the fish, to the use of modern monofilament gill nets, backed by refrigerator ships. The former support subsistence, the latter commercial fisheries. The methods can be grouped into those exploiting fish in the rapids, the main channel, the swamps and the flooded forest.

Twenty kilometres above Porto Velho lie the Teotonio rapids where a 10 m fall over a short distance creates a complex of fast and quieter water and pools as the water moves between the boulders. Many of the catfish species migrate up and down these rapids prior to spawning and a long oral

Table 5.4. Diets, expressed as volumes of food found in adult fish stomachs from samples of three species of fish from the Rio Machado (Brazil) floodplain. All three species are of the family Characidae

	Tambaqui (*Colossoma macropomum*)		Pacutoba *Mylossoma duriventris*		Piranha preta *Serrasalmus rhombeus*	
	High water (forest)	Low water (lakes)	High water (forest)	Low water (channel)	High water (forest)	Low water (lakes & channel)
Number of stomachs examined	96	27	96	114	157	97
Number empty	3	11	0	22	68	32
Fruits and seeds						
Hevea spruceana	3840	0	0	0	0	0
Astrocaryum jauri	1036	0	0	0	0	0
Neolabatia sp.	545	0	0	0	0	0
Pirahhea trifoliata	0	0	2990	0	0	0
Burdachia cf prismatocarpa	0	0	2870	0	0	0
Mabea sp.	0	0	1465	0	0	0
Other fruits and seeds	1130	5	6635	895	420	295
Percentage of total	94	14	95.5	22.3	7.7	11
Other plant material						
tree resin	0	0	0	0	100	0
flowers	0	0	0	1080	55	0
leaves	0	0	355	1245	10	175
detritus	0	0	0	0	0	5
Percentage of total	0	0	2.4	58	3	6.7
Animal material						
monkey faeces	355	0	250	0	0	0
fish	50	25	0	0	4447	1994
crabs	0	0	0	0	200	100
bird	0	0	0	0	100	0
porcupine spines	0	0	0	0	5	0
rat	0	0	0	0	0	100
mammal hair	0	0	0	0	50	0
lizard	0	0	0	0	25	0
zooplankton	12	0	0	0	0	5
mayfly larvae	0	5	0	0	0	0
cockroaches	0	1	5	0	0	0
beetles	0	0	45	15	25	0
caterpillars	0	0	5	0	0	0
ants	0	0	0	761	0	0
other invertebrates	0	0	0	15	0	0
Percentage of total	6	86	2	19.7	89.3	82.3

tradition exists to advise fishermen of where to catch particular species at particular times as they negotiate the cataract. Platforms are built of wood on the left bank (facing downstream) over a stretch of quieter water through which a big catfish, the dourada (*Brachyplatystoma flavicans*) moves. The water is stroked with curved hooks bound to a 5–8 m pole called *fisga*. Once a fish is gaffed, the hook comes free on a line attached to the pole and the fish is played until it tires and can be pulled out.

In the main channel, use of the gill nets also depends on knowledge of seasonal migrations. Some characin genera, *Brycon* and *Mylossoma* live in a clear water tributary, the Rio Machado, during the low water season, but as the flood rises, move downstream into the more turbid Rio Madeira to exploit the detritus and periphyton on the flooded forest trees in the main river valley. They move in a 10–14 day period when the water is a few metres below the peak flood. The fishermen watch for dolphin activity, which indicates a characin shoal, in the clear water tributary before the water mixes into the turbid Rio Madeira, where the fish disperse. They manoeuvre upstream of the shoal and place the 100–200 m gill net in a horseshoe pattern before beating the water with paddles to scare the fish, causing them to reverse direction and move into the net, which is then drawn tight with a rope threaded through the bottom of it. In May to July the same method is used for other species which are returning from feeding in the Rio Machado forests to move upstream in the main river.

In the swamps and the flooded forests, the methods depend much on individual skills. The swamps harbour cichlid fish which move along the edges of the floating mats of vegetation. At night they can be paralysed with a light and stabbed with a 1.5 m long, pronged spear. One voracious cichlid predator, *Cichla ocellaris*, is lured by movement through the water of a tassle of strips of red cloth or birds' feathers in which are embedded hooks. It mistakes the lure for its prey. The picarucu (*Arapaima*) which must return to the surface to breathe every few minutes is harpooned when it does so.

In the flooded forest—a refuge against large-scale fisheries where nets would be tangled—individuals can be caught from a knowledge of their diet. Seeds and fruit are released from the trees sporadically and are at a premium. They represent quite large but infrequent meals and are usually snapped up as they fall—their characteristic shapes and sizes making specific 'plops' thought to be recognized by the fish. These noises, particularly those of the jauri, a palm tree, (*Astrocaryum jauri*), can be simulated with metal ball bearings or nuts cast on lines. The fish are harpooned as they dart in to feed. Seeds can be used as bait also in simple rod and line fishing, with palm fruits or those of rubber trees (*Hevea*) or a cucurbit (*Cuffa*) the most successful baits.

5.9 MODIFICATION OF FLOODPLAIN ECOSYSTEMS

5.9.1 Wetland values

River floodplain ecosystems are notable parts of landscapes. To some peoples they represent home and figure prominently in their lives. To others—those who live on the surrounding uplands—they may represent a threat: from disease, or the fear of being lost or sucked down in them. Or they may be 'wasteland', useless for anything as they stand but potentially convertible by drainage to fertile agricultural land. The former fears are reflected in the housing in swamps of malevolent trolls, water witches and the like in traditional children's literature. Swamps have acquired a poor image in consequence, and nations have encouraged with alacrity the drainage of their floodplain swamps until the recent realization of their value as wetlands [517]. The value, reflected in increasing legislation to protect swamps, where before there was only legislation to finance the draining of them, rests in three areas: flood control, sediment and nutrient retention, and wildlife [365].

Swamplands are large sponges, with often great surface area. A small increase in water level results in the temporary storage of water which would otherwise rush downstream after a heavy storm to cause damage to human settlements. The effect is to spread out the flood peak over time, reduce its height, and minimize erosion of the downstream banks. It is estimated that for the Charles River catchment in Massachusetts, the loss of the 8422 acres—a small proportion—of wetland would result in average annual flood damage downstream of M\$17 [601]. In the spreading out of the water, some is also lost by evapo-transpiration, and this may also help alleviate downstream damage, though it also means a loss of water that might have been used for irrigation or domestic or industrial water. There is some controversy over the amounts lost by this process [365, 470, 746]. It seems that evapotranspiration from the swamp vegetation may actually be less per unit area than from an open water surface by as much as 40%. The vegetation reduces its own loss by the closing of stomata in hot, windy conditions, whilst also shading and cooling the water surface. Some people have thus argued that a wetland will save water. However, it will only save it if it would otherwise be spread out in a sheet as a lake behind a dam, for example. More water will be lost in total if the water spreads out in a swamp than if the water was allowed to move downstream in a direct channel.

5.9.2 Swamps and nutrient retention

Sediments and nutrients are retained in swamps, the former by the effects of reduced flow [593], the latter if the swamp is laying down peat which inevitably retains a proportion of the mineral nutrients originally contained within the plants. Nitrogen will also be removed from the water by denitrification if it is in the form of nitrates.

These properties may result in improvements in the chemical quality of the downstream water. Sediment might otherwise cause problems in the blocking of irrigation channels or the main channel if used for navigation, or might need to be expensively filtered from the water if it is to be used for industry or domestic supply. Excessive nitrogen and phosphorus can cause major problems in lakes and there are fears that increasing nitrate concentrations might cause health problems.

This retention has been eagerly recognized by civil engineers wishing to find cheap ways of treating effluents from towns and industry. Running the effluent into a riverine swamp to be 'treated' is apparently far less expensive than the building and operating of works for chemical treatment, and small 'artificial' swamps of reed, *Phragmites*, are being established at a number of sewage treatment works [267, 688]. Large natural swamps are also being viewed as potential treatment works [164, 420]. It is sensible therefore to look at the limitations of such treatment.

The capacity of a swamp to cope with sediment and chemicals is not infinite; only for the loss of nitrogen through denitrification does the adding of a substance not cause changes in the swamp. This is because the ultimate destination of the nitrogen is the atmosphere and not the swamp itself. The adding of sediment and the build up of peat will cause successional changes. The swamp surface will rise relative to that of the water until a floodplain grassland is produced which receives water for only part of the year and eventually not at all. Such successional changes, which eventually preclude the use of the ecosystem as a nutrient retainer, may be slow, however, and nutrient retention in the short term may be very successful. Some people think that succession can be prevented by harvesting of the vegetation and thus removal of some of the accumulating nutrients. There is a fallacy in this. Removal of the above-ground parts of the plants at the height of their growth when they contain most nutrients will kill them eventually, whilst removal when they are senescing will result in little harvesting of nutrients. These will have been, by then, translocated to the rhizome. And in any case the major sink for the nutrients (other than the atmosphere for nitrogen) is the sediment. The plants create the optimal environment for sedimentation but do not, when the vegetation is fully established, have any net uptake from year to year.

Swamp soils will retain metals, including toxic ones from industrial processes, and burial of these elements in their sediments may seem an attractive way of disposal. This should be counselled against, however. Plant roots may absorb such elements and mobilize them into food webs, or the build up of such substances may kill the swamp plants eventually or interfere with important bacteria in the sediments. Experiments with such disposal should certainly be encouraged—but with specially established artificial wetlands at sewage treatment works, not with natural wetlands where the properties of flood control, denitrification and natural sediment retention are too valuable to be jeopardized. Furthermore, natural floodplain swamplands have aesthetic and wildlife values which are important. In the United States, 20% of threatened or endangered plants and animals are associated with wetlands which figure prominently in lists of national parks and wildlife reserves. Is the cheap disposal of some industry's waste in the short term ever justified at the expense of the long-term loss of a habitat which is being increasingly lost through drainage?

5.9.3 Floodplain swamps and human diseases

The close links between cycles of flooding, high fish production, and movements of people and animals which support human settlement in floodplains are not without problems. Parasites which cause some of the most devastating human diseases have also fitted their life histories to this productive system. Hundreds of millions of people are exposed to malaria, schistosomiasis and filariasis, tens of millions to yellow fever, encephalitis, trypanosomiasis and various lung and liver flukes. The connection between them is that their agents of dispersal (vectors) are animals of wetlands. Of course these vectors often occur elsewhere—at lake margins and in irrigation ditches for example, but the extensiveness of floodplain swamps and the large populations of people dependent on them make such areas particular foci. Destruction of these areas is thus often advocated on health grounds.

Malaria is perhaps the most familiar of these diseases. It was not confined to the tropics until relatively recently, for 'paludism' or 'marsh miasma' was a risk of the European fenlands until the nineteenth century, and as late as 1827 people were fearful to enter the fens around the Wash in England because of the disease. Malaria is a protozoan parasite of red blood cells, and is carried by about 60 of the 400 or so species of the mosquito genus *Anopheles*. The female mosquito requires several blood meals to complete development of her eggs; otherwise she feeds on nectar, as does the male. In feeding she may transfer cells of the malaria parasite *Plasmodium* between human hosts. There are at least four important *Plasmodium* species, some, such as *P. falciparum* more dangerous than others (e.g. *P. vivax*), but all eventually proving fatal if not treated. *Anopheles* lay their eggs as floating rafts on almost any still water surface. Water among aquatic plants is ideal as long as it is not completely deoxygenated for there may be some cover from egg predators.

Control of malaria has long been sought because over 2 billion people are exposed to the disease, about 250 million are suffering from it at any one time and deaths from it number about 1.3 million each year. In 1956 the World Health Organization started a major campaign of spraying settlements with the inexpensive insecticide DDT to kill adult mosquitoes. For a time this was very successful but plans for spraying have often fallen into disarray for social or financial reasons, and resistance to DDT has developed in some *Anopheles* species. Curative or preventative drugs are available, but resistance of the parasite to many of these has developed and costs often preclude their widespread use. A third line of control is the potential development of immunization against the parasite and much research now concentrates on this. If successful, the fourth line—destruction of the swampland habitat by drainage, a method used in the nineteenth century in Europe—might be avoided. Indeed drainage and replacement of the swamp with irrigated agriculture and a plethora of canals may sometimes provide better breeding habitat for the mosquitoes than the original wetland.

Mosquitoes carry other diseases also. The genera *Aedes* and *Culex* transfer a wide variety of arbor, and other viruses, among which yellow fever and Japanese B encephalitis are well known and the former preventable by immunization. But others, such as Marburg virus, are presently incurable, though fortunately rare. More widespread (about 300 million current cases) is filariasis which is often called elephantiasis after one of its characteristic symptoms. Several mosquito genera—*Mansonia, Culex* and *Aedes*—carry the microfilariae, the larval dispersal stages, of the nematode worms *Wuchereria bancrofti* and *Brugia malayi*. The microfilariae are released into the human bloodstream from large (often several cm) adult worms which grow in the lymph tracts. These worms cause blockage and swelling of the tissues, much inconvenience, pain and eventually death. The microfilariae are produced in large numbers rhythmically each day and are present in the bloodstream at the time when the local mosquitoes are most active.

Mansonia, an important vector of *Brugia*, has larvae which, like those of other mosquito genera, breathe air, but not by spiracles held at the water surface as the larva attaches to the surface tension film. *Mansonia* larvae have a saw-edged siphon bearing a spiracle and cut their way into the roots of floating aquatic plants, like the water cabbage, *Pistia stratiotes*, and use the air supply contained in the lacunae of the plant. In this way they may be less exposed to predation than they would be at the water surface.

No less important than mosquitoes, as disease vectors in floodplains, are snails. Many genera of these specifically carry stages of flatworms (flukes, Trematoda), which cause debilitating and often fatal disease. Most widespread are the various *Schistosoma* species which cause bilharzia or schistosomiasis. The adult flukes occur in male-female pairs in the veins around the intestine or bladder of the human host.

An individual may carry only a few pairs or very many, each producing up to several hundred eggs per day. The eggs are provided with spines—the position and number varying with species—and burrow through the gut or bladder wall eventually to be voided with the faeces or urine. If, as often, this is in fresh water, the eggs will hatch to a stage, the miracidium, which infects snails of particular genera. Schistosomiasis can thus be controlled simply by provision of organized sanitation, though finance and custom may prevent this. As with many English public lavatories in the countryside, the great outdoors may personally be far preferable to an over-used pit. Also, the main source of schistosome eggs, the 10–24 year age-group and more particularly the 10–14 year olds, is the one most likely to make free in the open air.

After a period of development in the snail, the parasite is released as a cercarium—or rather as cercariae for a small cloud of them may make the water milky on their release. If it contacts human skin—a person wading or bathing in the water—a cercarium will burrow in, using enzymes and small teeth, taking a few minutes to do so. Inside the human host the cercariium matures to the adult fluke, moving to the blood system as it does so.

Schistosomiasis is a debilitating disease, affecting whole villages and undermining much will for an active life. It can be treated with drugs, the older antimony-based ones being almost as dangerous as the disease itself, the newer metrifonate, oxamniquine and proziquantel being safe, effective but relatively expensive. Control of the disease in the past has concentrated on killing the snails using copper compounds, which proved relatively ineffective, or the synthetic Bayluscide. Such approaches cannot work well on large swamps, though they are effective on a local scale—in village ponds, for example. In the past the draining or filling in of swamps has been recommended—and still is—to remove the snails' habitat but, apart from the intrinsic undesirability of doing this, it might actually increase the incidence of the disease. The irrigation schemes established in the Nile valley at Gezira following the damming of the floodplain by the Sennar Dam in 1924 led to an increase from 1% and 5% to 21% and 80% in the incidence of *Schistosoma haematobium* and *S. mansoni* respectively in the local populations. At present it seems likely that most of the 74 countries and 600 million people exposed to the disease will receive only local relief in the foreseeable future [875].

Schistosomiasis is only one, though the most widespread, of floodplain fluke diseases. Others include Busk's fluke (*Fasciolopsis buski*), affecting 15 million people, lung flukes (*Paragonimus*) and various liver flukes affecting 5 and 30 million people respectively. All of these have snail vectors and also a second swampland host.

For example, Busk's fluke—a large animal, 7–8 cm long, present as adults in human blood vessels—is carried in China and Thailand by the snail genera *Segmentina* and *Hippeutis*. The snails release not cercariae, but metacercariae which attach to water chestnut plants where they change to cercariae. Water chestnuts (*Trapa natans*) are floating rosette-plants with a flower which, although initially above water, droops as the fruit forms and dips underwater. The metacercariae attach to the fruits which are often collected and eaten uncooked by children who become infected. Extra reservoirs of infection in the Far East are domestic pigs in which the adult flukes also live. The pigs may be penned over ditches so that their excreta fertilize the water, in which fish are cultured for food. The fluke eggs passing in the faeces then have a very good chance of reaching the snail host.

Liver flukes include *Clonorchis sinensis*, the Chinese liver fluke whose vectors are snails of the genus *Bithynia* and a fish, the Chinese grass carp, (*Ctenopharyngodon idella*) under whose scales the cercariae develop. In Taiwan, *Opisthorchis felinus* similarly uses the common carp, *Cyprinus carpio*. In the Far East, fish is frequently eaten raw and the parasite is transferred. In Europe the sheep liver fluke, *Fasciola hepatica* may also infect humans, though not very seriously, through their eating raw watercress, its intermediary host between snail and mammal. Finally, in W. Africa and Asia, the usually fatal lung flukes are transferred to humans by the eating of raw freshwater crabs, on whose gills the metacercariae encyst.

5.10 DRAINAGE AND OTHER ALTERATIONS TO FLOODPLAIN ECOSYSTEMS

There have been very considerable changes to floodplain ecosystems in temperate regions. These changes are mostly drainage—to bring into cultivation sediments, which, because of the accumulated fertile silt and damp nature even when drained, often make extremely fertile soils—and for flood control—to allow settlement in the outer reaches of the river bed by confining it to its main channel. Often both of these aims are pursued together. Some of their consequences are illustrated by changes in the Florida Everglades.

5.10.1 The Florida Everglades

The Everglades is a complex of ecosystems, some of tree swamp, but mostly open, based on a shallow river almost as wide, up to 100 km, as it is long, 180 km. It flows slowly from the lowlands around L. Okeechobee to the sea at the south-western tip of Florida, USA (Fig. 5.9). The flow has never been rapid—only a few hundred metres per day—and the river is at most 30 cm deep over much of the area. The main freshwater community is of emergent aquatic plants, dominated by the saw-sedge, *Cladium jamaicensis*, which grows several metres tall, so that the description 'River of Grass' is embodied in the title of a well known book on the area by Marjory Stoneman Douglas [173].

The southern tip of Florida is floored by a porous limestone (Miami oolite) and is very flat. Running along the eastern coast is a ridge of rock some 6 m above mean sea level, and the west coast also bears a slightly wider but still narrow and subdued 'upland'. In the basin between the two lie the Everglades, with the land dipping only 7 m from near L. Okeechobee towards the sea. Rainfall is very seasonal in the area; most of the 200 cm or so falls in

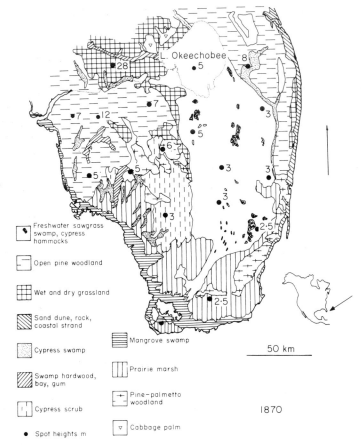

Freshwater sawgrass
swamp, cypress
hammocks

Open pine woodland

Wet and dry grassland

Sand dune, rock,
coastal strand

Cypress swamp

Swamp hardwood,
bay, gum

Cypress scrub

● Spot heights m

Mangrove swamp

Prairie marsh

Pine-palmetto
woodland

Cabbage palm

50 km

1870

Fig. 5.9. Natural vegetation of the Everglades as it was in 1870. (Based on Caulfield [106].)

June and July, sometimes in heavy showers associated with hurricanes which also occur in September. The winter months are dry and the natural water supply to the Everglades is then at its least.

To the south of L. Okeechobee was an area of several thousand hectares of peat, up to 4 m deep, laid down over several thousand years in swamps associated with the lake. This acted as a sponge, taking up much water in wet periods and releasing it steadily all the year round. The flow of water penetrated the Everglades along three main water-courses or sloughs, running south or south-westwards: the Shark River slough, the Lostman's River slough, and the Taylor slough. As the water reached the sea the saw-grass community merged with transitional vegetation and was then replaced by a dense, tangled forest of mangroves up to 30 m high. Mangrove forest is the sub-tropical and tropical equivalent of salt marsh and is subjected to much the same intertidal regime. The steady flow of fresh water confined the mangrove to the coast by stopping penetration of salt water inland and mangrove trees were not found beyond a few kilometres from the river mouths.

Dotted among the saw-grass vegetation on islands of oolite penetrating a few centimetres above the general basin level are woods of slash pine (*Pinus caribaea*) and palmetto (*Serenoa serrulata*) and sometimes clumps of hardwood trees. Small groves of other trees, swamp and pond cypresses (*Taxodium* spp.), occupy depressions in the oolite where peat has accumulated. The saw-grass does not undergo succession to drier forest because cool fires, begun by lightning at the start of the rainy season, have occasionally burnt the surface vegetation and litter, though left undamaged the deeper peat and rhizomes of the saw-grass [686]. The pine woodland is also fire-resistant and is replaced by hardwood only after a long period when fires have not intervened.

This complex of plant communities in turn supports a rich fauna, of which the birds, some 250 species of them, are best known, for the area is at the junction of several migration flyways [371]. Huge numbers were described by the naturalist John James Audubon in 1832 and it is still possible to see flocks of 100 000 ducks. Ibis, spoonbills, herons, pelicans, coots, plovers, gulls, terns, storks and cranes are also to be seen in abundance. Other vertebrates form a no less exciting collection. The alligator is perhaps most famous, but the American crocodile lives in the mangrove swamps, and snakes and turtles are common. There are 57 species of reptile and 17 of amphibia. Some 25 mammal species—opossums, raccoons, wildcat, otter, white-tailed deer, mountain lion, black bear, and, offshore, the manatee and dolphin—have been recorded, and half of these depend significantly on the freshwater communities. The fish (240 species) and invertebrate faunas are even richer.

Not surprisingly, the first threat, now much reduced, to the Everglades, came from poaching of the rich fauna. In the 1890s millions of feathers of egrets and other Florida birds formed the raw materials of a thriving millinery trade, and by 1930 100 000 alligator hides were being processed into leather each year in Florida tanneries. It has been estimated that the alligator population was reduced to only 1% of its original level by the 1960s, but it is now recovering owing to stringent protection laws [256]. The real danger for the Everglades' ecosystem, and also for the whole of southern Florida, now comes from interference with the natural drainage patterns.

As early as 1882 it was realized that the peatlands around L. Okeechobee would be extremely fertile land, if they could be drained of the standing water that covered them for 8 months of the year. A canal was dredged between the Caloosahatchee River and L. Okeechobee (Fig. 5.10). Sugar cane and winter vegetables thrived on the areas drained.

In 1925, and particularly in 1928, hurricanes were severe enough to cause flooding of water from L. Okeechobee into adjacent drained areas which had then been settled as fertile farmland. These areas had formed the natural

Chapter 5

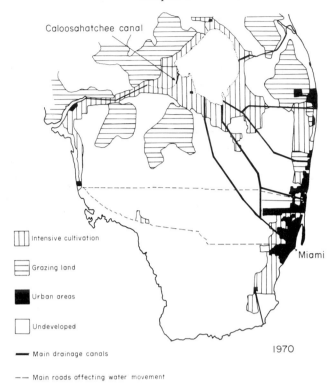

Caloosahatchee canal

Intensive cultivation

Grazing land

Urban areas

Undeveloped

Main drainage canals

Main roads affecting water movement

Miami

1970

Fig. 5.10. Florida Everglades in 1970. (Based on Caulfield [106].)

wetland of the lake, normally accommodating flood levels. In 1928, between 1500 and 2500 people were drowned in the floods. An embankment was constructed around the lake and a period of digging of further drainage canals began so that future potential flood water could be drained rapidly to the sea. The canals (Fig. 5.10) emerge at the coast among the built-up areas of the eastern oolite ridge and effectively divert the waterflow to the east from its original southward progress through the Everglades. Pressure for drainage has been stimulated not only by farming interests but also by the warm Florida climate. For several decades the area has been promoted as a retirement and holiday haven and the demand for building land, otherwise confined to the coastal ridges, has been aggressive.

The canal system incorporates large, shallow reservoirs called water conservation areas. These can be used for temporary storage of water released from L. Okeechobee. They are also necessary to delay the flow to the sea sufficiently for the groundwater aquifer, from which the coastal cities derive their fresh water supply, to be recharged. They have had to be built because the natural peat regulator south of L. Okeechobee, which bore exactly these functions, has been much reduced. Drainage results in rapid oxidation of the peat, which, once 4 m deep, now disappears at the rate of 2 cm per year.

A main road, the Tamiami Trail, skirts the southernmost of the reservoir areas, and water may be released through culverts under it to the remaining southern portion of the Everglades. The problem is that the water supply is now insufficient, a high proportion being diverted to the east coast, and the natural seasonal rhythm of supply is not necessarily maintained. Water is released to the Everglades largely when it is convenient for the drainage system. Between 1962 and 1965 none was released at all so that in 1968 it was necessary to pass legislation to guarantee at least a minimal supply for the Everglades National Park, a 500 km^2 remnant of the original 10 000 km^2. In 1970, plans were blocked just in time to prevent a large part of the adjoining Big Cypress swamp from being developed as an airport [766].

In the pristine Everglades, the seasonality of the water supply was something the ecosystem had adjusted to. In the dry season animals congregated in the deeper parts of the sloughs (ponds formed by the solution of the oolite) and breeding cycles were related to this. The concentration of invertebrates at the edge of the receding water, for example, provided the wood ibis, which is a wading bird seizing its prey after contact as it sweeps its bill through the water, with a rich food supply during the fledgling season. Nests of birds built on the swamp floor in the dry season may now be destroyed by unseasonal inputs of water. The wood stork which depends on a predictable supply of fish and crustacea for its fledglings has been reduced in numbers by 73% since the 1930s.

The overall lack of water is probably most crucial, however. The Everglades have always been subject to light fires, indeed the diversity of their vegetation depends on fire to prevent succession in the *Cladium* swamps. But the extreme drought now caused by diversion of the water supply, particularly in years of low rainfall such as October–April 1970–71 when only a third of the expected 35 cm fell, has led to especially destructive fires. These have bitten deep into the sub-surface peat, setting hundreds of hectares on fire at times, and damaging the vegetation perhaps irreversibly. Uncontrollabe fires such as these inevitably also threaten adjacent urban areas, as well as causing smoke pollution for long periods. The lack of water in the Everglades is also reducing the areas of water which persist through the dry season. Because fish and reptiles congregate in these pools this had led to heavier than normal predation and mass fish deaths due to deoxygenation. The lack of water is affecting the cities also. The groundwater aquifer is not being recharged rapidly enough to prevent sea water moving into the oolite and contaminating the drinking water wells. In recent years, new wells have had to be drilled further inland for east coast cities. Moreover, estuarine shrimps, the basis of a 20 million dollar industry, move into the freshwater of the Everglades for part of their life history. Interference with the hydrology of the system may put this profitable catch at risk.

5.10.2 **Drainage and river management in temperate regions**

The floodplains of most temperate rivers have been greatly altered in one way or another. Rivers now rarely expand over the plain, but are confined in deepened and often embanked channels sometimes perched above the surrounding plain. The surrounding soils, especially if peaty, have oxidized and shrunk on drainage, sometimes depressing the level by a metre or two. Even with embankment, there may still be some risk of flooding of the now agricultural or urban land of the plain, so the channel may be managed.

This means that its capacity to carry water may have been increased through dredging and deepening; it may have been straightened (canalization) to minimize siltation on bends, and its cross-section particularly in urban reaches may have been formed with concrete into a trapezoidal shape, which impedes flow least. Bank vegetation, overhanging trees, debris and aquatic plants may be regularly cut because these also may impede the flow and increase the risk of flooding [75, 591]. All of these measures decrease the variety of habitat for the channel ecosystems whilst the floodplain swamps may be reduced to the communities capable of growing in the ditches draining the land. Potential urban, industrial and agricultural pollution may also reduce water quality so that the river channel ecosystem is altered in so many ways that it may be difficult to link cause and effect.

The extent of river alteration is now very extensive. In the USA over 26 500 km of rivers are so managed (0.003 km km^{-2}) whilst in Britain a quarter (8500 km) of all main river lengths has been severely altered (canalized, dredged, piled) and a further 35 500 km are managed in a lesser way (removal of aquatic plants, and bankside trees and shrubs). This degree of management (0.06 km km^{-2}) is equally great in many parts of mainland Europe.

The effects are often to reduce the numbers, production and diversity of the bottom communities [37]. In the upper Rhine, the turtle *Emys orbicularis* has probably been eliminated and in one Swedish river, the smooth flow brought about by channelization has increased the incidence of biting black-flies (*Simulium*). One of the main reasons for the decline of otters in British lowland rivers is undoubtedly loss of their habitat, for cavities between bankside tree roots are preferred sites for the building of their nests.

A managed and canalized river is not pretty and there is considerable doubt that the costs to the community of the management, even if aesthetic and conservation considerations are ignored, are matched by even equal benefits. A current climate of opinion, somewhat alarmed at the consequences of intensive agriculture, has forced Water Authorities in Britain and equivalent bodies elsewhere to take a more sensitive approach [591]. The drainage engineer's original ideal of a completely straight, trapezoidal, smooth channel may be softened by the provision of rest stops—lengths of natural channel to maintain better habitat—by the keeping of bends as relief channels and by limited planting of trees at the edge of a shallower berm adjacent to the main dredged channel. This change is welcome but only cosmetic. When it becomes necessary to provide submerged lengths of pipe and to support sheets of concrete underwater to provide enough cover to maintain even a recreational fishery, then it seems clear that our policy of floodplain drainage may have been misguided for centuries [200a].

Areas of swamp or grazing marsh can be protected through compensation schemes to landowners, but in temperate regions very little such habitat is left and a future generation may hold that it has been irresponsibly wasted. It seems all the more poignant that past civilizations in South and Central America before the Spanish conquest found ways of cultivating swamps by building raised platforms of soil in them on which a range of crops was grown [16, 143, 144]. The platforms were fertilized by recycling the silt accumulating in the channels around the platforms and perhaps also by an obligatory growing of nitrogen-fixing legumes with the other crops. The method allowed much of the swamp to be preserved without drainage. These methods are currently being revived in Mexico [144] and may help prevent in the sub-tropics, some of the errors of temperate swamp management.

But against too sweeping a condemnation of the civil engineering approach to river management must be set two considerations. First, in the developed world such management has its origins on a small scale, several centuries in the past, when it must have appeared an entirely desirable practice. Once drainage and flood control have begun they cannot easily be ended without abandoning farm land or moving significant numbers of people. There is understandable resistance to this—maintenance of cultivated land represents for many people a coping with the vagaries of nature and perhaps subconsciously of themselves. And secondly, in the developing world, large numbers of people moving from the countryside to the cities to find work may frequently find the only places to build their shanties on the edges of floodplains in the dry season. Canalization may provide a less risky home for them. The problems of the floodplain wetlands ultimately arise from those of population, poverty and inequitable social and economic systems. The Pongolo R. case study in Chapter 1 illustrates the compromises that may need to be made.

5.11 LOWLAND RIVER CHANNELS

Whatever happens to the floodplain wetland, the river channel remains, except where the river has been dammed to form a lake. The pristine river channel will have been a quite varied habitat, with bends, fringing and some submerged vegetation, patches of silt and gravel. The animal communities of such a channel will have been modifications of those further upstream, with some organisms which colonized rocks and gravel upstream still present, but with a much greater emphasis on those which feed on the silt deposits. Such animals are able to tolerate lower oxygen concentrations in the warmer, more organic-laden water than those present in the more turbulent water upstream. The deposit feeders include a wide variety of invertebrates but particularly oligochaete worms, chironomid larvae and bivalve molluscs; insect larvae and Crustacea are less diverse perhaps than upstream. The communities of such sediments are discussed in Chapter 8.

Even without severe canalization, the river is likely now to be much modified as a result of human activities, particularly pollution. Historically, as a river valley has been developed, it has been organic matter from sewage which has been the initial pollutant, together with easily noticed pollutants like inorganic sediment from the washing of coal or mineral ores. Many subtropical rivers clearly show the effects of these, particularly in the dry season when flows are very low and dilution not very effective. I have stood on a river bridge over a heavily polluted river in Puna in India and watched what my Indian colleagues called the 'black flowers': circular clouds of ferrous sulphide, borne up to the water surface by decomposition gases in the anaerobic sediment, then gently subsiding to be replaced in a continual blooming.

There has then been added a variety of other obvious pollutants, for example dyes, bleach, and heat if industrialization has developed. A public outcry may follow either from the epidemics of cholera or typhoid which have followed sewage pollution of the water supply or the demise of the fish population. At this stage legislation has usually been passed to curb the worst of this obvious pollution and river quality has improved. Next, as industry and agriculture have become more sophisticated, the variety of pollutants has increased though controls have usually been set to limit the amounts of such substances that can be discharged. The developed world's rivers are generally at this last stage, the developing world's at the first or second. Only rivers in remote regions are generally unpolluted. In the final sections of this chapter I will look in more detail at those stages which often follow development of a river floodplain.

5.11.1 Pollution by organic matter

Input of organic matter is, of course, a normal feature of streams and rivers. The detritivores of streams (Chapter 4) and the sediment communities (Chapter 8) of slow-flowing rivers depend upon it for most of their energy. The main differences between natural organic input and pollution by organic matter, however, are that the former tends to be in large packets, like leaves, with a low surface to volume ratio, or is relatively refractory when finely divided, while the latter is usually soluble or finely divided and very labile. The bacteria which immediately colonize it need much oxygen to decompose it and organically-polluted water rapidly becomes deoxygenated. In rivers, loss of most invertebrates and fish follows and the remaining 'pollution community' comprises a mass of filamentous bacteria (including 'sewage fungus', *Sphaerotilus natans* and others), colourless flagellates, ciliate protozoons and anaerobic chemoautotrophic bacteria like *Beggiatoa*.

Progressively, as the organic matter is decomposed, this community may be replaced by one in which filamentous algae like *Cladophora* predominate. These are stimulated by the release of ammonium and phosphate from the decomposition, and support numerous chironomid larvae, which may cause a nuisance when they emerge as flies. If the stream bottom is muddy, numerically rich communities of oligochaete worms and chironomids may develop. These further process the organic matter indirectly by consuming the abundant bacteria, and as oxygen levels begin to rise again, a crustacean, *Asellus*, and other moderately tolerant invertebrates become abundant. Eventually, as the water again becomes fully oxygenated the 'clean water fauna', with a high species diversity, is able to return [385].

Some tidal rivers still receive raw sewage, and organic discharges from, for example, the food industry and intensive stock rearing units do occur, but organic pollution in the developed world is well understood. It is controlled through use of the same biological processes that break down organic matter in a river, but they are concentrated into the small area of a sewage treatment works rather than along several kilometres of waterway.

The deoxygenating ability, or oxygen demand of a sewage effluent, the product of a sewage treatment works, may be crudely assessed by measuring the rate at which oxygen disappears from it (diluted if necessary) in a sealed bottle kept for 5 days in darkness at $20°C$. The biological oxygen demand (BOD) of crude sewage is around 600 mg $O_2 l^{-1}$ 5 day^{-1}, and of unpolluted river water less than 5 mg $O_2 l^{-1}$ 5 day^{-1}. Good sewage treatment reduces the BOD of the discharged effluent to, at most, 30 mg $O_2 l^{-1}$ 5 day^{-1} and usually less than this.

5.11.2 Sewage treatment

The raw material pumped to a sewage treatment works is mostly water, with 1–2% of particulate and colloidal organic matter and a host of dissolved compounds. Domestic waste is only one component, and the bulk of this is bath water and kitchen water, rather than faeces and urine. In urban areas there is drainage water from the streets and some factory effluents, though these are strictly controlled in Britain by a consent system (see below) so that metal and other poisons do not inhibit the organisms harnessed at the works to remove organic matter.

After screening for objects like dead cats, the sewage has few particles larger than a few millimetres—passage through the sewerage pipes has generally broken up large lumps—but a macerator now completes the process. The 'foul water' is now greyish brown and turbid and has a BOD of

600 mg O_2 l^{-1} 5 day^{-1} still. It is led into large tanks, the primary sedimentation tanks where particulate organic matter settles out as a sludge, and the BOD of the overlying water decreases by 30–40%. The primary sludge is pumped to digesters, while the supernatant water enters the secondary stage.

The sludge digesters are enclosed tanks in which the sludge is heated to 30–35°C by methane produced by bacterial fermentation of itself. The methane is recycled to burners and may be used to generate electricity to power the works. Fermentation is carried out mainly by *Methanomonas* spp., which, being heterotrophic but anaerobic, cannot completely break down the organic matter to carbon dioxide and water. After several weeks' digestion the residual sludge should have an almost pleasant earthy smell. Its disposal is something of a problem, solved by compressing it to remove water or composting it with urban refuse on rubbish tips. It can also be given or sold to local farmers for fertilizer if it does not contain large quantities of heavy metals, or dumped at sea in the cases of very large works near the coast.

Meanwhile the supernatant water from primary sedimentation is usually treated in one of two ways—the trickling filter or the activated sludge processes. The former is the older method and removes a slightly lower proportion of the residual BOD than the latter, which, however, requires more control and is therefore more expensive. Both depend initially on the same principle—breakdown of organic matter by a similar community of bacteria and protozoa to that which responds to gross organic pollution of a river.

Trickling filters are typically circular beds of small rock fragments. The beds are up to 50 m diameter and about 2–2.5 m deep. Usually they are set on a valley slope below the primary sedimentation tanks so that water moves under gravity through pipes to the centre of the bed and out along arms extending radially from the centre. Gushing out through holes the water moves the arms by jet propulsion so that the bed is evenly sprayed. As a new bed is brought into use a film of bacteria and grazing protozoa develops on the rock fragments and converts the organic load into bacterial and protozoon cells and CO_2. The process may be aided by aerating the water before it enters the filter. Growth of the bacteria would soon clog, waterlog, and deoxygenate the filter were it not for grazers on the bacterial and protozoon film. These are a few species of oligochaetes and fly larvae, whose respiration removes yet more organic matter as CO_2, and a small proportion on emergence of the flies, sometimes to cause a nuisance in neighbouring housing areas.

The water trickles through the bed and out into channels. At this stage its BOD has been reduced to about 60 mg O_2 l^{-1} 5 day^{-1}, and it comprises a rich, inorganic solution containing phosphate, ammonium and nitrate ions, among others, with sloughed off bacteria, faeces of invertebrates, and residual refractory organic matter. The BOD is reduced to 20 mg O_2 l^{-1} 5 day^{-1} by settling of these solids in tanks to form a secondary sludge. This may ultimately be dried and disposed of to farmers or dumps, or digested with primary sludge.

The alternative treatment, the activated sludge process, is carried out in large tanks. Primary effluent is led into the tanks, seeded with a floc of particles of bacteria and protozoa, largely ciliates, kept back from the previous batch, and vigorously aerated and agitated. This promotes rapid growth of the floc (called activated sludge) and some rotifers and nematodes may graze it. The plant is usually run on a carefully regulated and monitored continuous flow system to maintain a high ratio of floc to incoming organic matter. The effluent is taken to ponds where the floc is settled then pumped to sludge digesters, while the supernatant may be the final effluent or may be further 'polished' by filtration or percolation through adsorptive activated carbon columns to remove dissolved organic matter. In tropical regions the sewage may simply be led into shallow lagoons where it is bacterially oxidized, aided by the photosynthetic oxygen production of dense populations of phytoplankton (usually Chlorophyta including *Chlorella*, *Scenedesmus*, and *Chlamydomonas*) which develop in the nutrient-rich medium.

From the point of view of BOD and turbidity, the standards of 20 mg $O_2 l^{-1}$ 5 day^{-1} and 30 mg l^{-1} suspended solids, set by a Royal Commission in the UK in 1913 for final effluents are easily achieved. If they are not, the fault lies usually in the management of the works, pumping machinery breakdown, or overloading of the works during heavy rainstorms when some diluted raw sewage has to be released to a river. Even good effluents, however, may still contain pathogenic gut bacteria and viruses such as those of poliomyelitis and infectious hepatitis, though these should be greatly diluted by the river flow and undetectable a little way downstream.

5.11.3 Pollution monitoring

The extent of organic pollution damage to the river community is usually assessed in one of two ways—by the presence or absence of indicator species, for example, sewage fungus, or of particular groups of species, genera or families of invertebrates in what are generally called biotic indices or by some mathematical measure of community diversity. River systems are so extensive and change so much from place to place because there are so many discharges to them that the need is for some simple, rapidly applicable measure which can be made by people without specialized taxonomic expertise. Measures of community diversity need no taxonomic expertise, but may be less sensitive than those which use indicator species. There is much debate about which method is most informative [341, 342] and one view is that none of the methods give any more reliable a measure than a simple counting of the number of species present [523].

The communities examined are usually those of invertebrates where a large sample can be easily obtained (compared, for example, with fish) and easily examined (compared with microorganisms). In parts of North America a method based on diversity indices and using the diatoms which will grow on glass slides suspended in the water is popular. Very large numbers of cells are counted on the slides and a graph plotted of number of species against their frequencies in the community. A truncated normal curve is obtained with the rarest species not detected and hence truncating the left-hand end of the curve. The peak of the curve and the spread of its right-hand tail are used to interpret the overall effects of the pollution. A low peak (few species of intermediate frequency) and a long tail (a small number of species (tolerant ones) of very high abundance) are considered to indicate a high degree of pollution, whereas a high peak and short tail indicate a pristine community. The method uses simple apparatus, integrates effects over a period (that of exposure of the slide) and can be standardized to a high degree. Counting of the slides is tedious, however.

In England and Wales, chemical measures and biotic indices using invertebrates find greater favour. A survey of rivers was made in 1980 [586] based on some simple chemical criteria and the National Water Council's Biological Score system. River lengths were classified (Table 5.5) from Class 1a—water of high quality, to Class 4—grossly polluted, on the basis of their dissolved oxygen concentration, BOD, ammonia concentration and their

Table 5.5. National Water Council river classification

Class	Dissolved O_2 (% saturation)	Biological oxygen demand (mg l^{-1})	Ammonia (mg l^{-1})	Notes and uses
1A	$\geqslant 80$	$\leqslant 3$	$\leqslant 0.4$	High amenity value, potable extraction, game fisheries
1B	$\geqslant 60$	$\leqslant 5$	$\leqslant 0.9$	Less high quality than 1A but similar uses
2	$\geqslant 40$	$\leqslant 9$		Potable after treatment. Coarse fishery. Moderate amenity. Should not show physical signs of pollution except perhaps humic colour and some foaming below weirs
3	$\geqslant 10$	$\leqslant 17$		Not likely to be anaerobic but fish absent or sporadic. Low grade industrial abstraction. Considerable potential if cleaned up
4	Likely to be anaerobic			Grossly polluted, likely to cause nuisance

ability to support fish. At the same time the invertebrates were sampled and the communities scored according to the system in Table 5.6. This allocates points to particular families of invertebrates present, with the highest points given to families least pollution-tolerant (e.g certain mayfly, stonefly, caddis-fly and beetle families) and least to oligochaetes and chironomids which normally live in organic sediments and hence can tolerate most easily an

Table 5.6. National Water Council biological scores system. Scores are summed for each group present, and groups are indicated by letters in parentheses

Families	Score
(a) Siphlonuridae, Heptageniidae, Leptophlebiidae, Ephemerellidae, Potamanthidae, Ephemeridae (mayflies) (b) Taeniopterygidae, Leuctridae, Capniidae, Perlodidae, Perlidae, Chloroperlidae (stoneflies) (c) Aphelocheiridae (beetles) (d) Phryganeidae, Molannidae, Beraeidae, Odontoceridae, Letpoceridae, Goeridae, Lepidostomatidae, Brachycentridae, Sericostomatidae (caddis-flies)	10
(a) Astacidae (crayfish) (b) Lestidae, Agriidae, Gomphidae, Cordulegasteridae Aeshnidae, Corduliidae, Libellulidae (dragonflies) (c) Psychomyiidae, Philopotamidae (net-spinning caddis-flies)	8
(a) Caenidae (mayflies) (b) Nemouridae (stoneflies) (c) Rhyacophilidae, Polycentropodidae, Limnephilidae (net spinning caddis-flies)	7
(a) Neritidae, Viviparidae, Ancylidae (snails) (b) Hydroptilidae (caddis-flies) (c) Unionidae (bivalve molluscs) (d) Corophiidae, Gammaridae (crustacea) (e) Platycnemididae, Coenagriidae (dragonflies)	6
(a) Mesovelidae, Hydrometridae, Gerridae, Nepidae, Naucoridae, Notonectidae, Pleidae, Corixidae (bugs) (b) Haliplidae, Hygrobiidae, Dytiscidae, Gyrinidae, Hydrophilidae, Clambidae, Helodidae, Dryopidae, Elminthidae, Crysomelidae, Curculionidae (beetles) (c) Hydropsychidae (caddis-flies) (d) Tipulidae, Simuliidae (dipteran flies) (e) Planariidae, Dendrocoelidae (triclads)	5
(a) Baetidae (mayflies) (b) Sialidae (alderfly) (c) Piscicolidae (leeches)	4
(a) Valvatidae, Hydrobiidae, Lymnaeidae, Physidae, Planorbidae, Sphaeriidae (snails, bivalves) (b) Glossiphoniidae, Hirudidae, Erpobdellidae (leeches) (c) Asellidae (Crustacea)	3
(a) Chironomidae (Diptera)	2
(a) Oligochaeta (whole class) (worms)	1

increased organic load. Scores for lightly polluted rivers may be greater than 100, and for heavily polluted ones, less than 10 or even zero. In general there was a reasonable correspondence between the chemical classification and biological score but the latter, in measuring the effect of pollutants over the period in which the community has developed, offers an advantage over the chemical methods. These may give different results depending on the time and date of sampling. A short discharge of pollutant may be missed by chemical sampling for it will soon be washed downriver. But in killing organisms its effect on the community will be detectable for some time. The 1980 survey of British rivers pointed to a general improvement overall with many stretches of river being moved upwards in their class, and only a few examples remain of rivers in Class 4 in the British Isles. This improvement, however, was not sustained between 1980 and 1985.

5.11.4 Current problems of river pollution

Whereas water pollution in the past often involved very obvious changes, it now involves the release of small amounts of substances which may have more subtle effects on behaviour or reproduction of organisms. These may be organic, such as pesticides or herbicides, or inorganic, such as some of the heavy metals (Chapter 3). These are now usually directed through a sewage treatment works where they can be precipitated out in the sludge. By a consent system the Water Authority can regulate the amount delivered to the treatment works so that it is insufficient to poison the bacteria in the digestion plant, but the final effluent often contains more metal than a natural drainage water except in areas of ore-bearing rock.

5.11.5 Heavy metals

Heavy metals are those with a specific gravity of 5 or more, and of the 40 or so such elements not normally present only as radioactive isotopes, manganese, iron, copper, zinc, molybdenum, cadmium, mercury and lead have attracted most notice. Some of them are required as trace elements by living organisms, though others (Cd, Hg, Pb) do not appear to be. At more than 'normal' concentrations even the trace elements are frequently toxic though the mode of action is often obscure. Selective inhibition of particular crucial enzymes seems one likely reason. Normal concentrations of these elements in unpolluted waters are around $1\ \mu g\ l^{-1}$, with zinc concentrations around $10\ \mu g\ l^{-1}$. Effluents from metal industries, and from the exposure of metal sulphides to air and water in mine waste dumps, may increase these concentrations by several orders of magnitude.

The bacteria and blue-green algae which colonize hot, volcanic springs may tolerate extremely high levels of these metals, and also of fluoride and arsenic (Chapter 3), and in 'normal' streams, polluted by heavy metals, highly resistant strains of algae may be selected for [864]. Dense populations of green algae (Chlorophyta) may frequently be seen in the water of recirculating fountains playing over the bronze statues of long dead worthies in European cities!

Fauna fare less well, however, and in Welsh mountain streams polluted by the waste from old lead mines, both zinc and lead levels may lead to a depauperate invertebrate fauna of insect larvae, *Tanypus nebulosus* and *Simulium latipes*, and some flatworms. Crustaceans and oligochaetes seem to be particularly susceptible, and fish are absent from such streams [417]. Salmonid fish tend to die at lower concentrations than coarse fish [418].

Death of specific organisms is not the only consequence of heavy metal pollution. There may be reduced growth rates of those that survive and accumulation of the metals in their bodies by factors of many thousands over the concentrations found in the environment. There may be, therefore, effects on amphibious but unresistant predators, such as water birds.

It would be convenient to be able to establish the concentrations at which each heavy metal causes physiological or behavioural changes or death in each species likely to be subjected to heavy metal pollution. As for herbicides (Section 5.4.1) LC_{50} values can be established. Unfortunately the LC_{50} varies not only with species, but with life cycle stage, concentrations in the water of other heavy metals, pH, oxygen, bicarbonate, organic matter and temperature. The situation is similar for other pollutants which may have subtle effects at very low concentrations such as pesticides, and organic waste products like PCBs (Chapter 3). A very large literature exists giving LC_{50} values (or other related measures of toxicity) for a limited range of test organisms. As has been suggested above for herbicides, the values may be quite worthless as means of deciding what amounts of a substance can safely be released to a natural community.

At best they give guidance on the upper limit that can be discharged, usually that which kills almost everything, and allow a Water Authority, working mostly on the basis of experience and inspired guesswork, to set a consent either for direct discharge to a river or through a sewage treatment works. The consent takes into account not only the concentration but also the total volume permitted to be released to a river. The consent must also allow for the period when natural flow in the river will be lowest (and concentration of pollutant therefore highest), for other effluents upstream and downstream, and, in deference to political realities, the economic state of the industry and area. Too stringent a standard may cost the industry so much in special extracting plant to remove the pollutant that it becomes unprofitable, and an appeals procedure discourages a pollution officer from allowing too great a safety margin. In the final analysis he hopes to maintain at least some fish populations and to minimize obvious visual effects of pollutant. The Water Authority has powers to prosecute (after warnings) if

the industry consistently exceeds the consent once the period in which it may
appeal is over.

5.11.6 Problems in pollution management

There are problems in the procedure in general. First, under the British
Control of Pollution Act (1974), discharges which are in accord with good
agricultural practice, which is defined by the Ministry of Agriculture,
Fisheries and Food, and thus through lobbying, by the industry itself, are
specifically excluded. There can be no prosecution therefore for any effects
of run-off of agricultural chemicals from field spraying, or of soil particles,
even if these cause measurable deterioration in the river community.
Secondly, there is no absolute standard to which a river ecosystem must be
held. The uses of the river are surveyed and the standards are set chemically
(not biologically) dependent on whether the water is largely for, for example,
fishery, or industrial or amenity use (Table 5.7). The bases of the chemical

Table 5.7. Selected water quality criteria used by Anglian Water for potable water supply
abstraction from rivers, fisheries and amenity and conservation. For use for these purposes 99%
(or 95% in the case of amenity waters) of regular determinations should fall lower than the value
given

	Potable water	Fisheries Salmonid	Fisheries Cyprinid	High amenity and conservation
Temperature ($^\circ$C)	30.0	22.0	28.0	25.0
Dissolved oxygen (mg l^{-1})	4.0	6.0	4.0	4.0
5 day B.O.D. (mg l^{-1})	9.0	6.0	9.0	9.0
Free and ionic ammonia (as N) (mg l^{-1})	1.6	1.25	2.5	1.5
Total oxidized nitrogen (NO$_2{}^+$ NO$_3$) as N (mg l^{-1})	22.6	90.0	90.0	
Anionic synthetic detergents (mg l^{-1})	0.4	0.4	0.4	0.2
Fluoride (mg l^{-1})	1.8	1.8	1.8	
Sulphate (mg l^{-1})	300			
Iron (mg l^{-1})	3.0	1.5	1.5	1.0
Manganese (mg l^{-1})	0.15	1.5	1.5	1.0
Zinc (mg l^{-1})	7.5	0.75	3.0	2.0
Copper (μg l^{-1})	75.0	170.0	170.0	110.0
Nickel (μg l^{-1})	150.0	600.0	600.0	400.0
Chromium (μg l^{-1})	75.0	150.0	750.0	500.0
Cadmium (μg l^{-1})	7.5	2.3	2.3	1.5
Mercury (μg l^{-1})	1.5	0.23	0.23	0.15
Lead (μg l^{-1})	75.0	60.0	750.0	500.0
Dissolved hydrocarbons (μg l^{-1})	200.0			
Total pesticides (μg l^{-1})	2500.0			

standards are flimsy in terms of understanding of their effects on natural ecosystems, and there are some notable omissions, for example standards for phosphate concentrations, which reflect a concern among the Water Authorities at the cost of meeting any standard that they might set of phosphate discharge from their own sewage treatment works.

Other countries, for example the remainder of the European Economic Community, prefer the setting of absolute standards of concentration in the final river water for all such waters. Such a system has the advantage that it cannot be easily manipulated to suit local political concerns, but also might mean that controls for high quality rivers, such as those in which salmon and trout breed, could be lessened. At present the British Water Authorities can insist on higher standards than those prescribed by the EEC as well as permit much lower ones, for example in areas of heavy industry.

6

Lakes, Pools and Other Standing Waters— some Basic Features of their Productivity

From the floodplain ecosystem of Chapter 5 it is only a small step to conceive of a hole or basin containing a volume of water relatively large compared with its annual inflow and outflow. Small and shallow basins are indeed formed within floodplain swamps by masses of emergent or semi-floating plants blocking the drainage of flood water back to the river. In these there is nonetheless likely to be a great influx and efflux of water and the nature of the lake will be greatly determined by that of the river. There will be much movement of suspended matter and relatively large changes in water level.

If a basin larger in area or depth is created however, perhaps by the movements of ice or rock or by deliberate damming, the nature of the lake will change. Proportionately less water will move in so levels will change slowly and with smaller amplitude. Suspended matter will be sedimented out at the river delta as water movement is slowed by the mass of that in the lake. The water, cleared of river-borne particulate matter, will support the growth of a suspended photosynthetic community, the phytoplankton, and a submerged plant community to greater depths than in the turbid river. The longevity of the mass of water in the lake may allow a physical and chemical structure to develop with characteristic horizontal and vertical patterns. Also sediment may accumulate on the bottom in an undisturbed, or only lightly disturbed, layer containing a history of the lake over time.

The nature of lake ecosystems in larger and larger basins will come to depend more on the water chemistry than does that of the lowland river. For, denied much energy in suspended material washed in from catchment or swamps, the lake ecosystem must function on what is synthesized within the water and hence on the dissolved nutrient supply. Eventually, the lake basin may fill with sediment and succeed to a swamp whilst its inflow rivers find channels across the original basin. Lakes, for this reason, although in the short term less frequently disturbed ecosystems than the annually disrupted river floodplains, are often, in the longer term, more temporary features of the landscape.

Variants on the lake theme are many. The water supply may be large in relation to the basin volume, as in a river-lake, yet creating a standing water ecosystem in temporary rock pools fed by rain water. These are washed out repeatedly in the rainy season but may persist for weeks or months over the dry season until they finally disappear. In other cases the basin may be large but the water supply so low that there is no outflow and water leaves only by evaporation. The catchment-derived water supply, in evaporating, leaves salts and the basin becomes an endorheic (internally drained) salt lake. In contrast most lakes are exorheic with some flow-through and no long-term salt accumulation.

6.1 EXORHEIC LAKES

The longest-lived lakes are those formed in continental trenches by separation of the plates of the Earth's crust—two series of lakes in the Rift Valley of East Africa (Fig. 6.1) are examples of this as also is Lake Baikal in the USSR. Basins formed in this way are very deep—they include the deepest in the world (L. Baikal 1741 m, L. Tanganyika 1435 m) with frequently steep sides. They are relatively long and thin and contain such masses of water that if emptied they might take tens of thousands of years to fill again at present rates. A useful statistic is the theoretical replacement time or hydraulic residence time—the mean volume (units, km^3) divided by the annual inflow rate (units, km^3 year^{-1})—or its reciprocal, the flushing or washout rate per year. For the upper layers of Lake Tanganyika, the residence time is probably several thousand years. Such lakes are old (Chapter 2), perhaps with several hundreds of thousands of years of continuous presence of water, though as climate has changed over such a period, so has the depth and volume of the lake.

More gentle geological (tectonic) movements may create shallower basins as land subsides in areas adjacent to those where more violent movements are taking place. Between the west and east arms of the East African rift valley lies the world's largest lake, L. Victoria (Fig. 6.1) covering an area of 69 000 km^2. It is about 80 m at its greatest depth and by the standards of smaller lakes formed in other ways, is not shallow. However, in comparison with the larger rift valley lakes it has extensive areas of fringing shallows covered by papyrus swamp. Lake basins formed in the manner of L. Victoria by gentle subsidence are probably not uncommon close to areas of major earth movements associated with ancient earthquakes and volcanic activity; Lough Neagh in N. Ireland, Great Britain's largest lake, is also an example.

Lava flows may dam rivers and extinct volcanic craters, their vents plugged by solidified lava, may harbour mountain-top lakes of singular beauty in near circular basins. Lake Kivu (see Fig. 6.1), formed by the damming of the upper Rutshuru R. by seven lava flows from the Virunga volcanoes within the last 20 000 years, is an example of the former, and there are crater lakes at the tops of the now generally inactive Virungas. Crater lakes are frequently isolated at the tops of mountains and the very small size and lack of development of their catchment areas, the crater rims, mean that

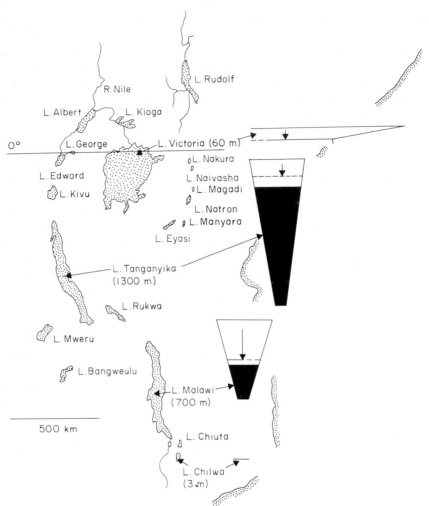

Fig. 6.1. Some major East African Lakes. For Lakes Victoria, Tanganyika, Malawi and Chilwa, scaled profiles are given, with arrows indicating the maximum depth to which the water column is mixed during the year. The darkened parts of the profiles indicate permanently unmixed and deoxygenated water. (Based partly on Beauchamp [42].)

little particulate matter is washed in and that a shortage of key nutrients may severely limit phytoplankton growth in very clear waters.

   Although tectonically produced basins may be old, the majority of the world's lake basins are relatively young. Most were produced in the recent glacial periods and hence date back only to 10 or 20 thousand years ago. As the ice melted back from the great land masses of North America and Eurasia, an uneven surface of hollows scraped by its movement was left in the rocks and debris. Such hollows might be large, for example those forming

Lakes Superior, Huron, Michigan, Erie and Ontario, the St. Lawrence Great Lakes which cover a total of nearly a quarter of a million km^2, or only a few ha in size. Many of the pockings of Canada, the N. American mid-west, Scandinavia and Russia were formed like this or through the melting of ice blocks calved from the retreating glacier and partly buried in washed-out rock and gravel debris. Such basins are called kettle holes; good examples are some of the Shropshire and Cheshire meres in England.

Other ways in which glaciers may form lake basins in upland areas tend to give steep-sided basins in infertile catchments. At the head of the glacier towards the mountain top, contact of the ice with the backing rock wall, and seasonal freezing and thawing, may pluck off rock fragments which grind out a basin under the ice as it moves downhill. Such are the corries, cwms and tarns of the Scottish, Welsh and English mountains. The ice moves rock fragments down usually pre-existing river valleys. If it melts back for a time at about the same rate as its forward progress from the cooler upland head, rock debris may be deposited as a pile or moraine at this point of dynamic equilibrium. When the ice has finally retreated, the moraine may form a dam retaining water behind it in the deepened valley. The main English Lake District lakes (Chapter 3), many Scottish lochs, the larger lakes of the European Alps and many in Norway were formed in this way.

There are many other geomorphological ways than ice action of forming lake basins. Landslides, sand dunes and shingle ridges may block streams; dissolving of limestone by rain water may end in the collapse of underground caverns to give a basin on the land surface. Equally heterogeneous are the lakes formed by biological action—damming by beavers, or the movements of great semi-floating rafts of papyrus which may block side streams on a floodplain. Especially there are the vast numbers of small, and smaller numbers of very large man-made reservoirs. Humans are probably now as important as ice was in forming lake basins.

6.2 THE ESSENTIAL FEATURES AND PARTS OF A LAKE

Essentially lakes (and their smaller versions, ponds) are bodies of water which exist for long enough for their waters to clear sufficiently of suspended particles to allow a suspended community of primary producers, the phytoplankton, to photosynthesize. Light penetration is then clearly an important feature of a lake water. Secondly, the size of the watermass may be great enough for winds to be unable to mix it completely at all times—it may thus acquire a structure. Thirdly, because the phytoplankton needs nutrients to grow, the nature of the catchment (Chapter 3) is also important, for the rate of supply of essential nutrients may influence the plankton production. Finally, the relative clarity of the water may allow a submerged plant community to develop over greater or lesser areas of the bottom dependent on the shape of the basin. This and any fringing swamp communities may contribute much to the productivity of the lake, by providing organic detritus to the sediment communities (benthos) or to the open water. The aquatic plants may also be important for the habitats they provide for larger animals (the nekton) which move between the fringes and the open water. These four aspects will now be examined in more detail, with a view to establishing a general scheme for the factors controlling production in lakes.

6.2.1 Light availability

The amount of radiation which emerges from the sun is considerable. Much of it is dissipated in space but the still large total of about $1350 \text{ J s}^{-1} \text{ m}^{-2}$ (watts) is received at the top of the earth's atmosphere. This radiation has wavelengths from about 0.15–3.2 μm (150–3200 nm) with the peak amount of energy at around 480 nm, but only the range between 400 nm and 700 nm is useful in photosynthesis (photosynthetically active radiation, PAR). This range also almost corresponds to the range detectable by the human eye so that it can be called 'light'. The rest of the radiation is either very short wave ultra-violet radiation or longer wave infra-red and other radiation. Molecules of O_2, O_3, H_2O and CO_2 in the atmosphere may directly absorb some of the radiation, including the light, and the energy is dissipated into heat.

Some radiation may be scattered from molecules or dust particles or cloud and be lost to space. These processes mostly reduce the amounts of infra-red radiation. Shorter wavelengths penetrate the atmosphere almost completely except for some of the ultra-violet radiation, so that of the total radiation received at the ground PAR constitutes about 50% compared with a proportion of about 38% in extraterrestrial radiation. The absolute value of the amount of light energy received at the earth's surface obviously varies with latitude, time of day, degree of cloud cover and dustiness of the local atmosphere. It is at most about $420 \text{ J s}^{-1} \text{ m}^{-2}$ (annual average) at the equator and about half as much (197) at the polar circles, with considerable seasonal change. Cloud cover may reduce these maxima by as much as 50%.

Further losses may occur through reflection at a lake surface and this is particularly important above about 40° latitude, where losses begin to rise from about 2–3%, to 6% at 60°, and 35% at 80°. Disturbance of the water surface by wind may reduce the reflection losses particularly at very high latitudes. The net effect of all of these factors is that the amount of PAR is much reduced from its potential maximum at all latitudes and the annual average received at a lake surface generally declines from equator to poles.

Once light penetrates into water, it finds itself in an environment much more efficient at absorbing and scattering it than the atmosphere was. There are four main components—the water itself, dissolved yellow organic substances (humic acids, etc), commonly called *gelbstoff*, suspended phyto-plankton and suspended inorganic and organic detritus. The latter two particulate categories may scatter as well as absorb the light and as scatter occurs in many directions there is an upward flux of light (a few per cent of the total) as well as a downward flux. The net effect is a downward flux, the amount of which dwindles as the light is attenuated (attenuation = absorption plus scattering).

Although at a given point in the water light is received, owing to scattering, from all angles, the behaviour of the dominant downward flux has proved adequate for a reasonable understanding of the behaviour of light in water. The absolute amount of energy received at a given point (the scalar irradiance) may be higher by a factor of up to about 2 than the downward flux, however.

For a given wavelength of light in a uniform mass of water, attenuation will be exponential and will follow the equation:

$$I = \frac{Io}{\exp k(z_1 - z_2)}$$

(previously introduced in Chapter 5) where I = the irradiance (amount of light energy received per unit area) at a depth z_2. Io is the irradiance at depth z_1 above z_2. e is the base of natural logarithms, and k is an exponential coefficient, which can be calculated from the gradient of the straight-line expansion of the equation:

$$k = \frac{1}{(z_1 - z_2)} \log_e \frac{Io}{I}$$

The units of k are depth^{-1} and it is correctly called the net downward attenuation coefficient, though conveniently it is referred to as the extinction coefficient. It describes the fraction of light energy that is converted to other forms per unit depth and its value depends on the wavelength and the nature of the water. In very clear waters values of less than 0.01 m^{-1} may be found (less than 1% of the energy absorbed per metre) whilst in the milky clay-laden waters of lakes at the feet of glaciers, values may be of 10 or more (the energy being almost totally converted within only a few centimetres). Values of k may be slightly higher with depth even in completely mixed water because of increased scattering that occurs with increased path length of the light in the water, but the deviation is usually not great.

Figure 6.2 shows attenuation of several wavebands (narrow bands of wavelength) in a lake, and illustrates how certain wavebands are more rapidly attenuated than others. The reasons for this are the differential capacities of water, gelbstoff, phytoplankton, and detritus to absorb at different wavelengths.

Water itself is a blue substance, absorbing strongly above 550 nm (orange and red) and below 400 nm (u.v.) but barely at all in the blue and green

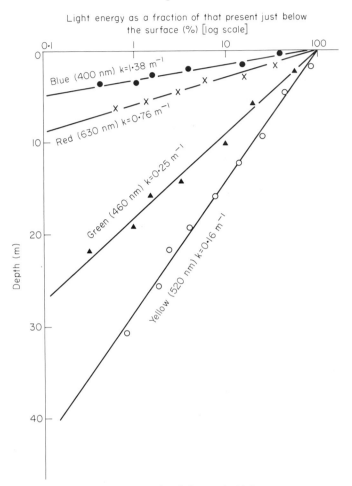

Fig. 6.2. Absorption of light of various wavebands in a typical lake.

wavelengths (400–500 nm). For this reason certain waters with very little dissolved organic material or suspended matter—the mid-oceans and lakes like Crater Lake, Oregon—look blue. This blue light, scattered out of the water to the eye, however, is usually absorbed in most lakes by the yellow substances, derived mostly from the soils of the catchment and sometimes very abundant, for example in peaty areas. Such lakes look brownish because these substances absorb least in the yellow and orange parts of the spectrum.

Phytoplankton possesses a range of pigments which specialize in absorption of particular wavebands (see Chapter 7). Chlorophylls absorb in the blue and red regions, carotenoids in the blue and green, and the blue and red bile pigments (phycoerythrin, phycocyanin, allophycocyanin) in the yellow and green regions. In a mixed phytoplankton community the yellow and green light is usually least absorbed.

Finally, inorganic suspended matter and detritus generally scatter light to a greater extent than they absorb it, though some detritus is brownish in colour and absorbs in the blue wavebands. Scattering is greatest at shorter wavelengths so the blue end of the spectrum is most affected.

Each lake water comprises a unique mixture (or set of mixtures for the composition varies with season) of these four components so it is impossible to give a general picture of how the spectrum of irradiance will change as the light is progressively attenuated with depth. However, in waters neither dominated by silt, gelbstoff or algae, but having measurable quantities of each present, the usual pattern might be of a light climate becoming progressively yellower and greener as light fades with depth. In peaty waters, red and orange light may come to dominate with depth, though all wavelengths will be rapidly extinguished, whilst in the clearest waters, bluish-green light will characterize the final fade out. Ultra-violet light is absorbed by the water very rapidly—in the first few cm—in all cases.

6.2.2 The euphotic zone

Photosynthesis can use all wavelengths in the spectrum of PAR and each quantum of light (photon) is approximately equal in value independent of wavelength. A measure of the total photon flux per unit area passing down the water column is valuable. The extinction coefficient for photon flux is greater in the surface waters as the more readily absorbed wavebands are taken out, and declines with depth as the more penetrative bands are left. However, the effect of increased scattering with depth may cancel this out giving a net effect which corresponds well with the exponential absorption equation.

The photon flux density is measured in $\mu mol\ m^{-2}\ s^{-1}$ or $\mu Einstein\ m^{-2}\ s^{-1}$ (1 μEinstein is about 1.5 μmol or 0.24 joules). Surface values may vary from zero in the polar winter to about 2400 $\mu mol\ m^{-2}\ s^{-1}$ on the equator under full sun. An overcast day in Britain might have 300 $\mu mol\ m^{-2}\ s^{-1}$. At values around 10 $\mu mol\ m^{-2}\ s^{-1}$ (very approximately) the energy absorbed in photosynthesis (gross photosynthesis) in an algal cell just balances the maintenance energy needs of the cell (respiration). Net photosynthesis (P_g-R) is then zero and no new production can occur. The depth in the water column at which photon flux density falls to around 10 $\mu mol\ m^{-2}\ s^{-1}$ is called the euphotic depth (z_{eu}) and the layer of water above it, in which net photosynthesis is theoretically possible, the euphotic zone.

As a rule of thumb, the euphotic depth for phytoplankton corresponds to that at which about 1% of the surface light still remains. (It lies higher in the water column for bulkier aquatic plants whose respiratory needs per unit weight are higher than those of microscopic algae). It can be calculated if the extinction coefficient of the photon attenuation (or more crudely that of 'white light' measured with a light meter sensitive to all wavelengths of *PAR*) is known:

$$z_{eu} = \frac{1}{k} \log_e \frac{100}{1} = \frac{4.6}{k}$$

Alternatively, a number of experiments in which the attenuation of different wavebands have been measured have shown that the equation:

$$z_{eu} = \frac{3.7}{k_{min}}$$

usually holds. k_{min} is the extinction coefficient of the most penetrative waveband—green in most lakes, red in peaty waters.

Values of k for photon extinction may be as low as $0.06\ m^{-1}$ for very clear water, e.g. Crater Lake, between 0.15 and $2.3\ m^{-1}$ for many lakes, and as much as $15\ m^{-1}$ for highly turbid silt-laden lakes such as those in semi-arid regions like South Africa and parts of Australia. Respectively these values would suggest euphotic zone depths of 77 m, 31–2 m and 0.3 m. In the middle and latter ranges, z_{eu} may be much less than the mean or maximum depths of the lake and net photosynthesis may thus be confined to a relatively thin surface layer. A theoretical maximum value for the depth of the euphotic zone around 200 m is imposed by the absorption properties of water itself.

A crude measure of the depth of the euphotic zone may be obtained by dangling a weighted, flat white disc (called a Secci disc after its inventor) into the water and recording the depth (z_s) at which it just disappears to an observer at the surface. This depth is a measure of the transparency of the water and although it varies with observer and surface conditions, it does bear an approximate relationship to the depth of the euphotic zone: $z_{eu} = 1.2$– 2.7 (mean 1.7) z_s.

6.2.3 Thermal stratification and the structure of water masses

A lake surface roiled by wind suggests a well mixed environment in which properties of the water like temperature and chemistry are much the same throughout. This is rarely true. A very shallow lake may be relatively uniform throughout the year and a deeper one also in a windy winter but in many circumstances temperature differences may develop down the water column. This may divide the lake into layers (stratification) which may not readily mix. In a large lake, the patterns of stratification may differ from one end to the other, or between the better mixed areas around inflows and the centre. This stratification may in turn eventually cause differences in water chemistry with depth.

Like all electromagnetic radiation, the longer wavelengths (above 700 nm), which impart most heat, are absorbed exponentially by the water.

The attenuation coefficients for such wavelengths are high (10–100 times greater than those for wavelengths in the photosynthetically active range) and heat radiation is effectively completely absorbed in the top metre or two of water. Theoretically this would create an exponential fall in temperature from the surface in a still water mass, but there is generally some wind disturbance. The wind mixes the heated surface layers downwards but may be insufficient to mix them to the bottom of the basin. This results in an upper, warmer, isothermal layer, a few metres or tens of metres deep, which, because water decreases in density with temperature above about 3.94°C (Chapter 2), floats on cooler, denser, usually also isothermal water. Traces of the idealized exponential temperature curve may be detectable in the lower layer but deep currents usually destroy them.

Between the upper layer, the epilimnion, and the lower layer, the hypolimnion, is a transitional zone called the metalimnion. In this zone a temperature gradient, the thermocline, of as much, or more than $1°C\ m^{-1}$ may be detected. This structure is called direct stratification and is often recorded in lakes in warmer climates, and in temperate lakes during the spring to autumn period. It may be very short-lived (a few days) on calm days in lakes of only a few metres depth so that a slight increase in wind action will mix the water completely again. On the other hand in many lakes it persists for all of the warmer part of the year. Even in lakes as shallow as 3–4 m it may be semi-permanent if the lake is very well sheltered by dense forest. Conversely, in somewhat deeper lakes in open landscapes swept by the wind, no persistent stratification may ever form.

Temperature ranges between epilimnion and hypolimnion may be from 20°C to 4°C in temperate continental lakes with severe winters and hot summers, for example in mid-western North America. It may be from 18°C to 10°C in maritime temperate lakes where the water mass does not cool down so much in winter, nor heat up so much in summer. Such conditions are typical of lakes in the English Lake District and the western Scottish lochs. In the tropics the range may be only a few degrees, from around 29°C to 25°C, but the stratification may be almost as stable as one supported by a much greater temperature range in temperate regions. This is because the change in water density per degree change in temperature is much greater at high water temperatures than it is at lower ones (Chapter 2).

The annual course of stratification is illustrated for three lakes, L. Victoria (East Africa), L. Windermere (UK) and Gull Lake (Michigan, USA), in Figs. 6.3–6.5 by means of depth–time diagrams.

To construct such diagrams, temperatures, obtained from a succession of depths by suitable remote recording thermometers or thermistors, are plotted against the dates on which they were obtained throughout the year. Isotherms are then drawn connecting points of equal temperature, much as contours are drawn along equal heights in making a map. The greater the vertical slope of the lines, the less is the temperature gradient, so that vertical lines indicate completely mixed (isothermal) water masses. The more the lines tend to the horizontal, the greater the temperature gradient in the water column, and the stronger the stratification.

A period of direct stratification ends when the water column is mixed from top to bottom. In temperate lakes this is usually caused by a combination of windier weather and cooling from the surface in autumn, as the sun's angle decreases. Eventually the density gradient becomes too small to be stable under the prevailing wind conditions. In L. Victoria (Fig. 6.3) which lies astride the equator, the surface fall in temperature is provided by the evaporative cooling of the Trade Winds which begin to blow just after the middle of the year. The two shorter periods of mixing in L. Victoria are probably not features of the main body of the lake. Regular measurements in this huge lake could only be made a few kilometres offshore, where temporary movements of permanently isothermal shallow water from the margins sometimes displaced the 'usual' water mass. This emphasizes that there may be horizontal as well as vertical differences in the structure of water masses.

The season of top to bottom mixing in temperate lakes may last from one period of direct summer stratification to the next such period (Fig. 6.4)—a condition known as monomixis (single mixing). This is also true of many tropical lakes, but not of many temperate continental lakes in areas with intensely cold winters. The maximum density of water at around 3.94°C (or in practical terms 4°C) leads, in these lakes, to a period of inverse stratification during winter with the warmest and densest water usually at 4°C at the bottom (though it may be cooler) and with an upward gradient of colder and less dense water to the surface where ice forms at 0°C.

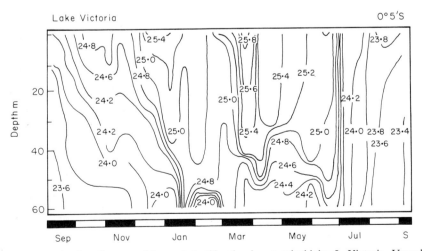

Fig. 6.3. Depth–time diagram of thermal stratification in a tropical lake, L. Victoria, Uganda. (Based on Talling [791].) Isotherms in Figs 6.3–6.6 are in °C.

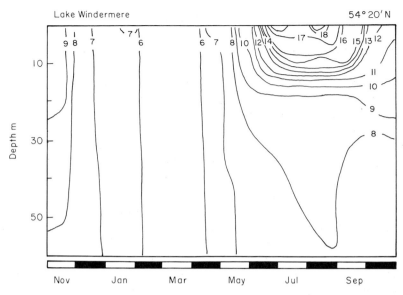

Fig. 6.4. Depth–time diagram of thermal stratification in a temperate lake experiencing a maritime climate, L. Windermere, English Lake District. (Based on Jenkin [399].)

The consequences of this property of water are immense for lakes in seasonally very cold regions. Ice and the snow which may accumulate on it act to some extent as insulators to further heat loss, so it is rare for a lake to freeze to the bottom. Fish and other organisms have therefore not had to evolve means of overwintering frozen solid.

The period of inverse stratification, where it occurs (Fig. 6.5), is broken by the progressive melting of the ice and warming of the surface water as the

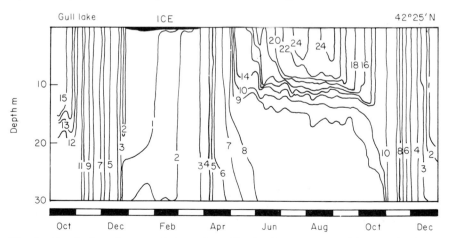

Fig. 6.5. Depth–time diagram of thermal stratification in a temperate lake experiencing a continental climate, Gull Lake, Michigan. (Based on Moss [558].)

sun's angle increases in spring. A period of spring mixing, at first at 4°C, afterwards at higher temperatures, precedes sufficient warming to create direct stratification in most sufficiently temperate lakes. This condition with two mixing and two intervening stratified periods is called dimixis. In polar regions, the summer is too short for this and after a period of mixing at a maximum temperature perhaps around 10°C, inverse stratification sets in again for the autumn and winter (Fig. 6.6), giving again a monomictic state, distinguished as cold monomixis from the temperate maritime and warm temperate conditions of warm monomixis.

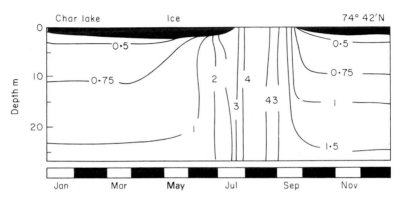

Fig. 6.6. Depth–time diagram of thermal stratification in a polar lake, Char Lake, Canadian Arctic. (Based on Schindler *et al.* [721].)

At the highest latitudes, within the polar circles and at high altitudes there are lakes which are permanently covered by ice, and never mix (amictic), and at the lowest, ones which probably mix but rarely (oligomictic) in their equatorial forest locations. Only in unusually cold spells is their direct stratification destroyed. At high altitudes in the tropics there are deep lakes which may mix very frequently with only brief periods of direct or indirect stratification. Figure 6.7 shows a suggested scheme [384] summarizing the most common thermal stratification patterns with latitude and altitude. It is a convenient generalization, and there is a high probability that a lake at a given altitude and latitude will behave accordingly, but local factors may upset this.

6.2.4 Key nutrients

Take a sample of water from almost any water body, dispense it into clean glass flasks and add, in some convenient replicated experimental design, a range of ions alone and in combination at concentrations of a few $mg\,l^{-1}$. Leave the flasks in good light for a week or two and the chances are very high that you will notice much greater growth of green or yellowish green algae in

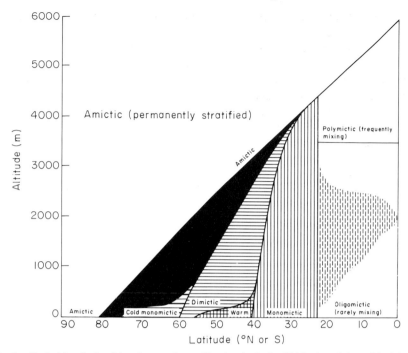

Fig. 6.7. Probable relationship of types of stratification (or lack of it) in deep lakes with altitude and latitude. Transitional regions between warm and cold monomictic types and dimictic lakes are shown. Also, in the tropics a region of mixed types, mainly variants of the warm monomictic type, is indicated at mid-altitudes. (Based on Hutchinson and Löffler [384].)

flasks to which phosphate has been added than in those to which it has not. For some waters it may have been necessary to add both phosphate and nitrate or ammonium ions, and in others, particularly some tropical ones, nitrogen compounds alone may suffice [557, 532, 670, 829, 830].

The experiment is even more convincing if done in larger containers set in a lake [732] or even on whole lakes [720]. In Canada a lake shaped roughly like a violin, though with a much narrower waist at the middle, was divided into two with a vinyl reinforced nylon curtain sealed into the mud and to the banks in the region of the waist [717]. Phosphate was added to one side but not to the other. Only on the side to which phosphate was added was there a dramatic increase in phytoplankton growth. The other side did not change from its previous state.

Phosphorus is, on average, the scarcest element in the earth's crust (the lithosphere) of those required absolutely for algal and higher plant growth (Chapter 3). Its compounds are also relatively insoluble and there is no reservoir of gaseous phosphorus compounds available in the atmosphere as there are of carbon and nitrogen. The case of nitrogen is more complex since despite the huge atmospheric supply (nitrogen constitutes nearly 80% of the atmosphere) elemental nitrogen is relatively unreactive and available as such only to a few organisms (nitrogen fixers). Nitrogen is converted to available compounds by atmospheric lighting sparks and perhaps ultraviolet radiation and by nitrogen fixers. These compounds (most importantly nitrate and ammonium) are very soluble and hence are readily transported into waterways. Nevertheless in some areas, particularly in East and Central Africa where rocks particularly rich in phosphate occur, and in areas where man enriches water with phosphate from sewage effluent, the *local* supply to need ratio of nitrogen may be less than the *local* ratio for phosphorus. A general conclusion from early work that nitrogen tends to be the most severely limiting nutrient in the tropics probably does not hold, for phosphorus has this role in at least some South American lakes [732]. As a generalization of wide validity, therefore, the extent to which the potential productivity of the open water phytoplankton in a water body is realized is set by the supply of available phosphorus compounds.

6.2.5 Nutrient 'limitation'

Phosphorus (and nitrogen where appropriate) have been referred to as 'limiting' nutrients. This is a graphic term, but can cause confusion. The stock of such nutrients in a water body does seem to set an upper limit to the average total algal crop which can exist at any one time (Fig. 6.8a). But this upper limit may not be reached because washout or grazers may be removing the crop as fast as it is being produced. Figure 6.8b shows some of the data in Fig. 6.8a plotted on linear scales and illustrates the great range of actual crop found at a given phosphorus concentration as a result of such processes.

Though the total potential crop may be set (limited) by the stock of P or N, the contributions of individual species within it may be set by other factors—diatom crops may be set by silicate, a particular flagellate species by stocks of a vitamin, or most of the phytoplankton community by nitrogen, whilst potential crops of nitrogen fixers are set by phosphorus or a trace element essential to N-fixation, like molybdenum.

In another sense, none of the individual cells of the crop may be physiologically short of ('limited' by) N or P. Mechanisms of continuous regeneration of N and P from bacterial decomposition of detritus or from grazing in the water body occur (Chapter 7). With supplies of these elements from the inflows also there may be just sufficient nutrient to meet the needs of the growing cells. Growth may be balanced with loss through grazing and nutrient regeneration from these lost cells. The algae may then show no evidence of individual nutrient deficiency ('limitation') whilst the total amount of the algal biomass at any one time is still set by the amounts of P or N in circulation.

A further confusion may also come from attempts to understand the role of light. In winter, light availability may certainly set the total amount of

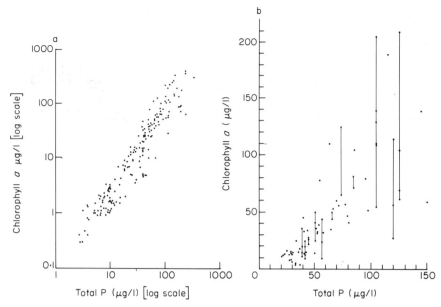

Fig. 6.8. Relationship between summer chlorophyll *a* concentration and total phosphorus concentration in lakes with data plotted (a) on log scales and (b) on linear scales. (Based on Shapiro [735].)

biomass sustainable in the water and no amount of fertilization with N and P will increase the crop. In summer, light may still be limiting if the algal crop becomes so dense (because of a very large availability of N and P) that it starts to shade itself. It is in the intermediate situation where the problems occur, when the availability of N and P is modest and when it is clearly demonstrable that light is severely attenuated with depth only a few metres below the water surface. It might then seem that light should set the crop size in much of the water column. However, in the epilimnion, or throughout the water column of an unstratified lake, cells are circulating through the light gradient and thus experience the higher surface light intensities for some part of their day. In such circumstances it seems that usually enough light energy is received for full use of the potential nutrient supply. Again this generalization is for the crop as a whole and may not apply to individual species or particular times of day, or where suspended silt or clay accounts for much of the light absorption in lakes of arid regions.

6.2.6 How the total phosphorus concentration of a lake is established

Phosphorus enters water bodies not only as inorganic phosphate ions (PO_4^{3-}, $H_2PO_4^-$, HPO_4^{2-}) but also in inorganic polymers, organic phosphorus compounds, in living microorganisms and dead detritus. Only some of these forms are immediately available for plant and algal growth, but others may become so through microbial activity in the water. The sum of all the forms of phosphorus, total phosphorus, P_{tot}, is a reasonable measure of the fertility of a water. A rather infertile lake may have only about 1 µg $P_{tot}\,l^{-1}$ while the most fertile may have 1000 µg $P_{tot}\,l^{-1}$ or more, with an unbroken continuum in between. P_{tot} is determined by a number of factors which can be quantitatively related in a general model, or equation [832].

The amount of a substance entering a water body per unit time, t, and per unit area of a lake is called the areal loading, L. There are usually several sources of loadings. They include subdivisions of the catchment area if it is not uniform (its land use or geology may vary considerably), and the atmosphere, including dry fallout and rainfall directly onto a lake. There may be loadings from the excretion and defaecation of visiting birds like gulls and other vertebrates like hippopotami, and, at certain times, release of the substance from the underlying sediment into the water. The latter is called an internal loading, the others are external. There will also be losses of the substance to the overflow and as particles settling to the bottom sediment.

Consider a lake receiving an element M from various catchment sources. Let M_w be the toal amount of element in the lake, M_{in} the total amount entering, M_{out} the total amount lost through the outflows and σ the fraction of M_w lost to the sediments during time t. The rate of change in M_w is then given by:

$$\frac{d(M_w)}{dt} = M_{in} - M_{out} - \sigma M_w \tag{a}$$

In v_{in} and v_{out} are the volumes of the inflow and outflow and $[m_{in}]$ and $[m_{out}]$ the respective *concentrations* of M:

$$\frac{d(M_w)}{dt} = v_{in}\,[m_w] - v_{out}\,[m_{out}] - \sigma M_w \tag{b}$$

Let V be the total volume of the lake and $[m_w]$ the concentration of M in it. Therefore, dividing (b) by V:

$$\frac{d[m_w]}{dt} = \frac{v_{in}[m_{in}]}{V} - \frac{v_{out}\,[m_{out}]}{V} - \sigma[m_w] \tag{c}$$

v_{out}/V equals ρ, the flushing rate (washout rate) or number of times the volume of the water body is replaced in time t, and $(v_{in}\,[m_{in}])V$ is the loading per unit volume of water body, called Q. If it is assumed (reasonably) that $[m_{out}] = [m_w]$, then:

$$\frac{d[m_w]}{dt} = Q - \rho[m_w] - \sigma[m_w] \tag{d}$$

Over several weeks the total concentration of the element M (though not necessarily of each of the sub-components that make up the total pool of M) often remains reasonably steady and in terms of order of magnitude it can be regarded as approximately constant over much longer periods. During each of the periods of relative stability:

$$\frac{d[m_w]}{dt} = o$$

and

$$Q - \rho[m_w] - \sigma[m_w] = o \qquad \text{(e)}$$

re-arranging this:

$$[m_w] = \frac{Q}{(\rho + \sigma)} \qquad \text{(f)}$$

Because water bodies have very varied volumes it is more useful to express $[m_w]$ in terms of loading per unit area. If the loading per unit area is L, and since $V = Az$, where A is the area of water body and z its mean depth:

$$Q = \frac{[m_{in}] v_{in}}{V} = \frac{[m_{in}] v_{in}}{Az} = \frac{L}{z} \qquad \text{(g)}$$

Thus (f) becomes:

$$[m_w] = \frac{L}{z(\sigma + \rho)} \qquad \text{(h)}$$

This is the basic model first used by R. A. Vollenweider. L can be expanded to equal the sum of the separate loadings from different sources; each can usually be measured by straightforward techniques. z and ρ can be readily obtained from survey and standard hydrological techniques though inflows do not usually mix uniformly and there may be areas of 'dead' unflushed water. To validate the model fully, however, σ must be measured and this is difficult. It is perhaps better to replace it in the model with R, the fraction of the loading that is retained in the lake water, not sedimented but ultimately destined to be washed from the lake through the overflows.

$$R = \frac{[m_w] V \rho}{LA} = \frac{[m_w] z \rho}{L} \qquad \text{(i)}$$

Because, from (h):

$$\frac{z[m_w]}{L} = \frac{1}{(\rho + \sigma)}$$

$$R = \frac{\rho}{(\sigma + \rho)} \quad \text{and} \quad \frac{R}{\rho} = \frac{1}{(\rho + \sigma)}$$

Therefore, from (h):

$$[m_w] = \frac{LR}{z\rho} \tag{j}$$

R can be obtained as the ratio of total efflux of *M* via the overflows (v_{out} [m_{out}]) divided by the total influx, *L.A.*

Despite its simplifying assumptions—that the lake is uniformly mixed and that the system is in a steady state with loadings and flushing rate not changing with time—this general model is important in that it allows prediction of [m_w] under varying conditions of water flow and changes in *L*. This not only helps to explain why water bodies have the characteristics that they do but allows the effects of man-made changes particularly in the nature of the loadings and in hydrological regime to be determined in advance. The natures of various sources of loading have been discussed in Chapter 3. The implications of sedimentation of a substance (σ) are discussed later.

6.2.7 Consequences of thermal stratification for water chemistry

The stratification of lakes often has large effects on the chemistry of the water. It may mean that a large proportion of the lake (in the hypolimnion) is almost completely isolated from the epilimnion, from the atmosphere and from the inflow water. Under inverse stratification the deeper layers will be similarly isolated.

Isolation of the hypolimnion from the epilimnion is often not complete because the water movements between the two layers may cause mixing upwards in the region of the thermocline. When wind blows strongly over the lake surface it creates a surface current in its direction which is balanced by a deeper current in a reverse direction along the top of the hypolimnion. As this current rubs along the top of the hypolimnion it may cause eddies which pare portions of hypolimnion water and mix them with the epilimnion. The effects of strong winds are then often to deepen the epilimnion by erosion of the hypolimnion—as happens completely at overturn.

Strong winds blowing in narrow basins may also set up internal waves along the interface between epilimnion and hypolimnion which continue long after the wind has abated. Initially the wind may pile up epilimnion water at the windward end as well as creating a return current. This pile of water deepens the windward thermocline by displacing an equivalent volume of hypolimnion water to the leeward end where the leeward thermocline is brought nearer the surface. When the wind ceases, the two layers may rock, or oscillate as the displaced waters in the two layers run back, overshoot and then run in the original direction. Such oscillations may take several days to die down completely and as the two layers move against one another, there may be mixing at their interface. Several such waves (seiches) may travel in different directions in a basin of complex shape and waves travelling in a circular motion around the edge of the lake may be imposed by the earth's rotation. For all this complex activity, however, it is still true to say that the epilimnion and hypolimnion of a lake may be essentially unmixing for long periods, particularly in continental climatic regimes.

Particularly because of its isolation from the atmosphere, the water of the hypolimnion begins to change in its dissolved gas chemistry. Gases in the atmosphere maintain dynamic equilibria with the same gases dissolved in waters, such that the concentrations of all but CO_2 in water freely mixing and exposed to the atmosphere are predictable almost entirely from the water temperature (Chapter 3). (Carbon dioxide is part of a series of equilibria involving carbonate, bicarbonate, hydroxyl and hydrogen ions. The concentrations of these as well as the atmospheric concentration of carbon dioxide also significantly determine its equilibrium concentration in waters.) Respiration and photosynthesis may lead to temporary departures from equilibrium of oxygen and carbon dioxide over a day even in well mixed, open water. There will, however, always be a tendency to return to equilibrium conditions.

This is not so in water isolated from the atmosphere in hypolimnia or under ice. Hypolimnia include the least well illuminated parts of a lake and inverse stratification takes place during cold periods when light intensity and day length are also low and short respectively. Photosynthetic oxygen production is small or zero yet respiration of bacteria associated with falling detritus or sediments and bottom living animals continues. Oxygen concentrations decrease and carbon dioxide concentrations increase under these conditions.

A continual rain of phytoplankton cells, detritus, zooplankton and fish faeces and corpses and associated bacteria falls through the metalimnion to the hypolimnion. The greater the epilimnetic production, the greater the supply of organic matter to the hypolimnion and the greater the demand on the irreplaceable (until overturn) oxygen reserves there. The effect on the concentration of dissolved oxygen in the water is not directly connected with the epilimnetic production, because the total amount of oxygen available depends partly on the hypolimnion temperature and partly on the hypolimnion volume. A given epilimnetic production may cause complete anaerobiosis of a hypolimnion in a lake only 10 m deep with half its volume below the thermocline and at a temperature of 10°C, but have a negligible effect on oxygen concentrations in a 100 m deep lake with 90% of its water at 4°C in the hypolimnion. In both cases, however, the rates of hypolimnetic oxygen depletion per m^2 of water could be similar.

Hypolimnetic oxygen concentration therefore is not a reliable guide to the productivity of a lake. In practice, however, many less productive lakes are situated in deep basins in rocky, upland catchments and for all these reasons have high hypolimnetic oxygen concentrations. In contrast many productive lakes have shallow basins in the subdued relief of fertile lowland catchments and greatly deoxygenated hypolimnia. Profiles of oxygen concentrations for three lakes in summer are shown in Fig. 6.9.

As particles fall through the hypolimnion, they are decomposed by the aerobic bacteria which have colonised them and the hypolimnion, provided some oxygen remains, may become enriched in ammonium, nitrate, phosphate, silicate and other ions. These substances may be returned to the surface waters at overturn. In a tropical lake they may then stimulate increased production. In a temperate lake the overturn usually occurs at a time when production is becoming limited by light and when the increased winter flow from the catchment area is restocking the lake with nutrients anyway.

In many lakes, however, decomposition is not complete before the particles reach the sediment so that elements contained in them may be locked away. One implication of this is that the maintenance of phytoplankton

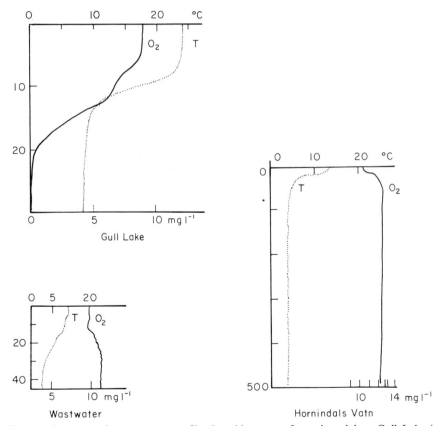

Fig. 6.9. Oxygen and temperature profiles in mid-summer from three lakes. Gull Lake is relatively fertile, but the other two are infertile. (Based on Moss [558], Macan [491], Hutchinson [379].) Vertical depth scales are in metres.

production in a lake is dependent on a continued supply of nutrients from the catchment area or atmosphere. Another is that a rich source of material accumulates for bacterial activity in the sediment.

6.2.8 Loss of phosphorus to the sediment

The former implication can be readily shown by use of Lund tubes [482, 483]. These are cylinders of butyl rubber 20 or 40 m in diameter, held floating at the rim and sealed into the sediments so that they isolate a body of water within a lake from the main source of nutrients, the catchment. When this was done in Blelham Tarn, a lake in the English Lake District, a reduction in phytoplankton crop followed very rapidly in the tubes (Fig. 6.10). Available

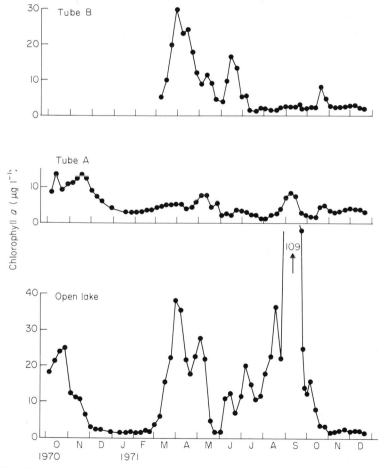

Fig. 6.10. Changes in chlorophyll *a* concentrations in Blelham Tarn (lower) and in two Lund tubes, A and B placed in the lake. The water in tube B was isolated from March 1971 onwards, that in A from mid 1970. (Based on Lund [482].)

nutrients, particularly phosphorus, were taken up into the algae and found their way rather quickly into the sediment. This is a very important result— it means that lakes may not accumulate nutrients in their waters, but are dependent for maintenance of their production on continued external supplies. There are instances, however, of lakes in which there may be a significant return of nutrients from the sediments and these will be dealt with later. For the moment we are dealing with a lake where sediments are accumulating particles which have passed through a still aerobic hypolimnion. What happens in the sediment?

6.2.9 Sediment and the oxidized microzone

In using organic detritus as their energy source, the many bacteria in the surface sediments absorb much oxygen which is replaced by diffusion. Below this surface layer, called the oxidized microzone, bacterial activity uses up oxygen faster than it can be replaced and the sediment becomes anaerobic only a few mm below the surface. The oxidized microzone is usually brown-red in colour because it includes a large quantity of oxidized iron compounds— largely oxide and hydroxide. Other substances are also present, especially ions like phosphate which are adsorbed and immobilized within the layer and largely prevented from diffusion into the overlying water.

   Below the oxidized microzone there is a great deal of bacterial activity for some centimetres. It is largely anaerobic, as different groups of bacteria, lacking access to oxygen and often unable to use it, use other electron acceptors to oxidize organic matter [414, 416, 554]. The reactions are those that also go on in deoxygenated swamp soils and were discussed in Chapter 5. Some of the products of these bacterial activities may diffuse through an oxidized microzone and accumulate in an oxygen depleted but still aerobic hypolimnion, for example CO_2, NH_4, MnII; others may be re-oxidized by aerobic bacteria or inorganic processes in the oxidized microzone: FeII, H_2S.

   If the hypolimnion becomes completely deoxygenated, the oxidized microzone may break down [554]. This is progressive as the oxygen concentration falls at the sediment surface and causes changing redox potential (Chapter 5).

   As the redox potential falls in the sediment surface microzone (Fig. 6.11), redox-sensitive substances can begin to escape from the sediment into the hypolimnion. FeIII is reduced to FeII between about $+300$ and $+400$ mV when there may still be 1–2 mg l^{-1} of oxygen dissolved in the adjacent water. When this happens the oxidized microzone starts to dissolve, allowing FeII and phosphate ions to move into the hypolimnion in quantity. At lower redox potentials, perhaps -100 mV, release of soluble sulphide ions occurs and hydrogen sulphide starts to accumulate.

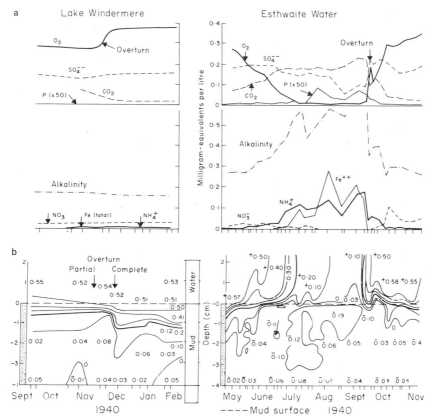

Fig. 6.11.(a) Changes in the chemistry of the water just (10–20 cm) above the sediment surface in Lake Windermere, which has an aerobic hypolimnion, and Esthwaite water, whose hypolimnion becomes anaerobic from July to September. (b) Redox potential (volts) measured in undisturbed cores taken from the sediment surface during the same period in the two lakes. The approximate lower limit of the oxidized microzone is at about +0.2 volts. The oxidized microzone remains intact in Windermere, but is destroyed in Esthwaite. (Based on Mortimer [554].)

The hypolimnetic water thus becomes greatly changed as it becomes deoxygenated. It steadily accumulates CO_2 (with a decrease in pH), MnII and NH_4^+ then FeII, phosphate and sulphide. Any nitrate present from previous aerobic decomposition is denitrified and as bacteria previously confined to the anaerobic sediment are able to colonize the water, methane may also accumulate.

During the stratified period, such hypolimnetic waters may become hostile to many eukaryotic organisms and fish are excluded from it at quite modest oxygen concentrations. At the concentrations found, FeII, NH_4 and H_2S may be toxic. Few bottom invertebrates can survive in anaerobic sediment overlain by anaerobic water, so that for most of the lake biota, the part of the lake capable of colonization is greatly reduced. Overturn results, of course, in a reversal of these changes as the hypolimnion water is reoxygenated. Under inverse stratification, similar deoxygenation and associated chemical changes may occur particularly if ice cover is prolonged. Winter kills of fish are then not uncommon in shallow lakes.

6.2.10 Aquatic plant communities and the shapes of basins

Thus far, light, nutrients and structure of the water mass have been considered as important general features of standing waters. The fourth aspect is the balance of open water and fringing and bottom communities in shallow waters. Shallow waters are likely to be more productive than deep ones because of the possibilities of recycling of regenerated nutrients from the bottom waters for continued phytoplankton growth in summer and because of the contributions made by extensive beds of aquatic plants.

Emergent, floating and submerged aquatic plants all contribute to the aquatic plant production of lakes. The emergent swamps are kept by the life forms of their plants to water about 2 m deep or less and the effects of wind on floating plants will confine populations of these to sheltered areas in inlets or among the swamps. The submerged plants penetrate to depths receiving enough light to allow net photosynthesis. This may be from only a few decimetres to perhaps a hundred metres or more depending on the water clarity. The submerged plants include vascular species which are usually not found below about 10–11 m, and bryophytes and charophytes which may go much deeper. One reason for the confinement of vascular plants to the shallower water may be that under pressures greater than about 1 atmosphere (about 10 m of water) the development of their lacunae (Chapter 5) is prevented [268] but others believe pressure to be unimportant [762]. Wave action at the edges of large lakes may also prevent colonization of the most shallow water, though communities of microscopic algae will generally be found attached to the rocks laid bare of sediment by the waves (see Chapter 8).

The submerged plants have considerable problems in obtaining sufficient light energy particularly since they are covered by efficiently absorbing dissolved and particulate screens in the water. They do have an advantage, nevertheless, in that they are rooted (or in the case of bryophytes and charophytes, have rhizoids) in a sediment generally much richer in nutrients than the overlying water, from which they can often also absorb nutrients via their leaves (Chapter 5) [163].

Some sediments—sands and gravels—are nutrient poor but finer sediments generally are not, and in lakes fed by catchments which provide few nutrients for the plankton, the rooted plants may be greatly advantaged. Data on plant production (and that of associated epiphytes) are much scarcer than those on phytoplankton production and few studies have considered all communities in the same lake (Table 6.1). Some general hypotheses might be expected to be true as data accumulate, however.

First, aquatic plant production might be expected to be less dependent on the nature of the pristine catchment area than that of phytoplankton. Infertile catchments, provided they are eroding to some extent, should contribute enough sediment to supply a quite dense crop if the nutrients are annually recycled between the growing parts and the storage tissues in roots and rhizomes. Secondly, the contribution made by aquatic plants should be greater in shallowly sloping basins than steep sided ones—the ratio of mean depth ($\bar{z}$) to maximum depth (z_{max}) may give a measure of this. Thirdly, a high degree of indentedness of the shoreline should favour aquatic plants by providing sheltered water. The shoreline development, the ratio of the

Table 6.1. Relative importance of different producing communities in lakes. (From Westlake *et al.* [853])

Lake	Mean depth (m)	Max. depth (m)	Area (km²)	Units	Aquatic plants	Photo-synthesis (% total) Epiphytes and bottom-living algae	Phyto-plankton
Lawrence L. (USA)	5.9	13	0.05	kg C lake⁻¹ year⁻¹	4400 (51)	1900 (22)	2150 (26)
Marion L. (Canada)	2.4	7	0.13	kg C lake⁻¹ year⁻¹	2340 (27)	5200 (61)	1040 (12)
Borax L. (USA)	0.7	1.5	0.43	kg C lake⁻¹ year⁻¹	511 (0.7)	28 000 (43)	36 500 (57)
L. Latnajaure (Swedish Lapland)	16.5	43	0.73		(20)	(20)	(60)
L. Mikolajskie (Poland)	11	28	5.0	MJ m⁻² year⁻¹	1.4 (12)	0.78 (6.5)	9.4 (82)
L. Myastro (USSR)	5.4	11	13	MJ m⁻² in May–Oct.	0.56 (7.4)	0.42 (5.5)	6.6 (87)
L. Chilwa (Malawi)	2	7	1400	MJ lake⁻¹ year⁻¹	4.2×10^{10} (63.6)		2.4×10^{10} (36.4)
Kiev Reservoir (USSR)	4	19	990	MJ m⁻² year⁻¹	0.74 (6.4)	5.7 (50)	5.1 (44)

shoreline length (l) of a lake to that (l_o) of a circle of the same area as the lake, is a convenient measure of this. And fourthly, the proportion of the toal production contributed by aquatic plants might be expected to decrease with fertilization of the lake by human activities, owing to competition for light (or possibly CO_2 because diffusion rates into bulky tissues are low) with the phytoplankton. Initially the absolute production of aquatic plants might increase with fertilization owing to leaf uptake.

In pristine lakes the proportion of production contributed by submerged aquatic plants might then be proportional to:

$$\frac{\bar{z}}{z_{max}} \cdot \frac{l}{l_o}$$

and in fertilized ones to;

$$\frac{\bar{z}}{z_{max}} \cdot \frac{l}{l_o} \cdot \frac{l}{[P_{tot}]}$$

where the total phosphorus concentration in the water is used as a measure of the effect of fertilization. The area covered by emergent and floating-leaved plants was predicted in a set of Swedish lakes by regression equations which included measures of lake area, total nitrogen and depth (for floating plants only) but the relationships have not been tested for submerged plants [752].

6.3 GENERAL MODELS OF LAKE PRODUCTION

Four main features of lakes which contribute to their productivity have now been discussed. What is their overall effect; can generalizations be made? Because many lakes have become more productive recently, posing problems for their management (see below) the matter is of wider than academic significance. Unfortunately most of the syntheses have related features of the lakes only to phytoplankton production, and the fringing and bottom plant communities have been ignored for lack of data.

6.3.1 Brylinsky's synthesis

Between 1964 and 1974 an attempt was made under the International Biological Programme to collect data on lake productivity from all over the world. Brylinsky and Mann [82] then attempted to draw generalities from these data and subsequently Brylinsky [81] refined the analysis.

Correlations among a variety of lake variables and photosynthetic production were made. Brylinsky reduced production rates measured by the O_2 light and dark bottles method (Chapter 5) by 15% to allow for respiration and make them more comparable with ^{14}C estimates (Chapter 5) which he considered to be measures of net production. Table 6.2. shows the amount of variance in photosynthesis explained by different combinations of those abiotic and biotic variables found to have greatest correlation. Among the abiotic variables, his conclusion was that latitude was particularly important and that nutrients had minor effects. The data available on nutrients, however, were relatively few and did not include the total amounts (e.g. P_{tot}, N_{tot}) in the water, only the immediately available forms. At a given time there may be little available because it has already been taken up into the plankton crop. The important effect of latitude was expressed not so much through light (Table 6.2, Equation 2) but through some other correlate, such as temperature or precipitation. This result was surprising in that the conventional wisdom was, and is, that nutrient supply is by far the most important determinant of production. Numerous bioassays of lake water in the laboratory suggest this (though many might be criticized for not exposing the nutrient-enriched water to light intensities as low as those in the water

Table 6.2. Relationships between phytoplankton primary productivity in a set of up to 93 lakes distributed across the world. Analysis was by mutiple regression and results are expressed as percentage of variation explained. In a perfect relationship 100% would be explained. (From Brylinsky [81])

Variables included and (percentage variance explained by each)	Number of lakes	Percentage of total variance explained
Abiotic variables		
Latitude (46.1), altitude (2.9)	93	49
Latitude (43.6), visible radiation (2.6) precipitation (11.3), mean depth (0.6) surface area (1.8), volume (0.3)	84	60.2
Latitude (42.1), precipitation (10.6) conductivity (9.7)	63	62.4
Latitude (44.1), precipitation (11.2) phosphate-P (1.7), nitrate-N (0.8)	54	57.8
Latitude (43.7), precipitation (9.9), conductivity (10.1), thermocline depth (6.2)	61	69.9
Abiotic and biotic variables		
Chlorophyll *a* (60.3), latitude (18.3)	27	78.6
Chlorophyll *a* (59.7), conductivity (4.4)	20	64.1
Chlorophyll *a* (60.1), nitrate-N (3.7)	25	63.8
Chlorophyll *a* (58.8), phosphate-P (2.1)	25	60.9
Chlorophyll *a* (60.1), latitude (17.6) precipitation (12.3), thermocline depth (9.2)	23	88.4

column). Experiments, either deliberate or unintentional, where lakes have become more productive due to nutrient addition in sewage effluent or agricultural drainage also support the crucial role of key nutrients.

When Brylinsky included mean chlorophyll *a* concentration in the water, a measure of phytoplankton biomass, in the correlations with productivity, it was found to be very significant, overshadowing the effect of latitude (see Table 6.2). Because algal biomass is an indirect measure of the total amount of potentially limiting (to crop size) nutrient in circulation, this may indicate a key role for nutrients. The pre-eminence of latitude in correlations using abiotic variables alone may then rest solely on the lack of an adequate measure of nutrient availability in the calculations. On the other hand, because algal biomass and algal productivity are not independent of one another, a circularity is introduced which technically invalidates the statistical techniques used.

6.3.2 Schindler's analysis

Following the original Brylinsky and Mann paper [82] D.W. Schindler [719] repeated the exercise using data gleaned from equally wide sources. Schindler's data included the energy related variables used by Brylinsky but also incorporated better measures of nutrient availability. Again O_2 release data were adjusted to make them comparable with ^{14}C uptake data. Variables tested included latitude, mean chlorophyll *a* concentration, mean total phosphorus concentration, P loading, N loading, mean depth, and residence time of the water.

Schindler found no significant correlation between phytoplankton productivity and latitude, nor between phosphorus loading and productivity. However, when he adjusted the loading for the effects of water residence time (see Section 6.2.6) to give an effective measure of total P concentration, he found highly significant correlations with productivity ($r = 0.59$) and with chlorophyll *a* concentration ($r = 0.88$). Schindler did not use data from lakes where he expected the key nutrient to be nitrogen (those with input ratios of N:P less than 5:1) and did not have sufficient sites to make separate correlations with nitrogen concentration. Again conventional wisdom suggests that such sites are scarce. Schindler argues cogently elsewhere [718] that because of the availability of atmospheric nitrogen and the widespread occurrence of N-fixing blue-green algae, that nitrogen limitation can only be temporary. It should be countermanded sooner or later by the development of crops of such algae to the potential set by phosphorus in the water.

Schindler's and Brylinsky's analyses are clearly at odds. In an attempt to reconcile them, Schindler suggests that the effects of latitude found by Brylinsky cannot be due to light. For a fiftyfold range of light energy available from the polar regions to the Equator, a thousandfold range in production is found. He suggests that the latitude effect may be related to the supply and efficiency of recycling (see Chapter 7) of the nutrients. Rain apparently increases in nutrient concentration with decreasing latitude and the higher temperatures at low latitudes may result in more rapid decomposition and recycling of scarce nutrients giving a more efficient use of the nutrients to support more photosynthetic production. This is perhaps supported by findings [531] that photosynthetic rates in pristine, saline East African lakes (Big Momela, Reshitani, Nakuru, Manyara, Magadi and Elmenteita) were around 30 g O_2 m^{-2} day^{-1} (11 g C m^{-2} day^{-1}) a value seldom found except where large quantities of nutrients are added by human activities.

6.3.3 Models incorporating other features

The role of communities other than the phytoplankton should not be ignored in generalizations about lake production though data are too few to incorporate them. However, indirectly they have been considered in equations attempting to predict fish biomass or fish yield (the harvest obtainable) from features of lakes. The links between fish production and physical and chemical features are far more distant than those between phytoplankton photosynthesis and nutrients or light so the relationship must be expected to be less secure.

One index (the Morpho-edaphic index [702]) relates fish yield to $\sqrt{TDS}/\bar{z}$ where TDS is the content of total dissolved solids, effectively an ultimate measure of nutrient availability in exorheic lakes. The inverse proportionality of mean depth ($\bar{z}$) may be a reflection also of nutrient availability and the recycling of nutrients regenerated by sedimenting material. It may also reflect the availability of shallow water and the associated aquatic plant communities. This index tends to describe the relationship reasonably in limited areas but the constant of proportionality varies greatly from region to region. It is a useful device for fishery managers (Chapter 9) in restricted areas but more difficult to use to throw light on fundamental principles. Melack [529], for example, found that for eight African lakes it was not related to fish yield whereas gross phytoplankton production was. On the other hand it was a reasonable predictor of fish yield in fifteen lakes in Southern India as also was gross production.

6.4 EUTROPHICATION AND ACIDIFICATION— CHANGES IN THE PRODUCTION OF PHYTOPLANKTON IN LAKES

There is a traditional belief that lakes become naturally more productive as they accumulate nutrients with time. This process is called natural eutrophication and it probably occurs very rarely, if at all (see Chapter 10).

The processes by which phosphate and nitrate are either washed downstream, locked into the sediment or denitrified there, seem to ensure that the productivity of a lake will reflect largely the contemporary nutrient supply to it and will increase or decrease in response to changes in this, though there may be some exceptions (see later).

In the last few decades, two man-induced processes have tended to cause major changes in phytoplankton production—artificial eutrophication and acidification. The former has increased, the latter decreased, the productivity.

6.4.1 Eutrophication

Artificial eutrophication (abbreviated here to eutrophication) usually is a problem. It is caused by the ingress of phosphorus in particular from sewage treatment works, discharge of raw sewage, (including that from large concentrations of farm animals, and from fish farms), and run-off of nitrates from arable catchments. It has briefly been dealt with in Chapter 5, when problems caused by increasing aquatic plant and filamentous algal growth were discussed. In lakes, the problem is reflected sometimes in increased aquatic plant growth at the margins but usually in increased phytoplankton growth. In extreme cases there may be eventual complete loss, through competition with the phytoplankton, of all submerged aquatic plants, and sometimes also the regression of marginal reedswamps [55, 563, 564].

Increased phytoplankton crops may cause amenity problems in terms of loss of clear water, and they are also commercially important. They make the filtration of water for the domestic supply more costly, and may cause tastes and odours in the water through secretion of various organic compounds. They may produce substances toxic to mammals [114, 337] and sometimes fish [353]. Fine organic matter from plankton crops passing through the waterworks' filters may support clogging communities of nematode worms, sponges, hydrozoans and insects in water distribution pipes. Dissolved organic matter secreted into the water by the algae may take chlorination more costly, or produce astringent-tasting chlorinated phenolic substances in the domestic water supply.

Organic matter sedimenting from the increased phytoplankton crops to the hypolimnion increases the rate of hypolimnetic deoxygenation. If the hypolimnion becomes anaerobic and rich in sulphides it may be unusable for water supply for a long period in summer, when, in many areas, water is short. A deoxygenated hypolimnion also excludes some fish groups, particularly the coregonids (white fish) and salmonids. They may then be unable to live in the lake at all or to pass through it to up-river spawning grounds because they are also intolerant of the high temperatures in the epilimnion. Such fish often depend on a cool, well oxygenated hypolimnia for their summer survival, and frequently support valuable commercial or recreational fisheries.

Fish tolerant of lower oxygen concentrations may increase in production as the coregonids and salmonids disappear (Fig. 6.12) but in extreme cases of eutrophication, usually shallow lakes, when dense algal growth out-competes the marginal aquatic plant beds, the major loss of structure in the ecosystem may have severe effects. There is a loss of spawning and living habitat for the fish, some of which attach their eggs to aquatic plants or their detritus. Large plant living invertebrates such as snails and damselfly and dragonfly nymphs are much reduced in numbers and this may lead to declines in the growth and survival of larger fish which depend on them for food. Fish-eating (e.g. bittern, heron) and plant-eating (swan, coot) bird populations may consequently be reduced in numbers so that the entire conservation value of a habitat declines. The loss of plants may also make bank edges more vulnerable to erosion by normal wave action or by boats so that banks may have to be expensively and unaesthetically protected by metal or wood piling. See Moss [563, 564] for a case study.

The degree to which eutrophication is considered a problem depends on place and people. A small lake fertilized by the village sewage in south-east Asia may be a major source of protein in the form of deoxygenation-tolerant fish (see Chapter 9) feeding on thick algal soups. Eutrophication here is perceived as a boon. A very modest degree of eutrophication by perhaps the effluent from a couple of holiday cottages in Canada or Sweden may be a

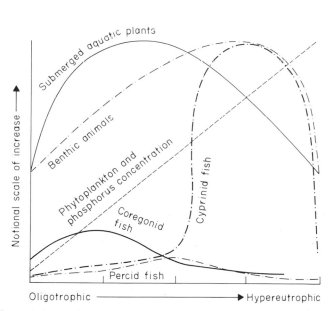

Fig. 6.12. General changes in north temperate lakes as they become eutrophicated. Changes in aquatic plants and benthic animals are considered in Chapter 8. In general, though yields of fish may increase, Cyprinids are less desirable as a commercial catch than coregonids. (Based on Hartmann [326].)

problem because it has decreased the transparency of the lake from 5 m to 3 m and has been associated with a reduction in lake trout catches.

In Great Britain the Water Authorities define eutrophication as a problem when it cannot be coped with by treating its symptoms within a drinking water reservoir in some chemical or mechanical way. More fundamental approaches to tackling the problem are not perceived as worthwhile unless they significantly reduce filtration costs [117]. The threshold at which this occurs is far above that which has sent governments in North America and mainland Europe hurrying to place anti-eutrophication legislation on the statute books. Table 6.3 shows a system proposed for Southern Ontario lakes in which the recreational and aesthetic values of the lakes are related to mean total phosphorus and phytoplankton chlorophyll *a* concentrations. In general, water supply problems might be expected at category 3 (30 µg l^{-1} total P and above). The extent of the eutrophication problem in the more populated areas of the world leaves few lakes in categories 1 and 2.

6.4.2 Solving the eutrophication problem

There are two approaches—treating the symptoms and removing the causes. The former includes raking out aquatic plants, poisoning algal growths, or altering the reservoir regime of mixing and circulation to favour more easily filterable algae or those less problematic in the production of tastes and odours (see Chapter 7). It is attractive in that it is inexpensive in the short term but it does nothing to tackle problems wider than those of producing drinking water or maintaining amenity. In the long-term it is usually cheaper to try to remove the causes of environmental problems. In the case of eutrophication this means limiting the supplies of the nutrients supporting the increased productivity.

The first question that then arises is which nutrient or nutrients to remove? To gain increased algal production, both nitrogen and phosphorus supplies must be increased because even if phosphorus is currently limiting the potential crop that can be supported, the supplies of nitrogen are generally not greatly in excess of algal need. To reduce the algal crop of a lake, however, should require reduction in only one nutrient. An analogy might be drawn with motor cars, which require oil, petrol and in most cases water to keep them running, but which stop if they run short of any one of these.

Phosphorus is the nutrient which can be most readily controlled. Nitrogen is not easily controlled—its compounds are too soluble, they enter waterways from many diffuse sources (every field seep) and in any case there is a potential supply from the atmosphere through nitrogen fixers even if those on the ground are removed. Phosphorus, on the other hand, is readily precipitated, enters mostly from a relatively few 'point' sources—large stock units and sewage treatment works—and has no atmospheric reserve.

Table 6.3. Maximum permissible average summer concentrations of chlorophyll a ($\mu g\,l^{-1}$) and total phosphorus ($\mu g P\,l^{-1}$)* in Southern Ontario lakes for different categories of use. (From Dillon and Rigler [167]). UK lakes falling into the various levels are also given

Level of water classi-fication	Use and appearance of water	Depth at which Secchi disc visible (m)	Chloro-phyll a level	Phos-phorus level	UK lakes falling into each level
1	Swimming in clear water, highly oxygenated hypolimnia suitable for game fishing	5	2	10	Scottish highland lochs, upland Welsh tarns, westernmost English Lake District lakes, Wastwater and Ennerdale and mountain tarns
2	Water recreation, preservation of game fishery not imperative	2–5	5	19	Some English Lake District lakes, e.g. Windermere
3	Aesthetic appearance (very clear water) for swimming not required. Emphasis on fairly productive coarse fisheries (bass, walleye, pickerel, pike, maskinonge, bluegill, yellow perch). Hypolimnetic oxygen depletion common and possibility of fish kills in winter by deoxygenation under ice	1–2	10	30	Esthwaite, some Shropshire Meres, Upton Broad
4	Suitable for coarse fishing but increased danger of fish kills in winter; hypolimnetic oxygen depletion very severe	Less than 1.5	25	56	Some Shropshire Meres, Martham Broad
5	Not applicable or included in the Ontario classification. Fish kills infrequent only because most British lakes do not freeze over for long in winter	Possibly much less than 1	Greater than 25	Greater than 60	Most Norfolk Broads, the London drinking water reservoirs and some Shropshire Meres

*Phosphorus and chlorophyll levels were related in these lakes by the significant regression: $\log_{10}[P] = (\log_{10}$ [chlorophyll a] + 1.14)/1.45

The next questions in planning restoration of a lake from eutrophication are then: what are the present supplies of phosphorus and how do these separately contribute to the concentration of total P in the water; how is the total P concentration related to the algal crop; what size of algal crop is wanted in the restored lake; is it possible to achieve a low enough phosphorus concentration to achieve that crop; how shall it be done; and what will be the complications?

6.4.3 What are the present supplies of phosphorus and do they contribute equally?

It is generally not difficult to identify the supplies of phosphorus though some may be cryptic. Survey of the catchment will reveal most sources and give an indication of land use. It will usually be too expensive to measure directly the amounts emanating from each source, but an approximate budget may be drawn up, as a desk-study, using data on phosphorus produced in sewage per person per year, catchment population, stock numbers, concentration of total phosphorus in drainage from land under various usages, and streamflow.

The loadings from various sources to the inflows of a lake can then be calculated and, from the amount of water entering the lake, a calculation can be made of what the total phosphorus concentration should be and compared with that which has been measured. Sometimes the calculated inflow concentration is much greater than that measured! This might be because of inevitable errors in estimating the phosphorus loading of a particular area from values derived from elsewhere. Alternatively, much of the total phosphorus draining from cropland may be in particulate form, may be readily sedimented in upstream wetlands or on the bed of the main river and may never reach the lake. Better correspondence is expected between predictions of total phosphorus concentration from the loadings of soluble phosphate compounds (inorganic and organic) and the measured total phosphorus entering the lake, although even then there may be losses of soluble phosphorus to the river bed during transit to the lake.

From the loadings (L), the mean depth (z), the flushing coefficient (ρ) and the retention coefficient, it is possible from the Vollenweider model (see above) to relate loading to mean concentration and therefore to calculate what the reduction in loading should be to attain a desired concentration in the lake. The problems of measuring true loading, let alone those of measuring the retention coefficient have meant that this approach has not been widely used (though see [167]).

6.4.4 Relationship of the phosphorus concentration to the algal crop

The concentration of phosphorus in the lake can be related readily to either mean or maximum chlorophyll *a* concentrations in the lake (see Fig. 6.8) and this relationship used as a basis for planning the reduction in loading which will give a desired chlorophyll *a* concentration. What this target should be, however, is a problem not easily solved for it concerns human perceptions.

Some limnologists divide temperate lakes into categories: ultra-oligotrophic, oligotrophic, mesotrophic, eutrophic, hypertrophic. These categories form a series of fertility and though it is not difficult to define the extremes, everyone has a different view of where the boundaries lie between categories. Ultra-oligotrophic lakes have very clear water, almost no phytoplankton, hypolimnia saturated with oxygen and coregonid and salmonid fish; hypereutrophic lakes have turbid water, dense algal growths, anaerobic hypolimnia (if they are deep enough) and cyprinid fish. Even these accounts are caricatures, however, readily to be disagreed with, and defining the intermediate categories of what is, in fact, a multivariate continuum is not really sensible.

However, lake restoration has political as well as scientific aspects, and politicians and administrators need definitions for the proscribing of law. A working party of the Organization for Economic and Cultural Development therefore gathered views on the definitions of the above categories in terms of phosphorus and chlorophyll *a* concentrations from a variety of people to obtain some consensus. It was then possible to draw a diagram (Fig. 6.13) which gives chlorophyll *a* and total phosphorus concentrations for particular combinations of inflow phosphorus concentration and T_w (water replacement time). Lake restoration attempts will generally wish to bring lakes from the hypereutrophic and eutrophic categories into the mesotrophic and oligotrophic groups.

6.4.5 Methods available for reducing total phosphorus concentrations

The best methods are those which reduce the phosphorus supply at its source. These include diversion of sewage effluent to the sea where it can be greatly diluted, and precipitation of it from the effluent (phosphate stripping) before it is discharged to the river. Alternatives for the presently used phosphate detergents might also be found. Less fundamental or effective are the precipitation of phosphate by adding chemicals to the lake, aeration of the hypolimnion, or removal of phosphorus-containing biomass.

Diversion of effluent is possible where a lake lies reasonably close to the sea. The pipelines necessary are expensive and the system must be such that the effluent itself does not constitute a large part of the water supply to the lake. Lake Washington is perhaps the best example of restoration by this method. Around the lake lie Seattle and its metropolitan suburbs. In 1955 a blue-green alga, *Oscillatoria rubescens* became prominent in the plankton, signalling a series of changes consequent on the progressive development of the area [191]. The lake was receiving sewage effluent (24 200 m³ per day) from about 70 000 people and the effluent was providing about 56% of the

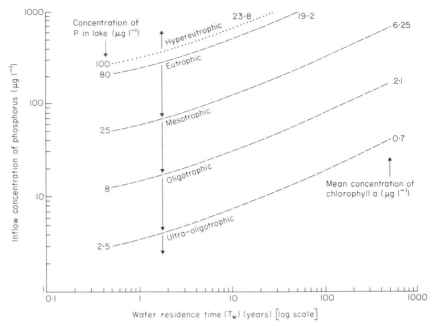

Fig. 6.13. Relationship between inflow phosphorus concentration, water residence time and the consequent most likely phosphorus and chlorophyll *a* concentrations in the lake. The longer the residence time, the more it is likely that phosphorus will be deposited in the sediment so that lower in-lake phosphorus and chlorophyll *a* concentrations will be obtained for a given inflow concentration. The terms 'hypereutrophic', etc, are placed arbitrarily on the basis of a consensus determined by the OECD. (Based on Vollenweider and Kerekes [833].)

total P load in the lake. In the early 1960s a scheme was conceived to divert the effluent to Puget Sound and by 1967 almost all of the effluent was piped to the sea. The transparency of the lake increased from about 1 m to 3 m and chlorophyll *a* concentrations decreased from 38 to about 5 µg l^{-1} [187]. The lake responded very quickly to diversion and since the early 1980s (Fig. 6.14) has improved even more [188]. These further improvements involve changes in the phytoplankton and its grazer communities and are discussed in Chapter 7.

Diversion is not always so practicable but all sewage treatment works can have phosphate stripping plant installed relatively cheaply. The effluent, after it has passed through the orthodox stages of removal of solids and biological oxidation of dissolved and fine organic matter (Chapter 5) is run into a tank and dosed with a suitable precipitant. Aluminium salts work well but have the disadvantage that disposal of the precipitate in the works' sludge as farm fertilizer may not be possible because aluminium salts are poisonous. Iron salts are not and ferrous ammonium sulphate is frequently the precipitant chosen. The costs of the process are largely in chemicals

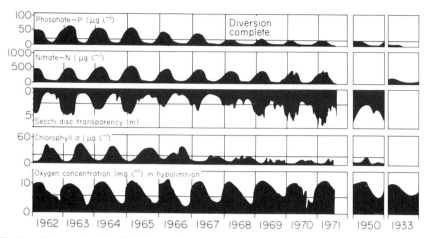

Fig 6.14. Changes in Lake Washington between 1962 and 1971. Data for two earlier years are placed at the right side to suggest the possible end point of the changes. The lines are smoothed to emphasize long-term trends. After Edmondson [188].

rather than installations and up to 95% of the phosphate can be removed. The availability of the process has allowed the US and Canadian legislatures to require that effluents in the area of the St. Lawrence Great Lakes should contain not more than 1 mg l^{-1} of $PO_4 - P$, compared with 10–20 mg l^{-1} previously.

Results of phosphate stripping have not been so dramatic as those for L. Washington. There still remains some phosphorus in the effluent, and the lake water may also contain a substantial background concentration resulting from agricultural activities in the catchment which are not readily controlled. In lakes with a long water residence time, also, it might take many years to obtain a marked improvement [165]. Many of the lakes for which stripping has been tested are shallow, extreme cases and in these a mechanism of return of phosphate to the water from the huge past stores in the sediment may take place (see below). Additional measures may be needed to restore these lakes completely, but phosphorus removal is essential to the success of the overall operation.

There has been much pressure to remove phosphorus from domestic detergents. It is true that such phosphorus comprises 40–50% of the total phosphorus in sewage effluent. However, the amount of phosphate remaining in the effluent in many places would still need removal by precipitation. The costs of precipitation increase greatly as the concentration decreases, so that removal of detergent phosphate would not reduce the costs of stripping proportionately. In some remote areas, however, removal of detergent phosphorus might obviate the need to install a stripping process at all for a considerable time; and because reduction of phosphorus concentrations in lakes is proving more difficult than at first thought, any reduction must be deemed ultimately sensible. Balanced against this are the facts that phosphate is a relatively cheap, effective, non-poisonous, non-corrosive component of detergents [183], whilst most of its competitors (e.g. silicate, borate, ethylene diamine tetra acetic acid, nitrilotriacetic acid, zeolites) fail on one or more of these counts [209, 272]. Nonetheless, legislation has been passed both in Europe and North America to reduce or remove the content of sodium tripolyphosphate used as a builder in detergents.

6.4.6 In-lake methods

Once phosphorus is actually in the lake, three main methods have been used to remove it. The first is to treat the lake with a solution of aluminium or ferrous salt to precipitate the phosphate; the method may also coagulate particulate matter. Results may be immediately very good with a clarification of the water [335] or ultimately negligible [229]. Of course the treatment must be repeated periodically so long as external inputs of phosphate continue.

A second method depends on the premise that entrainment of hypolimnion water contributes significant amounts of P to the epilimnion and involves hypolimnetic aeration [46, 54]. Air is bubbled into the hypolimnion in such a way that the hypolimnion is oxygenated without mixing with the surface waters during summer (Fig. 6.15). This technique at least increases the volume of water available to fish but again its use in reducing the phosphorus content of the lake in its productive upper layers is equivocal.

A third potential way of reducing the phosphorus content within the lake itself is to remove biomass which has accumulated in it [54, 374]. The problem with this is that compared with the sediment, the biomass accumulates very little of the total load. Clearly organisms like fish, at the top end of the food chains, accumulate only very small quantities and smaller organisms like invertebrates and algae which may accumulate a greater proportion, are uneconomic to remove from the water. Removal of aquatic plants has been suggested but they are bulky, comprise mostly water and air, have to be disposed of where they cannot pollute other watercourses in their decay, and accumulate phosphorus mostly from the sediments. In the latter the phosphorus reserves are very great and substantial improvement in a lake must not be expected by biomass removal for several to many years, and only then if external supplies are also reduced [85].

6.4.7 Complications for phosphorus control

By and large the above account of phosphorus control has been pessimistic. There are few cases where alone it has been dramatically successful. One of the reasons for this is that inevitably it has been tried on the most severe examples of lake eutrophication, where interactions with the phosphorus stores in the sediment may become important.

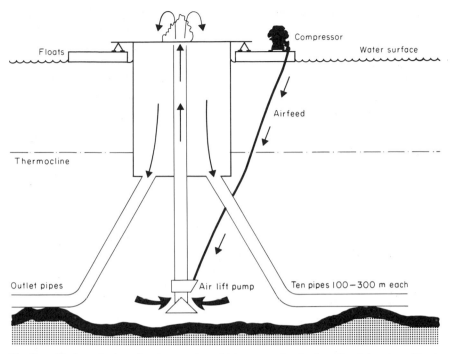

Fig. 6.15. Design of a hypolimnion aerator which oxygenates the hypolimnial water without mixing it with that of the epilimnion.

The Swedish Lake Trummen, for example, had become highly eutrophicated by 1981 after 30 years of discharge of sewage effluent into it [45, 53]. It was accumulating 8 mm year^{-1} of black, sulphurous mud and during ice cover in the winter its water column (2 m) became deoxygenated with consequent fish kills. The sewage effluent and some industrial discharges were diverted in 1958, but by 1968 the lake was still suffering high algal crops, low transparency, deoxygenation and fish kills. Much of the algal crop was being supported by phosphate released from the sediment. There was an internal load of 177 kg P per year compared with the external one which had been reduced to 3 kg P year^{-1}. In 1970 and 1972 a slurry of surface sediment and water was sucked from the lake. The sediment was settled in a lagoon and later disposed of as fertilizer whilst the water, after treatment with aluminium salts, was run back to the lake. Following sediment removal the water cleared, and winter deoxygenation no longer occurs.

In the discussion earlier in this chapter it was suggested that sediment acts as a phosphate accumulator, with very little of the phosphorus being released under aerobic conditions through the oxidized microzone. There has been much controversy over the role of sediments in the supply of phosphorus to overlying lake waters, particularly in shallow lakes where water mixes constantly to the bottom. There does appear to be a diffusion of some phosphate ions even through an oxidized microzone [352, 461, 604], though where comparative studies have been done (e.g. [423, 800]) this is small compared with the rate of release from deoxygenated sediment surfaces. However, in Lough Ennell in Ireland, Lennox [464] found that as much as 17% of the total phosphorus load could come from the aerobic marginal sediments. Drake and Heaney [179] found that at high pH ($>$ 10) significant amounts of phosphorus were released from the aerobic marginal sediments of a lake. But the high pH was created only over short periods by photosynthesis of large phytoplankton crops. Hence it is a consequential rather than a causative problem. In already eutrophicated lakes, the foraging of carp (*Cyprinus carpio*) for animals in the sediment may also mobilize large amounts of phosphorus [445]. Again this appears to be consequential as the size of fish involved presupposes an already high productivity.

Phosphorus supplies from the hypolimnion may mix sufficiently into the epilimnion in mid-summer to be significant in maintenance of the summer algal crops [119], but, using a range of examples, Schindler [719] found that the relationship between external phosphorus load and total phosphorus concentration in the water was the same for both stratified lakes and unstratified ones. He argued that in unstratified lakes, if internal loading from the sediment is significant, there should be a greater concentration of total P for a given external load. But there was not.

The problems seem to occur when the lake is well mixed but the sediment surface becomes anaerobic. In a shallow lake, like Trummen, wind circulation of water to the bottom is likely to be very efficient and oxygenated water must be expected always to be in contact with the sediment surface. Yet it is clear that in L. Trummen and in shallow eutrophicated lakes elsewhere [610, 613], that phosphate release does occur in quantity.

What appears to happen is that the supply of organic matter falling from the plankton to the sediment is both copious and highly labile, not having had time to decay in the short water column. It supports a degree of bacterial activity which cannot be sustained by diffusion of oxygen from the water to the sediment. Thus, despite overlying aerobic water, the sediment surface becomes anaerobic and the oxidized microzone breaks down. This allows both Fe^{2+} and PO_4^{3-} ions to escape. In the overlying aerobic water they should reprecipitate so that little net supply of phosphate should occur. However, in intensely reducing sediments with reduction of sulphate to sulphide by anaerobic bacteria, much of the iron is reprecipitated as sulphide, leaving phosphate to accumulate unhindered in the water. The phosphate can then help sustain a large summer algal community (usually limited for a time by nitrogen, so abundant is the released phosphate) which continues the supply of labile organic matter to the sediment and may maintain a self-sustaining system [569].

If the water retention time of the lake is short then much of the released phosphate may be washed out, together with some of the algal crop, if the external loading is reduced, so that the supply of organic matter and phosphorus to the surface sediment will diminish with time. The oxidized microzone should then re-form in summer. However, if the retention time is long and if winter stratification and deoxygenation also favours release there may be little net loss and no recovery.

To speed the process of sediment resealing, a technique has been invented called the Riplox process. It has been tried at L. Lillesjon near Varnamo in Sweden [54]. Diversion of sewage failed to restore it because of phosphate release from a sediment rich in P but not in Fe. The Riplox process injected, from a specially designed harrow, concentrated solutions of calcium nitrate, ferric chloride and lime into the sediment. The nitrate acts as a substrate for denitrifying bacteria which oxidize organic matter in the sediment, thus helping create conditions for the oxidized microzone to re-form. The lime adjusts the pH of the sediment to the optimum (7–7.5) for denitrification and the iron chloride helps in phosphate precipitation. Treatment was carried out in 1975 and the transparency of the lake was subsequently increased from 2.3 to 4.2 m; total phosphorus concentrations fell significantly. The treatment appears to have been successful but, perhaps for reasons of cost, has not been widely used elsewhere. It will only work, of course, where external nutrient supplies are also reduced but in such cases may give a more rapid recovery than simply leaving the sediments to reoxidize naturally. It may be cheaper than sediment removal, but the latter, in shallow lakes, is often needed anyway just to reinstate a reasonable depth of water.

6.5 ACIDIFICATION

Eutrophication, for the most part, is a severe problem of the populated lowlands and a lesser problem for the remoter upland lakes. In the uplands and remoter unpopulated regions, with their poorly weathered rocks and thin soils, it is generally acidification which is the more important problem [236]. It leads usually to a decline in productivity—sometimes called 'oligotrophi-cation'—of already unproductive lakes. Although this may aid the filtration of water for the domestic supply it may increase the corrosion of pipework, often of lead, in the cities to which the water is supplied. Acidification is also a problem because of its effect on the fish. The most obvious symptoms are the same as those affecting acidified rivers (Chapter 4) of aluminium and perhaps other heavy metal toxicity of adult fish, and interference with the hatching of fish eggs.

Lakes supplied with acidified water may become very clear, partly because mobilization of aluminium ions in the catchment soils leads to precipitation of phosphate which then stays in the soils [772]. Aluminium ions also readily flocculate particles, including phytoplankton, in the water of the lake [584]. In the very clear water, aquatic plant growth may spread further across the lake bed than previously, but the growth tends to be of acid tolerant mosses, *Sphagnum* spp., and of filamentous algae like *Mougeotia*. Aquatic vascular plants usually decline in numbers.

The diversity of all communities in the lake, including the phytoplankton, tends to decrease and an absence of fish predation leads to changes in the communities of invertebrate animals (see Chapter 7). The organic content of the sediment may increase, as leaf and other litter washed in is not decomposed, for populations of animals like snails and *Asellus* may disappear as pH drops below about 5 [11, 772].

Lakes affected by acidification are described from Ontario [327], New England [467], the Galloway region of Scotland [224, 225] and from the Brecon Beacons in Wales. In Norway an area of 13 000 km² is devoid of fish populations with lesser changes over a further 20 000 km² [733]. About 18 000 lakes in Sweden (a fifth of the total numbers of lakes greater than 1 ha) have pH < 5.5 at some times of the year and fish stocks have been affected in 9000 of them [407a].

Ultimately (Chapters 3 and 4) the atmospheric sources of hydrogen ions need to be reduced to solve the problem. Nonetheless, many catchments, acidified over several decades, have had their buffering capacities so reduced that high aluminium and H⁺ run off may occur for many years. In these cases, the catchments may be limed—an expensive business because of their size and remoteness—or lime may be added to the lakes themselves. Provided sufficient lime is added and the treatment is repeated as necessary, many of the changes which occur in acidified lakes can be reversed [204].

The problem is sometimes that the lime is readily washed out or, in lakes bearing high concentrations of humic compounds, it is readily precipitated and made inactive. In one such lake, Lilla Galtsjön in Sweden, an injection of 40 tonnes of 10% sodium carbonate solution into the sediments, with a sediment harrow, was made. It seems to have supplied a source of carbonate which slowly diffused out to neutralize the hydrogen ions over quite a long period [54].

6.6 VARIATIONS ON THE THEME—OTHER STANDING WATERS

So far this chapter has discussed the production and problems of exorheic lakes—those regarded as 'normal' by limnologists in Europe and North America. It ends by being less parochial. Temporary rainwater pools, meromictic lakes and endorheic salt lakes are three groups which set the exhorheic lakes in a wider context, and illustrate the continuum of variation in standing waters. In addition, for many parts of the world, endorheic lakes have a greater claim to be normal than exhorheic ones.

6.6.1 **Rainwater pools**

Temporary puddles of rain water form almost everywhere, though usually the rain soaks quickly into the ground and no particular ecosystem forms in them. Where the rain collects in depressions on rock, however, the pools may persist for weeks or months. They may then develop a particular community, which, although it is a very specialized one, illustrates some features of a simple standing water system. Such pools have been studied in Malawi in Central Africa where they form in depressions on the tops of rocky outcrops during the rainy season between November and March [510].

The depressions are generally bare at the start though perhaps with some fragmented mineral matter and the dried remains of sediment from a previous wet phase. On filling, the pools, perhaps only a metre squared or less in area and only a few cm deep, may be supplied with some nutrients leached from the previous sediment to supplement the meagre rain water supply. Phytoplankton may quickly develop [611]. This comes from inocula which are readily dispersed as dry cysts or spores by wind and is of motile algae like *Euglena* and *Chlamydomonas* which can use their flagella to remain suspended in the water. Other algae, generally Chlorophyta, may develop on the bottom or attached to the surface tension film (neuston). If the rainfall is heavy and storms follow each other quickly, the pools may be flushed out so rapidly that algal populations do not have time to form. However, when they do, they are often very large with biomasses, measured as their chlorophyll *a* content, of several hundred $\mu g \, l^{-1}$. This seems to be because droppings from crows and other birds are washed into the pools, and particularly because two small wild cats, the civet and genet, favour the depressions for their nightly defaecations [507, 508]. Phosphate and ammonium are leached from the dung to fertilize the water. Nitrate is generally scarce for it is readily denitrified by bacteria in the organic rich sediment formed from the dung. The larvae of flies dominate the invertebrate community, feeding on the sediment, and particularly its contained bacteria.

There are three prominent species: *Chironomus imicola* and *Polypedilum vanderplanki* which are chironomids and *Dasyhelea thompsoni* which is a ceratopogonid. All must complete their life histories to a stage where they are not killed if the pool dries out. This requires sufficient time and sufficient food.

Chironomus imicola can complete its life cycle from egg to adult in as little as 12 days, but it cannot tolerate drying out. It favours the larger pools for egg laying and disperses efficiently. The larger pools allow time for development of larger larvae and adults, which produce more eggs and can fly greater distances to deposit them in new pools as the old one dries out [509].

Polypedilum vanderplanki on the other hand is not so effective at dispersal but extremely tolerant of desiccation. Hinton [350] showed that the larvae could survive almost complete drying out (to only 3% of body weight) and, buried in sediments, they can survive baking to high temperature by the sun. Ninety-three per cent, even unprotected by sediment, survived 61°C for 14 hours. On re-wetting, the larvae resume feeding and growth in about an hour. *Polypedilum vanderplanki* can thus survive interrupted conditions as its pool dries between rainstorms and refills. Selection for a short life history has not been so crucial. It can complete it in 35–43 days of uninterrupted growth though with interruptions may take much longer, and can survive in much smaller pools than *Chironomus imicola*.

Dasyhelea thompsoni is intermediate in its features [89]. It does not lay eggs in large pools with a stable water level, where it might be forced into competition with *Chironomus imicola*, but lays them on drying edges where its larvae soon construct a cocoon against the bottom to protect them against drying out. It is not so tolerant of extreme desiccation as *Polypedilum vanderplanki*, but is a better invader of new pools, and within the pools favours the sediment richer in dung than does *Polypedilum*.

To the rain pool ecosystem can now be added a further component: the tadpoles of the Savanna ridge frog, *Ptychadena anchiaetae* [508]. The tadpoles, which hatch from eggs laid in the pools, may be very numerous (up to 1000 m^{-2}); they are not drought resistant but rely on completing their life history before the pool dries out. If they do not complete it, their dried carcasses may be later scavenged by *Dasyhelea* [508] and they may contribute further to the pool's nutrient supply. They may also be washed from the pool if it is flushed by heavy rain.

If they avoid these fates they may play an interesting role. Initially they feed on the sediments and through their excretion may mobilize nutrients more rapidly from them than would otherwise be the case [611]. In turn these nutrients support greater suspended algal populations, which the tadpoles also eat and which speed up the growth of *Chironomus imicola*, increasing its chances of maturing before the pool dries out. The algae are, in general, a food source richer in energy and nutrients per unit weight than the sediment. On the other hand tadpoles may also delay the final emergence of the adult flies by disruption of the habitat by their movements at a time when the insect must emerge through the water film.

This simple standing water system encapsulates a microcosm of the main features of larger lakes. These include the importance of nutrient supply from the catchment (in this case bird droppings and cat dung), the role of flushing in determining with nutrients the size of the suspended algal populations, the interaction between water and sediments, and the impact that a species at the upper end of a food web can have on the stages below it—a matter discussed in Chapter 7.

6.6.2 Meromictic lakes

Some lakes have almost permanently unmixed layers which usually become deoxygenated. Such meromictic lakes [379, 836] (as opposed to holomictic lakes which mix to the bottom at some time during the year) generally have a bottom layer of high density. This may be due to ingress of salts from the sea, from mineral springs, from concentration due to aridity over a long period, from freezing and thawing of surrounding soils or in very deep lakes from accumulation of salts following decomposition of falling detritus.

The Hemmelsdorfersee near Lübeck in Germany has a deep, 45 m hole which receives flood water from the sea every century of so (the last time in 1872). Subsequently fresh water forms a layer on top. Floodplain lakes close to the sea may also acquire such salt injections [130, 820]. Some Norwegian lakes (e.g. L. Tokke) have ancient salt layers which were trapped at the end of the last glaciation in basins which started as inlets of the sea. The basins were subsequently uplifted above sea level as the pressure of the ice decreased on the land surface. Concentrations of brine in Arctic lakes during the formation of surrounding permafrost give similar layers [619].

In the sub-tropics, the Dead Sea had a deep layer formed by ancient evaporative concentration, onto which less dense spring water had layered over several hundred years until the whole lake mixed in 1979 [770, 771]. The surface, spring-fed layers were characterized by relatively high ^{226}Ra concentrations and from the residual concentration in the monimolimnion (the unmixing layer) it was calculated that the last mixing took place 300 years ago.

In the tropics, Lake Kivu, a lake formed by the damming of the Rutshuru river in the late Pleistocene by lava flows from the Virunga volcanoes has a saline layer from about 70 m to its greatest depth (about 450 m). The salts are believed to have been leached from the lavas and the layer is anaerobic and rich in H_2S and methane. Fresh water has layered on the top of it. The deep lakes Malawi and Tanganyika [41, 161] also have anaerobic monimolimnia, though these are barely more dense than the surface water which, in major ion composition, they resemble. Only the top 200–300 m regularly mix through wind action, which presently seems unable to disturb the lower layers (down to 685 and 1470 m respectively). The constancy of major ion composition with depth, however, suggests that complete mixing has taken place relatively recently.

In many oligomictic tropical lakes the original hypolimnia may acquire denser and denser water as the products of decomposition accumulate in them. These lakes may then become biogenically meromictic (as opposed to ectogenically when the salt source is external, e.g. the sea or springs). Such lakes may then mix even less frequently—perhaps not for centuries. Examples may be some of the crater lakes in Indonesia and the Cameroons. Barombi Mbo is 110 m deep with no oxygen below about 20 m. That this stratification is very old is suggested by the presence in the lake of a fish which seems adapted to extensive movements into the anaerobic water to catch its prey, a midge larva of the genus *Chaoborus*. The fish, *Konia dikume*, [816] oozes blood when it is caught for it is heavily supplied with blood vessels. The blood has a high haemoglobin content—at 16.5 ± 0.9 g $(100$ ml$)^{-1}$ twice or three times as much as that of the other fish in the lake—and also rather more red blood cells of greater surface area per cell than the other fish [229]. Stratification in the lake is maintained by a temperature difference of only $1.3°C$ (29.3 to 28°C) between the top and bottom, and though chemical data are unavailable, the contrast in pH between upper (7.5) and lower (6.8) layers [299] suggests that the stratification is stabilized by meromixis.

When stratification is occasionally overturned in such deep tropical lakes, the mixing of the deeper, usually deoxygenated layers to the surface may cause mass deaths of fishes (e.g. Ranu Lamongan in Indonesia, [300]) and possibly, through escape of carbon dioxide, even of local human populations. Such an instance seems to have occurred in Lake Nios in N.W. Cameroon in August 1986 [496, 544] and at another lake some kilometres distant, L. Monoumi in 1984. In each case a gas without smell seems to have been released following heavy rainfall and to have blanketed a local village causing 1500 and 35 deaths respectively. The most likely contender is CO_2 which is heavier than air, rather than H_2, which is lighter or H_2S, which smells. All of these gases could have built up in the monimolimnion after release by underlying volcanic activity.

6.6.3 Endorheic lakes

It is a mistake to regard endorheic lakes as oddities [866]. They are very widespread (Fig. 6.16) though for much of the time they may be dry salt pans dominated by chlorides and sulphates or carbonates depending on the origin of the catchment rocks as marine or volcanic deposits respectively.

They form a particularly fascinating group. The diversity of their communities is very low. Some are far more concentrated than sea water and may support single-species communities, perhaps of the pink flagellate *Dunaliella salina*, or odd species of Halobacteria, fed upon by brine shrimps such as *Artemia salina*. *Dunaliella* is a potential source of carotene and glycerol for the food and pharmaceutical industries, *Artemia* is frequently cultured as a food for young fish in fish farms. In the less saline of the lakes, where some aquatic plants, particularly of the genus *Ruppia*, grow, there may be a potential source of genes for transfer of salt tolerance to crop plants. Salinization of soils following unwise irrigation of arid areas is an increasing problem.

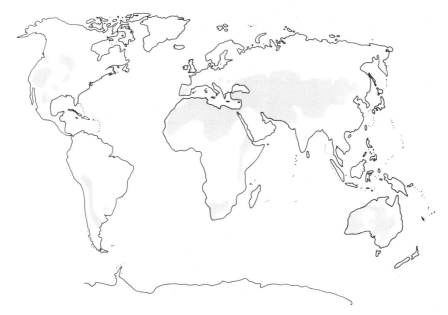

Fig. 6.16. Areas in which endorheic salt lakes are common. (Based on Williams [866].)

The more shallow endorheic lakes, when they have water in them, may be highly productive [797, 817] and in this way form a conclusion to this chapter which has essentially been about lake productivity. The most widely quoted case is that of L. Nakuru (Fig. 6.1) in Kenya, where for several years at a time a dense population, with as much as 16 000 μg l^{-1} of chlorophyll *a*, of a blue-green alga (*Spirulina platensis*) may supply food for up to a million and a half lesser flamingo (*Phoeniconaias minor*). These flamingos filter the algae from the water; other bird species filter out crustaceans or feed on detritus and algae in the bottom mud. The spectacle from the air of a blue-green lake trimmed at the edges with a dense pink fringe is one of the vivid cameos of African limnology.

Not all of the eastern rift endorheic lakes are like Nakuru—in fact few are. The particular combination of an intermediate salinity and intermediate depth seems to separate them from the deeper, less saline ones like L. Turkana dominated by a sparser plankton of species other than *Spirulina*, and the very shallow and very saline ones whose main producers are bottom-living diatoms (Table 6.4).

The particular interest of those like L. Nakuru is their very high productivity, and very high concentrations of soluble reactive and total phosphorus, and perhaps also of nitrogen compounds [378, 535, 796]. Of course it could be argued that these are simply the consequences of evaporative concentration. However, the inherent insolubility of phosphorus

Table 6.4. Some features of East African endorheic lakes. (Based on Tuite [817])

Lake	Area (km^2)	Max. depth (m)	Alkalinity (mEq l^{-1})	pH	Main primary producer
Nakuru	43	1.3	122–1440	9.8–11.0	*Spirulina*
Bogoria	33	8.5	480–800	9.8–10.3	*Spirulina*
Baringo	154	8	4.4–10.5	8.8–9.2	Other algae; *Spirulina* present
Turkana	8000	120	19.4–24.5	9.5–9.7	Other algae; *Spirulina* present
Magadi	95	0.6	420–3640	9.4–10.9	Benthic algae
Elmenteita	20	1.2	107–800		Benthic algae
Manyara	400	1.5	78–800	9.2–9.85	Benthic algae

compounds, and the fact that precipitants like iron should be equally concentrated make it likely that the explanation is more complex.

Perhaps the explanation is that the flamingos, *Spirulina* and sediment are part of a closed cycle in which the continual flamingo feeding and supply of excreta to the sediments prevent the formation of an oxidized microzone. Phosphorus may then be continually allowed to diffuse out from the sediment to support the algal growth. The source of nitrogen is a problem because sediments maintained in such an anaerobic state all of the time will contain large numbers of denitrifying bacteria. However, such conditions also favour nitrogen fixing bacteria and blue-green algae. The nitrogen excreted by the flamingo will also be converted from uric acid to ammonia rather than nitrate and will not be so vulnerable to denitrification. The high solubility of phosphorus compounds at very high pH values (> 10) which are found in carbonate-dominated endorheic lakes may also be part of the explanation. If this suggestion is correct, such lakes may provide exceptions to the principle that the fertility of pristine lakes depends on the nature of the catchment area at the time. Such exceptional lakes would have closed cycles, independent of the catchment once some threshold of phosphorus accumulation had been reached, and perhaps might be natural analogues in this respect to shallow, heavily eutrophicated lakes like L. Trummen.

The Plankton and Fish Communities of the Open Water

7.1 THE STRUCTURE OF THE PLANKTON COMMUNITY

Lake water sparkling in sunlight hides a miniscule waterscape which is closer to a slum than a paradise. It contains millions of organisms in every litre, some of which are photosynthetic, others of which feed on organic matter, both live and dead, dissolved and particulate. The water contains their excretions and secretions, faeces and corpses, intermixed with debris washed into suspension from the surrounding land. In this *mélange*, chemical and biological changes, both cyclic and irreversible, are taking place very rapidly.

Scaling one part of the plankton to human size and considering the rest relative to this will help indicate the structure of this community, the plankton. The rotifer *Keratella quadrata* (Fig. 7.1), a common small animal plankter, has a body about 125 µm long with spines half as long held out behind it, and is a convenient organism with which to scale up the rest of the plankton. In reality it is about half the size of a full stop on this page. If the body of *Keratella* is scaled to the size of a tall man then the rest of the plankton ranges in size from lentils and peanuts to large houses, or, if the fish which move through it be considered, to the size of 'whales' 30 miles (48 km) long! The water in which the plankton lives is viscous relative to such small objects. In the scaled up analogy, the viscosity of the fluid must also be increased, so the community must be imagined as suspended in light oil or glycerol.

7.2 PHYTOPLANKTON

The phytoplankton is the photosynthetic part of the plankton and the base of food webs in it. The individual phytoplankters are of many species, mostly oxygen-evolving and prokaryotic blue-green algae (Cyanophyta) and Prochlorophyta and eukaryotic algae. The eukaryote groups of greatest importance are the Cryptophyta (cryptophytes), Pyrrophyta (dinoflagellates), Chlorophyta (green algae), Euglenophyta (euglenoids), Bacillariophyta (diatoms),

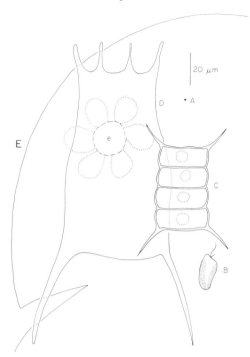

Fig. 7.1. Relative sizes of some major components of the plankton. (A) a bacterium; (B) *Cryptomonas*, a relatively small phytoplankter; (C) *Scenedesmus*, a moderately large phytoplankter; (D) *Keratella*, a small zooplankter; (E) outline of the head with eye (e), of *Daphnia*, a large zooplankter. The head constitutes about a quarter to a fifth of the total body size.

and Chrysophyta and Haptophyta (yellow-green, or golden-yellow algae). Some examples are shown in Fig. 7.2.

In stratified lakes where sufficient light may penetrate to metalimnion and hypolimnion layers with reduced redox potential or anaerobiosis, communities of photosynthetic bacteria (the purple and green sulphur bacteria) may be found. They form discrete layers only a few cm thick where their particular chemical needs are met (Fig. 7.3). These organisms use thiosulphate or hydrogen sulphide instead of water as a hydrogen donor for photosynthesis and hence do not evolve oxygen, but deposit granules of sulphur inside or outside the cells. Though such photosynthetic bacterial communities are common, most attention has been paid to the ubiquitous oxygen-evolving phytoplankters of the surface aerobic waters.

Taxonomically, and in size, the phytoplankton is very diverse (Fig. 7.2). In the scaled-up model the sizes of phytoplankters range from those of lentils and peanuts (1–5 μm) to those of footballs and water melons (up to 50 μm). When they occur as colonies they may be visible to the naked eye (several hundred μm) and would be scaled as heavy horses or elephants.

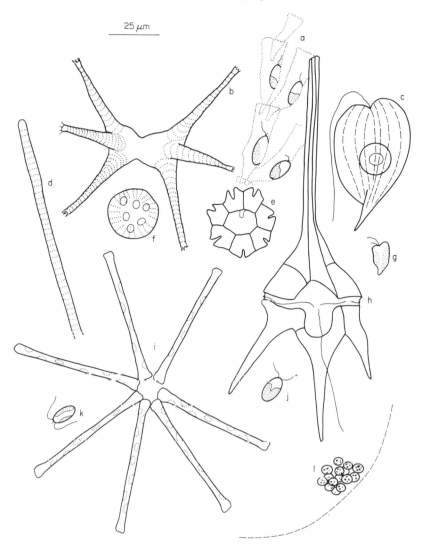

Fig. 7.2. Some typical phytoplankton algae, drawn to the same scale as Fig. 7.1. Cyanophyta (blue-green algae): (d) *Oscillatoria*, (l) *Microcystis*; Chrysophyta (yellow-green or golden algae): (a) *Dinobryon*; Chlorophyta (green algae): (e) *Pediastrum*, (b) *Staurastrum* (a member of a group called the desmids); (j) *Chlamydomonas*; Bacillariophyta (diatoms): (f) *Cyclotella*, (i) *Asterionella*; Euglenophyta (euglenoids): (c) *Phacus*; Cryptophyta (cryptomonads): (g) *Rhodomonas*; Pyrrophyta (dinoflagellates): (h) *Ceratium*; Haptophyta: (k) *Prymnesium*. *Microcystis* (l) is a very large alga, of which a diagram of the entire colony could occupy as much as this page. Only a few cells are shown.

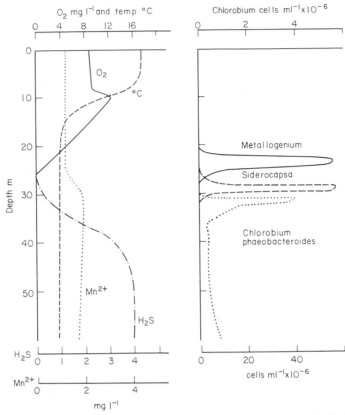

Fig. 7.3. Distribution of three bacterial species in the water column of L. Gek-Gel in September 1970. The lake is meromictic, having a deep and unmixing saline layer from 30–40 m down, which is permanently deoxygenated. *Chlorobium phaeobacteroides* is a photosynthetic bacterium, absorbing light efficiently between 450 and 470 nm (green) and using H_2S as a hydrogen donor. *Metallogenium* and *Siderocapsa* are chemosynthetic bacteria probably using the oxidation of Fe^{2+} and Mn^{2+} as sources of energy. (After Kuznetsov [440].)

A common misconception is that phytoplankters float, with similar densities to that of water. This is generally not so. Some which do are those blue-green algae, which are sometimes very abundant in highly fertile lakes, which have organelles called gas vesicles. They comprise masses of protein-bound prisms with conical ends, contain air and give positive buoyancy. Under some circumstances blue-green algae may truly float at particular depths in the water column which favour their growth and in other circumstances may form a paint-like scum or water-bloom at the lake surface (see later). Similar gas vesicles are used by the anaerobic photosynthetic bacteria to maintain themselves in the most appropriate part of the redox gradient (Fig. 7.3).

Excepting *Botryococcus braunii*, an alga which may remain positively buoyant by storing large quantities of oil, all other phytoplankters are more dense than water. In the case of diatoms, which have cell walls of silica, they may be considerably more so, having a specific gravity of as much as 0.25 in excess of that of water [558]. These phytoplankters are kept suspended in the water largely by wind-generated currents. Some species have flagella and movement of these may help counteract the tendency to sink.

Non-flagellated species have evolved cell or colony shapes which decrease the rate of sinking, or delay alignment in the water in a position which would aid sinking. Flat plates, needle shapes with curved ends, and the possession of spines and projections all seem to be advantageous. However, too easy an acceptance of shape as adaptive should be avoided. Envelopes of mucilage, invisible unless the cells are mounted in Indian ink, may be thick enough to give a spiny cell an effectively spherical shape, and may even lubricate passage through the water.

Why have phytoplankton cells not all evolved positive buoyancy, when sinking, with its potential for loss from the euphotic zone, so clearly has disadvantages? The answer is that it also has advantages, which must, on balance, outweigh the disadvantages.

Phytoplankters need a supply of inorganic nutrients from the water. These they absorb from the water layer, a few micrometres thick, immediately in contact with the cell wall or membrane. Molecular forces tend to preserve this layer intact and it soon becomes depleted of nutrients which are not rapidly replaced by diffusion alone. Continuous movement of the cell through the water, as it sinks, and is retrieved by upwardly directly turbulence, sloughs away the depleted nutrient shell. It is replaced by a supply of undepleted water to the cell surface [380, 576].

The phytoplankton was long assumed to comprise organisms which satisfied their energy needs through photosynthesis and absorbed only inorganic nutrients. This is far from the truth. Many require at least one preformed organic compound for their growth. Of those algae tested until now, 70% require one or more of the three vitamins cyanocobalamin (B_{12}), thiamine and biotin. B_{12} is most commonly needed, perhaps sometimes as a readily soluble source of cobalt which it contains, but often as a complete organic moiety, the co-factor of a necessary enzyme. The test of independence of organic compounds is the ability of the alga to grow in a pure culture through a series of many sub-cultures in entirely inorganic media in the laboratory. This has been shown for a relatively few species, mostly blue-green algae, diatoms and desmids. Many algae are able to take up simple organic compounds in the dark or light, but the concentrations of such compounds are normally low in natural waters and bacteria successfully compete for them. It is probable that dependence on organic compounds is less marked in the open water phytoplankton than in benthic algal communities (Chapter 8). But some plankters of highly organic sewage oxidation ponds (usually green algae and euglenoids) may depend as greatly on organic uptake for their energy needs as on photosynthesis.

The smaller phytoplankters (up to 10 μm) may occur in very large numbers: 10^6 per ml is a characteristic upper figure (though subject to wide variation) compared with about 10^3–10^5 for the larger phytoplankters. On the scaled up model the smaller species would appear as a population of objects the size of tennis balls spaced at distances of about 8 feet (2.5 m) in three dimensions. A population of the larger algae can be imagined as a similar constellation of water melons 30 yards (27 m) apart from each other. Even considering that several species may simultaneously be forming large populations in the water, there is clearly a lot of space between the individuals. This might explain why, despite their ubiquity, parasitic fungi (frequently chytrids) and parasitic protozoa only cause epidemics when algal population densities are very large. Successful infestation of a host cell requires an encounter between the parasite (size about 5 μm—golf ball size) and the host in an environment where hosts are well spaced out and both host and parasite are continually moved by turbulence.

The phytoplankters have sometimes been divided into ultraplankton (< 5 μm or thereabouts), nannoplankton (5–60 μm) and net plankton (> 60 μm). Lately another term, picoplankton, has been coined to describe both prokaryotic blue-green algal cells and tiny eukaryote algae of bacterial size (about 1 μm) which may carry out a substantial part of the photosynthesis of the plankton community in the oceans and in infertile lakes [777]. Such cells are readily detected by the use of fluorescence microscopy—with orthodox microscopy they may be difficult to pick out against the larger numbers of heterotrophic bacteria. Phytoplankters come in a continuum of sizes, however, and the only real use of naming different size categories is to point out the fact that much early work, which used samples taken by relatively large-meshed nets towed through the water, undoubtedly failed to sample most of the phytoplankton.

The very wide size range of phytoplankters (from lentils to elephants in the model) is remarkable, particularly because in a nutrient-scarce medium (Chapter 6), small bodies with high surface to volume ratios should be able to compete more effectively for nutrients. Large cells also sink faster and hence are more vulnerable to loss from the epilimnion in stratified lakes. There *are* large phytoplankters, however, and they have persisted in the plankton for a very long time. There must be an advantage to large size and this seems to be that big cells are less readily eaten by filter feeding zooplankters which are mechanically unable to manipulate large cells or colonies into their mouths.

Phytoplankters are also remarkable for their almost complete abandonment of sexual reproduction. They live vulnerable lives, most of them are readily grazed, all may be washed out of their habitat by incoming floods, many may be lost to the sediments by sinking. Such conditions have favoured selection of the most rapid means of reproduction of many individuals to replace those lost from the population. Hence simple cell division with generation times of only hours or a few days is how the phytoplankton reproduces.

A simple equation describes the conditions for increase in a population of a phytoplankton species. It will increase its number (N) if its growth rate (b) is greater than the sum of its rates of loss by sinking and trapping in the sediments (v), grazing (g), and flushing out of the water body (w):

$$\frac{dN}{dt} = bN - (vN + gN + wN)$$

The existence of many thousand planktonic species—with several hundred in almost all lakes—means that this equation has, in thousands of different ways, been successfully balanced by each. The next sections will look at the components of the equation.

7.2.1 Photosynthesis and growth of phytoplankton

Photosynthesis—the fixation of light energy into energy-rich chemical bonds, and growth—the synthesis of new cell material, are different processes. Although the latter requires the former, photosynthesis is not necessarily followed by growth if the necessary nutrients are not available for synthesis of cell materials.

The methods by which photosynthesis can be measured were discussed in Chapter 5. When samples of water from a well-mixed water column are placed in bottles and resuspended at a series of depths in the water column, and the photosynthetic rates measured by the oxygen release method, a characteristic curve of gross photosynthesis with depth is frequently obtained (Fig. 7.4). It often has a low value at the surface, thought to be due to inhibition of photosynthesis by ultra-violet light, and a peak at some depth below the surface. The extent of this peak is set when the energy-absorbing pigments in the cells are unable to abstract any more radiation (at light saturation, I_k). The peak is a function of the light saturated rate of photosynthesis per unit of biomass (P_{max}) and also of the total biomass (n) of photosynthesizing cells present. Usually biomass is expressed as chlorophyll *a*. Finally the curve of photosynthesis with depth declines more or less exponentially below the peak. The depth to which photosynthesis extends depends on the surface intensity (I_o) and the attenuation coefficients for light absorption in the water (k) usually expressed as the minimal of these (k_{min}) (Chapter 6). As the biomass increases it becomes a major contributor to the attentuation and the curve is displaced upwards. Such a curve can be described by an equation for the total photosynthesis per unit area of lake

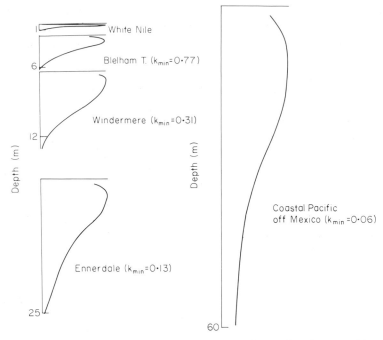

Fig. 7.4. Examples of gross photosynthesis–depth profiles in a series of waters of increasing turbidity as indicated by the value of the minimum extinction coefficient in each case. (Based on Talling [793].)

(Σa), which Talling [790] has determined to be approximately:

$$\Sigma a = n P_{max} \cdot \frac{1}{1.33\, k_{min}} \cdot \frac{I_o}{I_k}$$

The value of 1.33 was determined empirically and the term I_o/I_k approximately describes the extent of the bulge in the curve in the water column—the lower the intensity of light saturation, the deeper the peak. I_k varies with different algae—it is often low for blue-green algae but higher for green algae, and the value of P_{max} tends to increase with temperature with an approximate doubling for every $10°C$ within the range at which the algae can live. A similar photosynthesis—depth curve can be obtained even in stratified water columns (where the algae and photosynthetic bacteria may be layered or where there may be mass movements of flagellated or vesiculate algae) if the values of photosynthesis measured are expressed not per unit volume of water but per unit biomass of algae.

The effects of the algae themselves in absorbing light can be described by plotting, for a given lake, the minimum attenuation coefficient on a number of occasions against the chlorophyll *a* concentration. The intercept on the *k* axis (biomass = 0) gives the attenuation due to the background properties of the water and the slope of the line the attenuation per unit chlorophyll *a*.

The value of the latter varies with the nature of the algal biomass (it is lower for populations of large cells than for those of smaller ones) but often falls into a range from about 0.008 to 0.021 ln units m^{-1} (μg l^{-1})$^{-1}$. A crop of 100 μg l^{-1} of chlorophyll *a* in a lake of background attenuance, $k_{min} = 0.3$ m^{-1}, could then reduce the potential euphotic zone (Chapter 6) from 12.3 m to between 3.4 and 0.17 m. At high concentrations the algae can then induce a state of shelf-shading where light availability becomes limiting to the rate of photosynthesis through the activities of the algae themselves. This seems to begin at covers of algae in the water column of about 300 mg m^{-2}.

Mean values of gross phytosynthesis in g O$_2$ m^{-2} day^{-1} include for example 21 for Loch Leven, 15.6 for a bay in L.Neagh and 57 for some tropical African lakes. Converted to yearly values on the basis of regular measurements and to units of carbon, the rather fertile lakes Leven and Neagh had annual productivities of 785 g C m^{-2} year^{-1} and 1500–1800 g C m^{-2} year^{-1} [404].

The actual growth of algae, however, is likely to have been much less, for these are values of gross photosynthesis or production. Net production is less by the amount of respiration of the algal cells and even then may be expressed as carbohydrate or fat storage in the cells, rather than by true growth, the production of new ones.

7.2.2 Net production and growth

Net production may be small if respiration rates are high but there is a problem in measuring algal respiration in water which contains other respiring organisms—animals and bacteria. There are several circumstances in which algal respiration rates are likely to be high relative to photosynthesis. First, in early spring in a temperate lake when the water column is vigorously mixing, the cells will also be circulating between the (albeit poorly) illuminated upper layers and the layers below the euphotic zone (z_{eu}). If the mixing depth (z_m) is large relative to z_{eu} they may spend long periods unable to photosynthesize but inevitably respiring. Until, on average, the gross photosynthesis made by the cells whilst they are in the upper layers exceeds the respiration whilst they circulate throughout the whole water column, no net production will be possible. This phenomenon seems to determine the incidence of growth in spring. For example, in the deeper northern basin of L.Windermere, growth starts about 2–3 weeks later than in the shallower southern basin where z_m is much less.

Secondly, as the algal crop grows and becomes self-shading, the photosynthetic rate per cell, determined by light, may be decreased but the respiration rate, determined by temperature does not. And thirdly, in lakes

ıigh concentrations of suspended silt particles may effectively
,siderably whilst not affecting z_m.
respiratory demands can be met, there may still be no growth
some net production. If nitrogen or phosphorus is scarce, the cells
y be unable to synthesize proteins and nucleic acids. They may then be
ıorced to respire synthesized carbon compounds, to store them or to secrete
them to the water. In one study only about one-fourteenth to one-thirtieth of
the carbon fixed in photosynthesis was incorporated into new cell material
[671]. In general, in freshwater lakes, N or P is likely to be scarce and the
degree of scarcity may determine the rate of growth as well as the potential
maximum yield of algae (see Chapter 6).

7.2.3 Nutrient uptake and growth rates of phytoplankton

Phytoplankters require about 20 elements for growth but only C, N and P are
likely to limit growth rates on any general basis. All three are present in the
water at lower concentrations than are required in the cell, so that active,
energy-requiring uptake mechanisms, involving transport enzymes are needed
to concentrate them into the cells. The efficiency at which these mechanisms
operate differs between species and can be approximately measured as the
half-saturation constant (K_s) for uptake rate (μ) in an equation [545]:

$$\mu = \frac{\mu_{max} S}{K_s + S}$$

The half saturation constant is the concentration (S) of enzyme substrate—
in this case the nutrient—at which half the maximum rate, μ_{max} (when the
enzymes are saturated to full capacity) can be achieved. It might be expected
that values of K_s will be lower for species growing in infertile lakes than those
in fertile ones. Coexistence of two species competing for the same nutrients
may also be possible if the first is more adept at taking up one nutrient and
the other a second nutrient. For example, Titman [809] studied the
relationship between two diatoms, *Asterionella formosa* and *Cyclotella
meneghiniana* both of which grew in Lake Michigan, with *Asterionella* more
abundant in the open lake and *Cyclotella* near shore. Both species require
both silicate and phosphate. Using a continuous culture apparatus in which
a nutrient solution of constant composition can be supplied, Titman
determined $K_s(P)$ for *Asterionella* to be 0.04 μmol PO_4 and for *Cyclotella*,
0.25 μmol PO_4.

This suggests that if phosphate is scarce *Asterionella* will tend to compete
favourably with *Cyclotella*. On the other hand if silicate is scarce the reverse
is true, for $K_s(Si)$ was 3.9 μmol SiO_3 for *Asterionella*, and 1.4 μmol SiO_3 for
Cyclotella. For each species in turn growth rates will be similar when:

$$\frac{\mu_{max}S(Si)}{S(Si) + K_s(Si)} = \frac{\mu_{max}S(P)}{S(P) + K_s(P)}$$

Hence both nutrients are in balanced supply for growth when:

$$\frac{S(\text{Si})}{S(\text{P})} = \frac{K_s(\text{Si})}{K_s(\text{P})}$$

For *Asterionella* this ratio is 3.9:0.04 or 97, so that at Si:P molar ratios greater than 97 in the water, *Asterionella* will be phosphorus limited. For *Cyclotella* the ratio is 5.6 and above and below this the diatom will be phosphorus and silicate limited respectively. When the two species are present in the water together, both will be phosphorus limited when the ratio is greater than 97, and *Asterionella* will tend to survive rather than *Cyclotella*. When ratios are below 5.6, both are silicate limited but, being more efficient at silicate uptake, *Cyclotella* will predominate. But between 5.6 and 97, each diatom's growth rate is limited by a different nutrient and they should be able to grow together. In Lake Michigan, the Si:P ratio is between 200 and 500 in the open lake but is between 1 and 10 near the shore. The general predominance of *Asterionella* offshore and *Cyclotella* inshore is consistent with the laboratory findings [806].

Some time ago, Redfield [665] suggested that cells which had a balanced nutrient supply would have a ratio (by atoms) of carbon:nitrogen:phosphorus of 106:16:1 (or by weight 42:7:1) and that severe departure from this ratio indicated a (physiological) nutrient limitation. A demonstration of the Redfield ratio in a phytoplankton population might suggest that it is not being limited by nutrients (note—limitation of growth not yield, see Chapter 6). In the open ocean, Goldman *et al.* [281] found that the phytoplankton did in fact have ratios of C:N:P close to the Redfield ratio and that their maximal growth rates were at this ratio.

In fresh water, departures from the Redfield ratio are common, as are properties of the cells which indicate a physiological nutrient limitation, suggesting that scarcity of N or P is keeping growth rates below the maximum. A variety of indicators may be used to demonstrate this [221, 336, 841]. For nitrogen, high rates of ammonium uptake in the dark and increases in the ratio of carotenoids to chlorophylls are suggestive of nitrogen deficiency. The chlorophylls are nitrogen-containing compounds of which the cell may be unable to synthesize sufficient if nitrogen is scarce. For phosphorus an increase in the activity of acid phosphatase enzymes (which can break down organic phosphorus compounds), a high rate of ^{32}P uptake and an absence of free phosphate in the cells suggest phosphorus scarcity.

The departures from the Redfield ratio in lakes compared with the ocean are of interest in that it is suggested that a very efficient system of remineralization of N and P takes place through grazing and bacterial activity in the ocean. This maintains the availability of N and P in the correct ratio and sufficient always to allow high growth rates. In fresh waters, there seems no reason why a similar system should not have developed. It may have done and may be indicated by discovery of Redfield ratios in pristine lakes whose nutrient loadings have been unaltered by human activities—the situation which pertains, because of its huge extent, in the mid-ocean. In most fresh waters, however, alteration of the loading rates of N and P may have disrupted such a system. First of all increased nutrient concentrations have favoured large algae with low surface to volume ratios which may not be readily grazed so that their contained nutrients may not be readily regenerated, and secondly the imbalance of P and N loadings may itself create scarcities. Heavy phosphorus loading may induce nitrogen scarcity for example. The implication is that the behaviours of many of the freshwater phytoplankton communities studied are artefacts of human influence. The regeneration and cycling of N and P will be considered later in this chapter after some information on bacteria and zooplankton has been given. For the moment the theme of growth in phytoplankton will be developed with reference to the spatial distribution of phytoplankton communities.

7.2.4 Distribution of freshwater phytoplankton

Land vegetation differs obviously from place to place, both in its appearance—herb-dominated grasslands, shrubby savannahs, forests—and in its species composition, depending on the local climate, soils and other factors. The same is true of phytoplankton communities though the differences are not so obvious.

'Appearance' for phytoplankton includes features of size, organization into filaments or colonies, possession of flagella or buoyancy vesicles. The environmental features which might be associated with differences among these include the availability of nutrients and the degree of mixing or stratification, among others. There will always be exceptions to generalizations, but the phytoplankton of an infertile lake is likely to be of small organisms with high surface to volume ratios. These are readily grazed so that scarce nutrients might be rapidly recycled by excretion from the grazers before the cells are lost to the sediment and their nutrients removed from use. Unicells are more likely to dominate than colonies or filaments. Fertile waters, on the other hand, may be able to sustain greater proportions of larger organisms. Nutrient uptake may be less of a problem and consequently the organisms may invest more energy in devices to avoid grazing, the loss of nutrient to the sediments being met by sufficient supplies from the catchment or even return from the sediment (Chapter 6).

Table 7.1 gives some general associations of algae in waters of increasing fertility but many variants on the theme are to be expected. One feature might be picked out as an example for further discussion. This is the correspondence of a diverse desmid plankton with very infertile conditions.

Table 7.1. Some general associations of phytoplankton communities with waters of low and high fertility. (Based on Hutchinson [380])

Infertile waters ('oligotrophic')

(a) Dominated by a large variety of desmids (e.g. *Staurodesmus* and some *Staurastrum* species (small unicells) with some colonial green algae (*Sphaerocystis*) and thin-walled (*Rhizosolenia*) or small (*Tabellaria*) diatoms. Often in slightly acid waters, low in alkalinity and Ca

(b) *Dominated by small diatoms* (*Cyclotella, Tabellaria*) and some filamentous diatoms (*Melosira* spp.) in neutral or slightly alkaline waters. Chrysophyta may be major components perhaps together with very small (< 3 μm) flagellate and coccoid blue-green algae

Waters of intermediate fertility ('mesotrophic'–'eutrophic')

(a) Dominated by dinoflagellates (*Peridinium, Ceratium*) and diatoms (*Cyclotella, Stephanodiscus, Asterionella formosa*), with desmids but of the genera *Staurastrum, Closterium* and *Cosmarium*, other green algae (*Scenedesmus, Pediastrum* and perhaps some filamentous blue-green algae either fixing N_2 (*Anabaena*) or not (*Oscillatoria*)

Highly fertilized waters ('hypertrophic')

(a) Dominated by *Oscillatoria* with diatoms in lakes

(b) Dominated by small Chlorococcales (green algae) in small ponds

(c) Dominated by *Spirulina* or *Dunaliella* in saline lakes

(d) Dominated by Euglenophyta in small ponds heavily fertilized with organic matter. The Euglenophyta appear to be confined to ammonium as a nitrogen source. Ammonium is a major breakdown product of organic decomposition

---

7.2.5 The desmid plankton

Desmids are a group of the Chlorophyta (green algae), often said to be calciphobic. The confinement of many of their species to waters of low calcium content has often been assumed to be causative. Grown in laboratory culture, however, they will often tolerate high concentrations of calcium and most of the other major ions which generally increase in concentration from the least to the most fertile water. The exception appears to be bicarbonate, whose concentrations are linked with pH, HCO_3 and CO_2 in natural waters.

Carbon dioxide associates with water to form a series of equilibria:

$$CO_2 + H_2O \rightleftharpoons H_2CO_3 \rightleftharpoons HCO_3^- + H^+ \rightleftharpoons CO_3^{--} + 2H^+$$

and the proportions of the total inorganic carbon concentration ($CO_2 + H_2CO_3^- + CO_3^{--}$) which are accounted for by each form depends on pH. Below pH 4.5 almost all is in the form of H_2CO_3 and CO_2; above about 8.4–9.0, increasing proportions of CO_3^{--} can exist. Bicarbonate dominates in the middle range. Absorption of CO_2 or HCO_3^- for photosynthesis (CO_3^{--} is generally not able to be used) causes the equilibria to move to the left. The concentration of $[H^+]$ thus decreases and the pH rises. This resets the equilibrium proportions of CO_2, HCO_3, and CO_3 to favour progressively HCO_3^- and CO_3^{--}. At high pH free CO_2 is very scarce.

In a series of experiments using a variety of desmid species (on the grounds that confinement to only one or two held the risk of aberrant behaviour from any particular strain), Moss [559] showed that several desmids from infertile waters ceased growth at pH values between 8 and 9 with a mean around 8.4. Other algae, including some desmids found in fertile lakes, would continue to grow at higher pH values (some beyond 9.5). In the media used, pH 8.4 was associated with an alkalinity (a measure of bicarbonate) of about 2.5 mEq l^{-1}. A survey of the literature revealed a confinement of diverse desmid floras to alkalinity below about 1.5 mEq l^{-1} in nature. It seemed likely that these desmids of infertile waters were unable to use CO_2 at low concentrations or to use HCO_3^- as a carbon source for photosynthesis.

Talling [794] showed that phytoplankters could be ranked according to their ability to take CO_2 at lower and lower concentrations (and thus higher and higher pH values). Desmids of infertile lakes were able to grow in fertile lake water if the pH was reduced to below 8.4 [560]. These findings might suggest why many desmids do not compete readily with other algae in alkaline waters, but leaves the question of why the fertile lake species do not displace the desmids from the infertile ones. After all, the former may take up either CO_2 or HCO_3^- and hence should be at no disadvantage in an acid lake. Indeed many of the fertile lake species can be found, though they are scarce, in low-alkalinity water but for some reason they do not compete effectively with the desmids. One reason for this may lie in the supplies of phosphorus and nitrogen. If the rates of supply of these are low, appropriately low growth rates might allow a species to cope without becoming physiologically nutrient limited. A genetically determined high growth rate, on the other hand, which might be sustainable and advantageous in a fertile lake, might be a disadvantage in an infertile one. The maximum potential growth rates of the desmid species tested turned out, when corrected for the effects of cell volume, to be lower than those of the species of fertile lakes.

7.2.6 Mixing, stratification and washout

As well as by the effects of water chemistry, phytoplankton growth is influenced by mixing and stratification and the washout rate. The latter can be demonstrated in the Norfolk Broads, a series of small, shallow riverine lakes and the rivers which connect them. These waters can be arranged in a gradient from rapidly flushed (say once every 2 weeks or less) to poorly flushed (once or twice per year). In the most flushed parts, the phytoplankton is dominated by centric (radially symmetrical) diatoms (*Stephanodiscus, Cyclotella, Melosira*), which have heavy cells requiring a high degree of turbulence to maintain them in suspension. Where lowland rivers develop phytoplankton it is also usually diatom dominated. As the flushing rate decreases, small, bilaterally symmetrical pennate diatoms like *Synedra* and *Diatoma* join the centric diatoms, together with filamentous blue-green species (*Oscillatoria*). Finally, in the least flushed areas, colonial blue-green algae (*Aphanothece*) dominate with only pennate diatoms. The blue-green algae are often quite large and do not, for the most part in this case, have gas vesicles. They might be expected to sink fairly rapidly though the short water columns are well mixed at all times. The association of these species with the least flushed areas might be associated with their relatively lower growth rates which cannot cope with high washout rates. It is impossible to be certain, however, for many other environmental factors are also correlated with low flushing. These include reduced nitrogen inputs, slightly higher pH, potentially greater development of grazer populations and perhaps greater deoxygenation rates through respiration during night-time in summer.

In the Broadland, however, there is at least an association with flushing which matches associations in deeper lakes with mixing (high turbulence) or stratification (low turbulence). Diatoms are associated with mixing, blue-green algae, particularly gas vesiculate forms, and large motile flagellates with stratification. This may be reflected in the seasonal pattern of development of the phytoplankton with a single lake (see later).

Gas vesiculated and flagellated algae tend to be common in stratified lakes at a time, mid-summer, when new nutrient supplies to the lake may be at their lowest for the year. Light is not scarce but algal crops have accumulated making large demands on whatever nutrients might be regenerated by grazing and decomposition within the eplimnion. Daily movements up and down by the flagellates may give access to greater nutrient supplies at the top of the hypolimnion [705]. This seems to be the case with a colonial flagellate, *Volvox* in Lake Cahora Bassa in Mozambique [755] and may be one of the reasons for the success of a very common, but quite large dinoflagellate, *Ceratium hirundinella* in many temperate lakes. Vesiculate blue-green algae also may migrate to gain access to nutrients, and the paint-like scums or blooms which are sometimes seen at the surfaces of fertile waters may reflect breakdown of this mechanism.

7.2.7 Blue-green algal blooms

The genera concerned in algal blooms include *Aphanizomenon, Anabaena, Microcystis* and *Oscillatoria* which are able to regulate their buoyancy with gas vesicles. The protein walls of the vesicles are permeable to atmospheric gases but not to water, and contain a gas mixture in equilibrium with that dissolved in the surrounding water. Some vesicles are weaker than others and can be collapsed by application of external pressure (about 7 atmospheres) to the water in which the cells are suspended. If the vesicles are removed from the cells (this can be done by ultra centrifugation following mechanical breakage of the cell wall and membrane) rather more pressure is needed to collapse them (about 11 atmospheres). This is because in the cell they are already subjected to considerable internal osmotic pressure.

Since they may occupy up to 30% of the cell volume, gas vesicles can make the cells less dense than water, so that they float upwards. This relative movement, like sinking, has advantages in promoting nutrient uptake. However, as the cells rise into regions of higher light intensity in the water column, their photosynthetic fixation of carbon dioxide increases and the concentration of soluble organic substances in the cell rises. This is accentuated if available phosphorus and nitrogen supplies are low and new cell growth is prevented. As the internal concentration of soluble organic substances rises, so also does the internal turgor pressure. Some of the vesicles, the weaker ones, collapse. The cell may then become denser than water and starts to sink. This mechanism prevents the cells moving into the highest light intensities at the water surface which, in summer, may be lethal for these organisms [672].

As the cells sink into less well-illuminated parts of the water column, conditions are such that the rate of gas vesicle formation is greater than that of cell synthesis and division. The concentration of gas vesicles in the cell therefore increases, the cells become positively buoyant again, and start to float upwards. This cycle keeps them in water strata most suitable for their growth.

The cycle may also gain them access to nutrients at the top of the hypolimnion. Blue-green algae are tolerant of low oxygen concentrations perhaps because they first evolved in the anaerobic waters of the Precambrian about 2 billion years ago. Indeed some can use H_2S as well as H_2O as a hydrogen donor for photosynthesis (Stewart and Pearson [775]).

Ganf and Oliver [252] showed that the development of vesiculate blue-green algal communities in Mt. Bold reservoir in Australia, at the time of thermal stratification, was linked with their access to nutrients in the hypolimnion. Nutrient supplies in the epilimnion would not support growth of any of a variety of algal species tested. They placed water samples from the lake in dialysis bags (which allow free movement of small nutrient molecules through their walls) and showed that algae in the water would not grow if the bags were suspended near the surface (0.2 m) or at 10 m, which was below the lower limit of the euphotic zone. However, if the bags were circulated on a daily basis between 0.2 m and 10 m, abundant growth was obtained as the algae met their light requirements at one place in the water column and their nutrient needs at the other.

If the water is extremely fertile the phytoplankton crop as a whole may, in summer, build up to such a concentration that the euphotic zone becomes much less deep than the epilimnion. The blue-green algal cells then may spend much of their time at very low light intensities and the differential rates of gas vesicle synthesis and cell production lead to formation of cells so buoyant that they rise to the surface very rapidly on calm days. The mechanism by which increased turgor pressure bursts the weaker vesicles is unable to operate, for photosynthesis is inhibited at the high extreme surface light intensity and the cells are trapped at the surface forming the characteristic scum. This may cause odour problems as it decays or if it is later windrowed at the lake edge.

At least that is one explanation of the formation of algal blooms and in support of it, it is claimed that cells from the surface aggregations are not viable if attempts are made to grow them in culture [672]. Other reports claim that this is not so, that the cells are viable and that surface bloom formation is a device by which the algae cope with high pH and a shortage of CO_2 in very fertile waters packed with algae. Increasing the pH of suspensions of the algae caused movement to the surface and closest access to the supply of atmospheric CO_2 [56, 617]. It may be that there are different sorts of blooms with different explanations in different places. On some lakes the blooms are dramatic indeed, forming, for example, thick porridges at the edge of the Hartbeespoort Dam in South Africa and clogging the cooling water intakes of boat motors [685].

7.2.8 Phytoplankton communities and drinking water

In reservoirs fed by fertile water, large growths of algae cause problems. They are best resolved by reducing nutrient loading (Chapter 6). If this is not possible it may be feasible to manipulate the reservoir water to favour species of algae which are more desirable (if still undesirable) than others. Blue-green algae are a particular problem as they may be toxic and produce particularly noxious ('pigpen') tastes if they decompose on the filters used for clearing the water. Diatoms may also produce 'fishy' odours but are more readily filtered out. Algicidal chemicals (e.g. $CuSO_4$) can be used but are costly and undesirable.

Reduction of blue-green algal growths is often possible through judicious mixing of the water. This may be done by bringing the inflow river water into a storage reservoir through jets placed near the bottom. The jets are angled so that the upward movement of the water causes the maximum turbulence [441, 768]. Aeration may also be successful and helps prevent formation of anaerobic hypolimia, whose sulphide content may make the water unusable at a time when supply is short and demand great. It is possible that intermittent mixing and periods of stabilization may not only favour certain more desirable species. It may also prevent build up of very large crops of them by changing the environment frequently enough to prevent any particular group reaching its maximum potential [673].

7.3 MICROCONSUMERS OF THE PHYTOPLANKTON— BACTERIA AND PROTOZOA

Heterotrophy is the sole source of energy for most bacteria in lakes and some evidence suggests that large proportions of the organic matter produced by the algae may be used by the bacterioplankton. The bacterioplankters are about the size of lentils in the analogy drawn on p. 205, though with shapes varying from the familiar rods and cocci to filaments and branched (prosthecate) forms. They are suspended freely in the water as single cells or small colonies and commonly are studded onto a nucleus of dead organic detritus or other organisms. It is difficult to know at what population densities they occur because methods of study are in their infancy. The most promising current methods of estimating their numbers or biomass use counting, after the staining of centrifugates or filtrates of water with live stains such as acridine orange. Determination of adenosine triphosphate (ATP) [657] after separation of the bacterial size fraction by screening and filtering is also useful.

The numbers or biomass measured of bacteria are not good measures of their activity for the turnover of the populations may be very rapid. Moreover there is probably much specialization among the strains present. Some may metabolize mucose sugar polymers or one or more of the many carbohydrate polymers found in algal cell walls; others may break down the chitins of crustacean exoskeletons and yet others dissolved organic compounds secreted by phytoplankton.

Glycollate and peptides are among substances found secreted [226]. They are detected in experiments where inorganic ^{14}C has been supplied to phytoplankton in bottles, and combined organic ^{14}C is ultimately detected in the water. The secretion could be artefactual—phytoplankton doubtless behave somewhat abnormally in glass bottles—but the phenomenon is so widespread as to make this unlikely to be a complete explanation. Secretion may amount to over half of total carbon uptake and may have some advantage to the organisms, for example in chelating scarcely soluble nutrients such as trace elements. However, it seems to happen to a greater extent in infertile than in fertile lakes and hence seems likely to be a function of the mismatching of energy and nutrient supplies.

The activity of the heterotrophic bacteria is difficult to study but can be measured by separation of the bacterial cells from the water using fine filters and then measurement of the incorporation of tritiated (H^3) thymidine into the cells. Thymidine is a component of DNA and the method must be calibrated against some measure of actual cell production. Using this method in L.Michigan, Scavia, Laird and Fahnenstiel [714] showed that secondary bacterial uptake of carbon accounted for more than the current fixation of carbon by photosynthesis. The bacteria were presumably drawing on previously produced reserves in the water.

Alternatively the activity can be illustrated by the turnover rate of specific substances in the water [6, 172]. The concentrations of such substances (e.g. glucose, acetate, glycollic acid) are often too low for convenient and precise chemical analysis. The life of a given molecule may be only seconds and the entire pool of some dissolved substances may be turned over every few minutes in summer.

7.4 PROTOZOA AND FUNGI

Over 20 years ago, working with inshore sea water, Johannes [407] suggested that protozoans might have an important role. This was in grazing bacterial cells and detritus formed from dead algal cells and other sources and in regenerating their contained nutrients at much higher rates than bacteria or the larger zooplankters alone could do. More recently, also for the sea, Azam *et al.* [19] have suggested that the bacteria and smallest phytoplankton cells are fed upon largely by ciliate and flagellate protozoa. In doing so they form larger aggregates which may be more easily ingested by the zooplankton grazers (see p. 226) than freely suspended cells only 1 μm or so in size. This microbial complex is thought to have a major role in regenerating nutrients for the phytoplankton. The position is still controversial in marine circles, but suggests that freshwater protozoans may be similarly important, especially in infertile water particularly favouring growth of small algae and of heterotrophic bacteria. Ciliates are certainly common in fresh waters (see, for example, [215, 339]) as are flagellates and amoebae, and *inter alia* have been shown to eat small diatoms [87], blue-green algae [879] and bacteria [647]. Even the distinction between autotrophs and phagotrophs (those that ingest solid particles) is not clear in many algal groups. Some chrysophytes, for example, can themselves feed on bacteria as well as photosynthesize [50, 51, 206].

Phytoplankters are also attacked by internal parasites, both fungal (largely the chytrids [883]), and protozoan. Many of these are specific to single species of algae or groups of them, and there is usually a degree of infestation in any population. Such infected cells also become part of the detrital aggregates of organic matter, bacteria and protozoa which are common in many lakes and which with the phytoplankton form the food of the herbivorous zooplankters.

7.5 ZOOPLANKTON

The rotifers and crustaceans are the major groups of freshwater zooplankton (Fig. 7.5) other than the protozoa. Freshwater jellyfish, carnivorous on other zooplankters, some flatworms, gastrotrichs and mites do occur, but not commonly. The rotifers are man-sized to horse-sized on our scaled-up model (see p. 205) and are mostly suspension feeders. Their name comes from the rhythmically beating, apparently rotating 'wheel' of cilia close to the mouth. The cilia direct water with its suspended fine particles into the gut. Some rotifers may have more complicated food gathering mechanisms. One grasps the flagellum of phytoplankton cells like *Cryptomonas*, and tears the cell open to release its ingestible contents. Rotifers feed on particles from about 1–

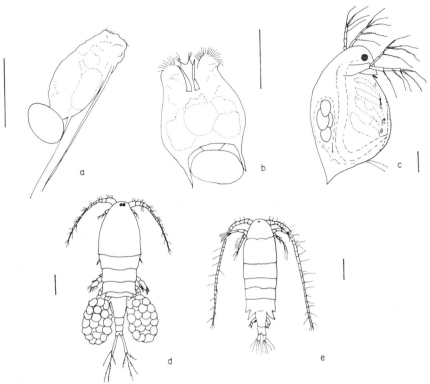

Fig. 7.5. Some representative zooplankters. In all cases the length of the scale lines represents 100 μm. Rotifera: (a) *Filina*, (b) *Brachionus*. The corona of cilia can be seen in each case and at the rear, a single egg. Crustacea (Cladocera): (c) *Daphnia*. The filtering limbs are enclosed by the carapace which also contains the egg pouch containing a few eggs. Crustacea (Copepoda): (d) *Cyclops*. The egg sacs are paired and the antennae, the lower pair of appendages on the head, are not branched; (e) *Diaptomus*, a calanoid copepod, in which the antennae are branched. When the animal is carrying eggs these are contained in a single egg sac, in contrast to the paired sacs of the cyclopoid copepods.

20 μm in size, a range shared by the filter feeding Cladocera (Crustacea) which can also take food a little larger: up to 50 μm or more in size.

The crustacean zooplankters include the Cladocera which have a carapace which covers the body in most genera. The group includes the well known herbivorous genera *Daphnia* and *Bosmina* and also some carnivores on smaller zooplankton, *Leptodora* and *Polyphemus* which are raptorial, meaning that they actively grasp their prey. The small particle feeders have thoracic limbs provided with hairs (setae) on which are closely spaced (a few μm) setules which retain small particles as they beat, and eventually convey food to their mouths. There is some controversy about whether the action is a simple filtering one or one in which the creation of small currents pushes particles of food towards the mouth. A correlation between the sizes of particles taken and the spacing of the setules supports the former case [263, 290, 344] but theoretical considerations suggest that the water is too viscous to allow simple filtration [253]. Direct impaction of particles on the limbs may be important and the organisms also have an ability to reject unsuitable food—that which is toxic or too large to ingest. A claw on the lower abdomen can be used to prise out unsuitable food from the feeding groove between the limbs along which food must pass to reach the mouth.

Cladocera move through the water more actively than the rotifers, using a rowing action of the large, branched second antennae which gives them their common name, water fleas. On the scaled model (p. 205), cladocerans, up to a few millimetres in length in reality, would be as tall as church steeples in some cases.

The third important group of zooplankters, also crustacean, is the Copepoda, whose adult members are usually a little larger than Cladocera. They may be small-particle feeding (mostly the calanoid copepods, like *Diaptomus*) or raptorial, the cyclopoid copepods, which include *Cyclops*. The prey of these may be smaller zooplankters, larger colonies, or masses of phytoplankton. Overall the copepods can tackle a wider range of bigger food particles ($5-100 \mu m$) than the non-raptorial Cladocera and rotifers. It seems likely that the calanoid copepods do not filter particles but actively select from those that are brought into the mouth region by the movements of the limbs [618, 630].

The size of particles taken by zooplankters depends on the size of the organisms—this has elegantly been shown by feeding plastic beads of known size to them—but there is clearly some separation among the groups in terms of the range eaten, and also in terms of the grazing rate. This can be measured as the volume of water swept for particles per animal per day. Again it depends on a number of factors but is generally low for rotifers, ten to 100 times higher in copepods and often even higher than that in cladocerans.

Life histories vary among the zooplankton also. Although the cytological mechanisms differ, both rotifers and Cladocera are parthenogenetic. Females produce broods of eggs asexually which hatch into more females. This allows rapid replacement of populations especially vulnerable to predation (see later). The eggs are born in sacs by rotifers and in pouches deep in the carapace in Cladocera and young are released which resemble the adults and soon grow large enough to reproduce. The rate of egg production is high, for a new generation of rotifers is produced in only a few days and each female produces up to 25 young in her lifetime of 1–3 weeks. Cladocerans take longer per generation (1–4 weeks), but each have a longer life expectancy (up to 12 weeks or so) and may produce up to 700 young per lifetime.

The genetic advantages of sexual reproduction are not lost in these parthenogenetic animals because most produce males during times of food shortage or other inclement conditions. Eggs are then fertilized, become thick walled and do not hatch out for some time. In some Cladocera they are held in thickened egg pouches called ephippia. Eventually the eggs hatch to form a new season's parthenogenetic females.

The Copepods are quite different. Each generation is sexual, and before the new mature adults are formed, there are eleven successive moults in the life history. The first six, after the egg hatches, are of juveniles called nauplii, which look quite different from the adults. The next five, the copepodites, do look like the sixth copepodite stage which is the reproductive adult. In terms of their longevity and fecundity the copepods are similar to the cladocerans.

The zooplankters thus include a diversity of form and activity in their communities and much of this is ultimately related to the major effects that predation, among themselves, or by vertebrates, has upon their numbers (see later). They are also much more heterogeneously distributed than the phytoplankters. In a mixed water body the latter are usually randomly distributed and so may be the small rotifers. The crustacean zooplankters move actively, may shoal both vertically and horizontally and often go through diurnal vertical migrations, reaching the water surface by night and moving down by day. They are tricky to sample quantitatively.

Nets are used, for the concentration of zooplanters (up to one to two per ml for the larger ones) is not great enough for a sample dipped from the water to contain sufficient for counting. However, some of the larger, faster moving zooplankters—the copepods and mysids—may be able to detect the shock wave that precedes a net as it is drawn through the water and avoid being caught. The mysids are a group of shrimp-like animals, including the genera *Mysis* and *Neomysis* which are omnivores or predators on other zooplankters, relatively large (up to 2 cm), and which have sexual life histories.

7.5.1 Grazing

The herbivorous zooplankters feed on the phytoplankton, on bacteria and aggregates of detritus and microorganisms. This simple statement embraces an enormous literature, for zooplankton feeding has attracted much attention [630].

Comparison of the mean herbivorous zooplankton biomass or production against those of the phytoplankton shows a general correlation. For example, McCauley and Kalff [503] determined that in over twenty Canadian lakes: Log_{10} zooplankton biomass $= 0.5 \log_{10}$ phytoplankton biomass $+ 1.8$. The puzzling feature of a greater biomass of consumer than consumed is explained by the greater turnover rate (a measure of productivity) of the phytoplankton.

The ratio of zooplankton to phytoplankton biomasses or productivities tends to decrease as the phytoplankton biomass increases, and this hints at the complexity of the relationship. Not all the phytoplankton is readily available to the grazers and the proportion of those that are inedible—usually the larger forms and often the blue-green algae—tends to increase with increasing lake fertility. A better correlation is thus found between zooplankton biomass and biomass of the smaller (usually less than 30 μm) phytoplankters.

The implication of these correlations is that zooplankton crops are set by phytoplankton production, much in the same way that the latter is set by the key nutrients. A comparison [753] of the same species, *Daphnia hyalina* in an infertile lake (Buttermere) and a fertile lake (Esthwaite) in the English Lake District, for example, showed that the maximum populations were six times as great in the latter, with higher birth rates and rates of increase. The animal was forced in winter to form resting eggs in Buttermere, but survived as adults in Esthwaite.

In Lake Constance also [446], *Daphnia* may at times be food-limited. The rate of egg production in this species depends on food concentration (measured as carbon in particles < 50 μm in size) with no eggs produced with food at < 0.2 mg C l^{-1} and a rise in production to about 0.7 mg C l^{-1} when egg production increases no further. Lake Constance has less than 0.7 mg C l^{-1} for much of the year. Other workers (for example, [546]), basing their conclusions on an apparently full gut at all times in the animals, feel that temperature and other factors rather than food availability are likely to be more important.

There may be a danger in extrapolating from lake to lake and from species to species for, particularly in the latter case, marked selectivity of food occurs. Some zooplankters are very fussy eaters. Cladocera, for example, along with rotifers can take bacteria-sized particles (< 1 μm) whereas calanoid copepods cannot. And within the size ranges acceptable, a preference may be shown for different sorts of food. Tests with bacteria, a yeast, and three small algae showed that among three rotifer species, *Keratella cochlearis* would take all but preferred *Chlamydomonas* (an alga) whilst two *Polyarthra* species would take only *Chlamydomonas* and *Euglena* (both algae) [56]. A cladoceran, *Bosmina longirostris*, also preferred *Chlamydomonas* and other workers showed that it exerted up to a 13.7 times preference for *Chlamydomonas* over *Aerobacter* (a bacterium), whilst another cladoceran, *Daphnia rosea* showed no preference at all [162].

Some animals prefer live food (e.g. *Bosmina longirostris* and *Diaptomus spatulocrenatus*), others dead detritus (*Keratella cochlearis*), and others (*Conochilus dossuarius*) have no preference [767]. Bacteria may be more readily taken, particularly by the larger animals, if they are attached to detritus than if they are free-living [728]. Among the algae, the assimilability of different species, and hence their ability to support growth and egg production, varies greatly and is not just a function of size. For *Daphnia* species in Lake Washington, Infante and Litt [393] found the greatest egg production and biomass production with the flagellate *Cryptomonas erosa* and a small diatom, *Stephanodiscus hantzschii*, a middle range with larger diatoms, *Asterionella formosa* and *Melosira italica* and the lowest production with the small celled *Chlorella* and a thinner variant, *tenuissima*, of *Melosira italica*.

The blue-green algae in general seem unfavourable. Arnold [18] has shown that indefinite maintenance of a *Daphnia* species is not possible on blue-green algae, but there is considerable variation in the results of different workers. Schindler [722] found the blue-green algae not the most readily assimilated, but not the least either. Part of the problem may be due to the size and handleability of the larger blue-green algae [646]; but sometimes they may be directly toxic, reducing feeding rates even if given as separated cells rather than as colonies [447].

On the other hand there is also evidence that filaments of blue-green algae are too large or awkward (too stiff) for ingestion. Some cladocerans may not have wide enough gaps between the edges of the carapace that cloaks their limbs to bring the filaments or colonies onto the limbs. If the filaments can reach the feeding groove, rejecting the filaments with the abdominal claw may pose heavy energy demands [645]. Other more palatable foods may be rejected at the same time so that growth of the animal is further decreased [392]. This phenomenon may give an advantage to small animals (which cannot take in the large blue-green algae at all so do not have to reject them from the feeding apparatus) [842]. It may be one reason why small species are often predominant in mid-summer in fertile lakes [275, 276] but there are other reasons for this, connected with fish predation (see later).

Much of the literature on feeding is based on laboratory experiments. Studies in enclosures in lakes give clues as to how selective grazing may help, with nutrient effects, to determine the composition of phytoplankton communities. Again the results may be very different in different places.

Porter [642] enclosed 500 litre samples of lake water in large polyethylene bags which she sealed and resuspended for several days in the lake. From some of the bags she had removed the larger zooplankters (crustaceans) by previously filtering the water through a 125 μm mesh net. In others she increased the zooplankton population severalfold by adding animals caught with a net from the open water. The major effect of grazers, which included small particle feeding herbivores (*Daphnia galeata mendotae* and *Diaptomus minutus*) and raptorial herbivores (*Cyclops scutifer*) was to suppress populations of small flagellates and nanoplankters and of large diatoms. Populations of large colonial green algae were not affected or even increased. They were ingested but not digested and may even have benefited from passage through guts. Phosphate released there by digestion of other species was taken up by the colonial green alga *Sphaerocystis schroeteri* as it passed through undigested and emerged growing healthily on copepod faecal pellets [643].

In a hypereutrophic Scandinavian lake, Schoenberg and Carlson [727] set up plastic enclosures 1.4 m in diameter by 2.5 m deep. They first asphyxiated any existing fish and zooplankters by adding solid CO_2 and then added to some enclosures *Daphnia galeata mendotae* and to others *Bosmina longirostris*.

The latter is a small animal unable to handle large filaments or colonies. In its enclosures the total algal crop was increased, higher phosphorus concentrations were maintained in the water, the pH was forced upwards by CO_2 uptake and the proportion of blue-green algae in the phytoplankton community was increased. The *Daphnia*, however, kept the water clear, the phosphorus concentrations and pH low, and reduced the blue-green algal component of the algal biomass. Phosphorus was kept low perhaps by sinking of phosphorus-rich faeces to the sediment. *Daphnia* may have directly grazed the blue-green algae (*Microcystis*) or, by grazing the other species more readily than *Bosmina*, may have created conditions in the water (lower pH, higher transparency, lower phosphorus availability) less favourable to blue-green algal growth. The potential impacts of grazing on the phytoplankton are thus varied and not easily predictable yet. They depend very much on the species, both of animals and algae, that are present. They may even depend on the state of a particular alga as the interesting relationship between *Daphnia* and *Aphanizomenon* seems to show.

Aphanizomenon is a blue-green algal genus which has stiff filaments, which sometimes bundle together in parallel to form 'flakes' which are visible to the naked eye. Often there is an association between such flakes and an abundance of *Daphnia* [250, 356, 485]. *Daphnia* can feed on single filament *Aphanizomenon* and even sometimes on small colonies. Larger colonies are not ingested and interfere slightly with feeding on small green algae such as *Anhkistrodesmus*, which *Daphnia* assimilates more efficiently (35%) than *Aphanizomenon* (11%). It appears that at high population densities (say 10–15 l^{-1}) *Daphnia* may cause selection in favour of flake *Aphanizomenon* by removing potential algal competitors with it for nutrients.

In L.George, Uganda, the major zooplankter, *Thermocyclops hyalinus* assimilates between 35 and 58% of ingested *Microcystis*, the major phytoplankter present, which is a blue-green algal species [551]. Reports of inabilities of temperature zooplankters to assimilate blue-green algae may reflect the present state of flux in many temperate lakes due to nutrient enrichment. This has stimulated growth of perhaps previously less common blue-green algae, for which the zooplankters may not yet have evolved suitable means of coping. L.George is not a recently polluted lake, but has had a naturally large crop of blue-green algae presumably for a very long time.

7.5.2 Feeding and grazing rates of zooplankton

Ingestion, or feeding rate depends on the volume of water handled per unit of time (the grazing rate) and on the food concentration. Burns and Rigler [84] studied this with *Daphnia rosea* feeding on a yeast (*Rhodotorula glutinis*) isolated in culture from lake water. The yeast was grown in a medium containing radioactive (^{32}P) phosphate until it was uniformly labelled. *Daphnia* were allowed to feed in yeast suspensions for 2–5 minutes, after which they were anaesthetized in saturated CO_2 solution, which prevents defaecation on disturbance of the animal. The radioactivity incorporated into the animals was measured and grazing rate (G) calculated as:

$$G \text{ (ml animal}^{-1} \text{ hour}^{-1}) = \frac{\text{Radioactivity per animal}}{\text{Radioactivity per ml of yeast suspension}} \cdot \frac{60}{\text{time (min) of feeding}}$$

Ingestion rate could then be calculated as grazing rate times the number of yeast cells per ml as cells animal^{-1} hour^{-1}. Grazing rate decreased almost exponentially as food concentration increased from about 1.5–2.0 ml hour^{-1} at 25 000 cells ml^{-1} to about 0.2 ml hour^{-1} at 500 000 cells ml^{-1}. Increasing body length and temperature increases up to about 20°C were associated with increased filtering rates. Increasing food concentrations were met by increasing ingestion rates, up to a maximum of about 100 000 cells ml^{-1} where a plateau was reached. In general grazing rate seems to decline above a threshold level as food concentration increases so that there is an upper level of feeding rate dependent on species and conditions.

These studies were carried out under ideal laboratory conditions, with a food source known to be acceptable to *D.rosea*. When studies were carried out in natural lake water from Heart Lake, Ontario, where *D.rosea* is common, filtering rates were smaller by a factor of 2–3 in mid-June to mid-October, than those predicted from the laboratory studies. Possible reasons for this include a lack of particles of filterable size or interference with feeding by large particles. Removal of animals from a lake to measure their filtering rates may itself affect the rates measured. This has led Haney [313] to design a sampler which will enclose *in situ* a volume of lake water and simultaneously will allow labelling of the particulate matter with radioactive yeast or algae.

Measures of the grazing rates of complete communities are scarce, though it is clear that grazing rates for rotifers are of the order of a fraction of an ml per day, whilst *Daphnia* species may handle 5–30 ml day^{-1} and *Diaptomus* about 35 ml per day. This means that the latter two genera, if present in reasonably high numbers (say 10–50 l^{-1}) may process substantial proportions of their habitat every day [450, 451] and this suggests a major impact of grazing if these groups are abundant.

One attempt has been made to derive overall equations linking grazing and feeding rates to the major factors which influence them by using multiple regression analysis (Peters and Downing [630]). Size of animal and food concentrations were the more important variables influencing the rates. It is generally accepted that grazing rates increase by some power of the body length, for example, Porter *et al.* [644] derived an equation $G = 0.54 \text{ length}^{1.55}$

for a group of Cladocera. The power to which length is raised may be greater for some species.

7.6 FISH IN THE OPEN WATER COMMUNITY

Fishes are very diverse. They include omnivores like the roach (*Rutilus rutilus*), eating a range of foods from invertebrates to plants, and species such as *Ramphochromis macrophthalmus* of Lake Malawi, part of whose diet of fish includes the eyes of prey too large to be tackled in their entirety (Chapter 2). Size ranges from the 12 mm of a Philippine goby to the (sometimes) 5 m of the European catfish, or wels (*Siluris glanis*).

The greatest number of freshwater fish species lives in the Tropics. This may reflect partly greater growth rates and shorter generation times in warm waters, which may permit faster evolution and partly the long history as permanent waterbodies of many deep, tropical lakes. The fish fauna of islands such as Great Britain is depauperate. Glaciation eliminated pre-existing lakes and the period afterwards when the waterways of Britain were connected with those of mainland Europe, before sea levels rose to isolate the islands, was short. The fish faunas of continental North America and Asia are less rich than those of the Tropics but much richer than those of temperate islands. Table 7.2 lists some major freshwater fish orders with some indication of their distribution. There are few species which breed, feed, grow and die entirely in the open water—most use the bottom and edge habitats (Chapter 8) at some time and, except where there has been a long time for evolutionary specialization, many fish species can live both in rivers and lakes.

7.6.1 Predation on the zooplankton and fish production

Probably almost all freshwater fish feed on zooplankters at some stage in their life history. This may only be for a few weeks immediately after the fry have used up their yolk sacs, for a year or so with switches to benthic or fish food as the fish grow, or in the case of many shoaling open water species, for their lifetime. This suggests two generalizations. First, that for fish production as a whole there will be no simple relationship with the zooplankton community and its production. But secondly, that for zooplanktivorous fish some relationship should exist albeit with the complications afforded by the natural history of two very diverse animal groups.

Two examples will illustrate this, one of them [323, 324] showing how production of the small mouth yellowfish, *Barbus aeneus* in Lake Le Roux, South Africa is linked with zooplankton production and through that with other aspects of the ecosystem, the second illustrating the strong impact of predation on zooplankton populations.

Table 7.2. Some major orders of freshwater fish

Order	Freshwater spp./total spp.	British spp.	Temperate US spp.	Tropical African spp.	Notes
Lepidosireniformes	5/5	0	0	4	Lung-fish; air breathing
Polypteriformes	11/11	0	0	6	Bichirs
Acipenseriformes	15/25	1	14	0	Sturgeons: primitive, partly cartilaginous skeleton: anadromous*
Semionotiformes	7/7	0	7	0	Gars
Mormyriformes	101/101	0	0	101	Elephant fish: snout is often elongated and proboscis-like
Clupeiformes	25/292	2	5+	2+	Herrings, shads: often plankton feeders with long gill rakers
Salmoniformes	80/508	14	70	0	Pike, salmonids ciscoes, coregonids, grayling, smelts: believed ancestral to many other orders; often anadromous
Cypriniformes	3000/3000	20	209	180	Cyprinids ('coarse fish') with Siluriformes; constitute 73% of freshwater fish: carps etc.
Siluriformes	1950/2000	1	37	345	Catfish: sensory barbels on head; no scales
Atheriniformes	500/827	0	186	0	Killifish, guppies
Perciformes	950/6880	11	55	700+	Extremely diverse, perches centrarchids, cichlids; spiny fins
Anguilliformes	15/603	1	?10	0	Eels, often catadromous†
Total	6559/14 259	50	583	1338	

Total freshwater fish = 6851, Total fish 18 818

* Breed in fresh waters, but spend part of life cycle in sea.
† Reverse of anadromous.

7.6.2 The small mouth yellowfish in Lake Le Roux

The small mouth yellowfish is endemic to the Orange River system of South Africa, which drains a relatively arid area, subject, even in the absence of human activity, to much erosion in the wetter months. The river, and the lakes formed by damming along it, including Lake Le Roux, are thus relatively turbid with suspended fine inorganic clay and silt. Depending on the particular weather conditions in a year, Lake Le Roux may be more or less turbid, with secchi disc transparency varying (between 1977 and 1984) from only 80 cm to less than 30 cm. The silt absorbs much light that would otherwise be used in photosynthesis and the phytoplankton crops and their dependent zooplankton stocks are correspondingly low in turbid years (Fig. 7.6). The yellowfish population (measured as catch per unit effort of fishing) is correspondingly low in these years as also is its growth rate (Fig. 7.6). This latter undoubtedly depends on the availability of suitable 'forage' zooplankters but also perhaps on temperature. The turbid years are also ones in which heat radiation is intercepted higher in the water column, and in which water temperature is lower because more heat is radiated back to the atmosphere instead of being mixed down into the water column.

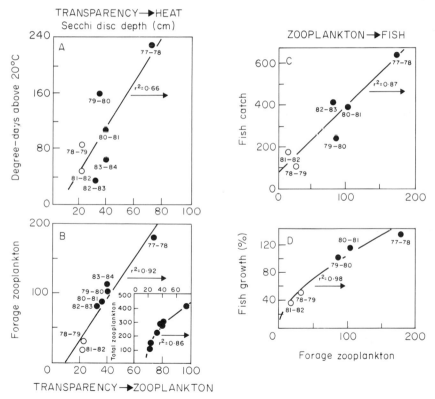

Fig. 7.6. Relationships between: summer heating and transparency (the greater the penetration of light and long wave radiation the warmer the water; zooplankton available to fish and transparency (the greater the transparency, the greater the phytoplankton and consequently the zooplankton production); fish catch and fish growth and available zooplankton, in Lake Le Roux between 1977 and 1984. (After Hart [324].)

7.6.3 **Predation by fish**

From data such as those in Fig. 7.6 it is not possible to assess the impact of fish predation on the composition and size of the zooplankton populations. However, a detailed analysis of the population dynamics of individual zooplankton species can reveal much information. An example is the study of *Daphnia galeata mendotae* carried out in Base Line Lake, Michigan, by Hall [309]. The method, however, is applicable to any zooplankter, and depends ultimately on finding the birth rates of the animal as the year progresses.

The rate of change of numbers at any instant, t, is given by:

$$N_t = N_o\, e^{rt}$$

where N_t and N_o are the numbers at the end and start of the period, e is the base of natural logarithms and r is the intrinsic rate of natural increase. r equals $(b-d)$ the difference between instantaneous birth and death rates. Death may arise from predation, parasitism, and washout.

If by regular sampling in the field, the numbers of animals can be established, estimates of r can be obtained for the periods $t \to t + 1$, between samplings, from integration of the above equation:

$$r = 1/t(\log_e N_{t+1} - \log_e N_t)$$

r, however, is a composite rate, and to understand the dynamics of the population b and d must be estimated. b, the instantaneous birth rate, is defined by:

$$N'_{t+1} = N'_t\, e^{bt}$$

N' represents a potential population size in the absence of deaths and cannot be estimated. Hence b cannot be directly measured. However, a finite approximation to birth rate, B, can be defined as:

$$B = \frac{\text{Number of newborn (during interval } t \to t+1)}{\text{Population size at } t}$$

$$B = \frac{N'_{t+1} - N'_t}{N'_t}$$

and because

$$b = 1/t \log_e \frac{N'_{t+1}}{N'_t}$$

and for

$$t = 1, \quad b = \log_e \left(\frac{N'_{t+1}}{N'_t} - \frac{N'_t}{N'_t} + 1 \right)$$

so

$$b = \log_e \left(1 + \frac{N'_{t+1} - N'_t}{N'_t} \right) = \log_e (1 + B)$$

B can be independently estimated since it is equal to the number of newborn per individual per day. Those about to be born are carried as eggs or embryos by the female until they are released when the female moults.

$$B = \frac{\text{Number of reproductively mature adults (N}_A)}{\text{Total population (N}_t)} \times$$

$$\frac{\text{Number of eggs and embryos carried per adult (brood size, } E)}{\text{Number of days for an egg to mature from production to release } (D)}$$

N_A, E and N_t are readily estimated from sampling of the natural population. D depends largely on temperature and is measured in laboratory cultures at a range of temperatures. In *D.g.mendotae* it ranged from 2 days at 25°C to 20.2 days at 4°C. From the lake temperature at the time of sampling, an appropriate D value can be selected for calculation of B.

b and *r* can be used to calculate *d*, the instantaneous death rate ($r = b - d$). Figure 7.7 gives these data for *D.g.mendotae* in Base Line Lake, together

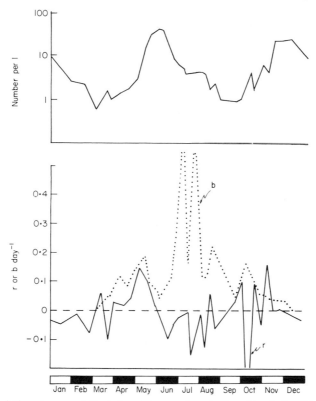

Fig. 7.7. Population changes of *Daphnia galeata mendotae* in Base Line Lake, Michigan (after Hall [309]). Upper graph shows changes in the total population, and the lower one the observed intrinsic rate of increase (*r*) and the calculated birth rate (*b*). The difference between the graphs gives *d*, the instantaneous death rate.

with changes in population. From March to early June the population increased and death rates were low. In mid-summer the population remained steady at a low level, but birth rates (and therefore production) were very high. Death rates were then also high. There was no evidence that large losses were occurring through parasitism or washout and the main cause of loss seems to have followed fish predation after young zooplanktivorous fish hatched in May and June. Although the population was reproducing at such a rate that it could double itself, on average, every 4 days, predation more or less accounted for all of the production and kept the population at a low level.

The theory given above assumes that all of the mortality is of adults but this is probably not the case. The eggs being carried by the adults at the time they were eaten are also eaten, so that the method overestimates birth rates and therefore death rates also. Eggs carried may not always develop and newly hatched *Daphnia* might also be eaten by invertebrate predators. Threlkeld [805] has demonstrated potentially important effects of egg mortality in some populations and Prepas and Rigler [652] also give a sophisticated criticism of the principles of the basic method used by Hall, which derives from work originally by Edmondson [186]. Other refinements are given in [99], [189] and [621] but in comparison with the errors inherent in sampling natural populations of animals which are usually patchily distributed, the refinements may not always justify the extra time needed for extra measurements and observations. The problem is in knowing in advance whether they will or not!

7.6.4 Predation and the composition of zooplankton communities

There is clearly an important set of relationships upon which the size of zooplankton populations and the production of fish depend. There are complexities in these relationships, however, for the nature of the zooplankton community may be shaped by the balance of invertebrate and vertebrate predation on it.

The first clues to this came from Hrbacek *et al.* [372] and Brooks and Dodson [77]. The effects of fish predation have been shown clearly when fish have been newly introduced into lakes where previously they did not live. Fig. 7.8 shows the marked shift in zooplankton species and sizes which followed introduction of the planktivorous fish *Alosa aestivalis* into Crystal Lake in Connecticut, which previously lacked such a planktivore. The zooplankton community was changed from one of *Epischura*, *Daphnia* and *Mesocyclops*, which are all usually more than 1 mm in size to one of *Ceriodaphnia*, *Tropocyclops* and *Bosmina* which are all rather smaller than 1 mm. *Cyclops*, at just under 1 mm, persisted in both situations.

This work led Brooks and Dodson [77] to state their size–efficiency hypothesis to explain why crustacean zooplankton communities at a given time during the year tend to be of rather uniform size range. They believed

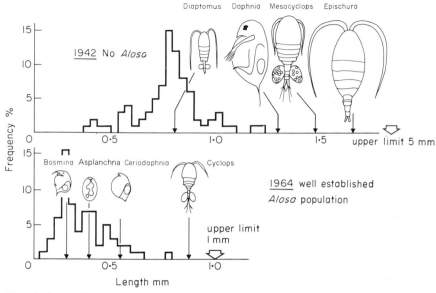

Fig. 7.8. Composition of the mainly crustacean zooplankton of Crystal Lake, Connecticut before (1942) and after (1964) introduction of a planktivorous fish, *Alosa aestivalis*. Planktivorous fish had previously not been present. Specimens are drawn to scale and represent mean size for each species. The arrows indicate the size of the smallest mature instar of each species. The effect of the fish has been to replace a community of large species with one of much smaller organisms. (From Brooks and Dodson [77].)

that all zooplankters competed for the available particulate matter in the water but that the larger zooplankters competed for it more effectively. Small animals were thus excluded by starvation if large ones were present. Vertebrate predators (largely fish but also amphibians like salamanders and even wading birds in shallow pools [170]), however, select the larger Crustacea (cladocerans, calanoid copepods), and, depending on the intensity of predation, smaller zooplankters (smaller Crustaceans, rotifers) could coexist up to the state of the complete elimination of the large forms where predation was intense.

There seems little doubt that fish do select the largest prey that they can catch and ingest. The larger the prey, the greater is the return of energy for that invested in catching it. They must see their prey to eat it and may also need to see it against the light background of the water surface. There is some evidence that objects smaller than about 1 mm in size are not readily seen by fish, so rotifers will escape and also small Cladocera like *Bosmina* and the nauplii and early copepodites of copepods.

Among the larger animals which are potentially available, the Cladocera seem most vulnerable. They move relatively slowly and probably do not have sensory mechanisms capable of detecting the shock wave of an approaching fish. In contrast the copepods have sensory hairs on their antennae and, with a flick of their abdomens can move away from the line of attack by the fish with great speed. The association of a community of small Cladocera, rotifers and copepods with fish predation has been frequently shown (e.g. [484]). Selection of the larger animals can also be shown by comparison of fish gut contents with the availability of prey in the lake [49] (Fig. 7.9).

The reasons for exclusion of the smaller animals in the absence of fish predation are less well founded. In general smaller Cladocera, for example, remove particles from water no less efficiently (measured per unit volume of animal) than larger ones, though if food is scarce, the larger filtering areas of the bigger animals may give them some advantage. Studies in which a large *Daphnia* species (*D.pulex*) has been introduced to enclosures in a lake [825] have demonstrated that it reduced the numbers of copepods, rotifers and a small cladoceran, *Bosmina longirostris*. Others, however (see, for example, [169]), have not shown any competitive advantage of the larger animals in feeding. Gliwicz [275] suggests that interference by larger algae may put the larger Cladocera at a disadvantage because energy is required to reject filaments which are easily excluded by the narrow gap of the carapace from the smaller.

Smaller animals, however, seem to be more vulnerable to invertebrate predation. They are eaten by some of the larger raptorial zooplankters: *Cyclops, Leptodora, Polyphemus,* mysid shrimps [454] and insect larvae such

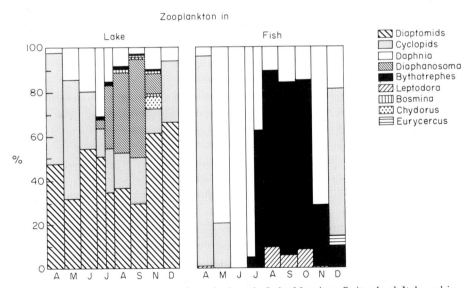

Fig. 7.9. Percentage composition of zooplankton in Lake Maggiore, Switzerland–Italy and in the stomachs of zooplanktivorous whitefish (*Coregonus* sp.). Diaptomids and Cyclopids are copepods, *Daphnia, Bythrotrephes, Leptodora* and *Eurycercus* are large Cladocera. The others are small Cladocera. (From de Bernardi and Giussani [49].)

as *Chaoborus*. In the presence of fish they may increase in numbers because their invertebrate predators, being in general large, are taken by the fish. Nonetheless invertebrate predation may be considerable even in the presence of fish [452].

7.6.5 Predator avoidance by the zooplankton

There are a number of ways in which the zooplankton have evolved mechanisms to minimize the risk of being eaten. Cladocera frequently produce individuals that are smaller, thinner and more translucent during the summer when fish feeding is intense. These forms may be less readily seen by the fish [396] but have the disadvantage that they have smaller brood chambers and produce fewer eggs. Within a lake where predation is concentrated, for example at the margins where fish may find refuges against their own predators, the larger, or sometimes just different, forms may persist offshore and the smaller ones survive more readily inshore. Green [298] has shown this for *Daphnia lumholtzi* in L.Albert, whilst a similar contrast has been shown by Zaret [883] for *Ceriodaphnia cornuta* in Gatun Lake, Panama.

The smaller, thinner animals often produced apparently in response to fish predation (though there is evidence that physical factors such as temperature and turbulence may also stimulate increase in their numbers [331, 338, 380]) often have protruberances like long spines or extensions (helmets) of the head in Cladocera. The phenomenon is called cyclomorphosis for the occurrence of these forms is often seasonal. The protruberances may be devices to minimize predation by invertebrates for even the rotifers and smaller Cladocera produce them. Fish suck in their food and the shape of it seems not to matter very much; but invertebrates must grasp and bite so the food must be conveniently handleable. The pattern of bite damage to the prey (all of which is usually not eaten) is often consistent [168] (Fig. 7.10) suggesting a particular way of attack, and the production of spined or distorted forms may interfere with the handling mechanism of the predators [331, 430, 599]. The exuberant forms may be genetically different and selected from a genetically heterogeneous population by differential predation, or they may be developmentally induced by particular substances emitted by the predator. The rotifers *Keratella tropica* and *Brachionus* spp. produce longer spines when their populations are declining and their predator, another rotifer, *Asplanchna*, is present [271, 297].

There are also behaviours which bring the prey into contact with the predator as infrequently as possible. Vertical migration has long been known in zooplankters, with the animals remaining at depth, in the dark, by day and moving to the surface to feed at night. Many explanations have been put forward for this. It might be most economical, for example, to grow at low temperatures which minimize respiration rates and to spend time in the warmer surface waters only when it is necessary to feed.

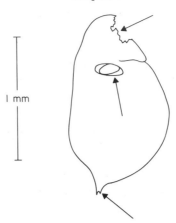

I mm

Fig. 7.10. Typical damage to *Daphnia galeata mendotae* attacked by the copepod, *Diaptomus shoshone*. The antennae are usually broken and the tail spine and head bitten. (From Dodson [168].)

However, zooplankters of the same species may show patterns of movement (presence or absence or the extent of it) which are more closely related to the risk of fish predation in the surface waters [96, 885]. Vertical migrations seem to be stimulated by the presence of some light so the animals often move to the surface at a time when there is still sufficient light for fish to see them and feed on them. In one lake a *Daphnia* species was thus vulnerable only for about an hour at twilight in summer, when fish moved from the lake edges to feed on them [848]. In this way a coexistence between quite large animals and fish can persist with the help of the refuge of the dark deeper waters.

7.6.6 Consensus

Despite the fascinating complexity of relationships between zooplankters and their predators, one important generalization emerges. This is that the larger Cladocera, particularly *Daphnia* spp. are very vulnerable to fish predation, whilst at the same time usually being the most effective grazers on the phytoplankton. The presence of a large stock of zooplanktivorous fish should thus be associated with a dearth of grazing and development of dense algal populations. Husbandry of the zooplankton population by reduction of fish predators thus offers possibilities for reducing algal growth in eutrophicated lakes where only limited or no nutrient control is possible. The technique is called biomanipulation and is discussed later. For the moment an interpretation of why domestic garden ponds with lots of goldfish are often bright green with algae may be obvious.

7.7 FUNCTIONING OF THE OPEN WATER COMMUNITY

The plankton community is a very dynamic one. Not only are the relative positions of all of its particles, live or dead, changing from second to second, but also dozens of chemical changes are going on simultaneously. We can divide the plankton into a series of 'compartments' for the purposes of examining its activities. These are phytoplankton, microorganisms (bacteria and fungi), detritus, Protozoa, zooplankton, dissolved substances and fish. A particular phosphate ion or carbon atom may, in summer, find itself shuttled through several such compartments in a few minutes. Figure 7.11 shows some of the main pathways between compartments.

Some of the processes illustrated in Fig. 7.11—phytoplankton nutrient uptake, grazing, fish predation—have already been considered. Others are very imperfectly understood. They happen very rapidly and steady states, amenable to analysis by the relatively imperfect methods available, are short-lived. The movements of phosphorus and nitrogen form a good means of illustrating some of these processes. The general picture is one of frequent re-use of each element within the system, with inevitable losses to the sediment and overflow more or less matched (if the lake is in a quasi stable condition) by the various inputs to the lake.

7.7.1 Cycling of phosphorus in the plankton

Some aspects of the phosphorus cycle within the plankton community are shown in Fig. 7.12. Phosphorus enters the cycle from the catchment area (see Chapters 3 and 6) or by release from the sediment (see Chapter 6). It leaves it when it is washed through the outflows, or is incorporated permanently into sediments as detritus or precipitates. During the plankton cycle between these events, there seem to be at least two main means of regeneration of inorganic phosphate: by zooplankton and by microorganisms.

The relative importance accorded to these must vary with the lake concerned. In L.George (Uganda), for instance, it is only by rapid recycling of nitrogen and phosphorus through the zooplankton that dense blooms of blue-green algae can be maintained [254]. When released by zooplankters N and P are taken up so rapidly by the phytoplankton that their excretion can only be detected if the phytoplankton is experimentally removed.

Calculations based on the plankton of Heart Lake, Ontario by Peters and Rigler [631] suggest that about a quarter of the particulate matter in the epilimnion (expressed as its phosphorus content, particulate phosphorus) is ingested by zooplankton each day in summer. Of this, about half is assimilated and during the approximately steady state of the zooplankton population in summer an equivalent amount is daily excreted. Thus 0.5×25 or 12.5% of the total phosphorus in the plankton is regenerated daily by excretion and a

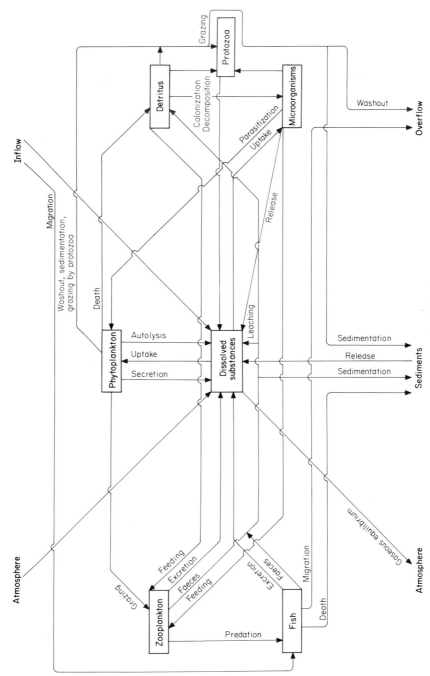

Fig. 7.11. Some important relationships between the major compartments in the plankton system.

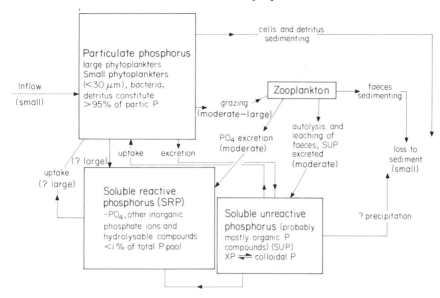

Fig. 7.12. Cycling of phosphorus during a day in summer in a temperate lake. There is relatively little through-flow of water so losses to the overflow are negligible. Values are given as percentages of the particulate pool turned over per day and are approximate. Compiled from various sources discussed in the text.

similar amount is daily turned into faeces which enter the detritus compartment. Here more of the phosphorus may be released by microbial activity.

There has been increasing recent interest in the role of the microorganisms in regenerating phosphate. The microbial complexes of small phytoplankton, protozoa, bacteria and detritus may be important particularly in infertile lakes[797a, b]. Zooplankton excretion has consequently been less emphasized recently[504]. It seems that phosphate is taken up very rapidly by the smallest organisms and that the larger phytoplankters might re-acquire it only slowly following excretion of it, or of organic phosphates, from the microbial complexes [135]. Some have phosphatases on the outsides of their cells capable of releasing phosphate from the organic phosphorus compounds.

Lean [458] has demonstrated some of the potential complexity of movements of phosphorus through the microorganisms. He added radioactive inorganic phosphate to water from Heart Lake, Ontario. Within a minute, half of it had been taken up by the smaller phytoplankton and microorganisms. Within an hour an equilibrium had been established in which only about 0.2% of the dissolved phosphorus was present as dissolved inorganic phosphate and 98.5% was in particulate form. By filtering the water through molecular gel filters two further sets of phosphorus compounds were detected. One, with a molecular weight of about 250, called XP, constituted 0.13% of the total and one with molecular weights greater than 5×10^6 constituted 1.16% of the total and was called 'colloidal phosphorus'. XP was released from the particulate compartments probably by bacterial rather than algal secretion, but certainly not through death and decay as the process was too rapid (less than 3 minutes). XP reacted with 'colloidal phosphorus' with release of inorganic phosphates, but XP did not itself readily hydrolyse to inorganic PO_4.

The turnover time for the passage of inorganic PO_4 ion through the particulate, XP and colloidal P compartments is likely to be only a few minutes in summer (Rigler [682]). Some of the colloidal P ceases to react with XP after a few days and hence its contained phosphorus becomes unavailable to the plankton.

The overall picture of phosphorus movement is thus one of very rapid cycling and reuse of inorganic phosphate by the phytoplankton and bacteria. Organic phosphorus compounds may not always be available directly to the phytoplankton, but must be metabolized first by bacteria other than those which produce and secrete it. Zooplankton also cycle phosphorus with a rather longer turnover time (days rather than hours) and may also be prime agents, through sedimentation of their faeces, of loss of phosphorus to the sediments. This seems to occur at the rate of a few per cent of the total phosphorus pool per day and implies again (see Chapter 6) that maintenance of plankton production relies on continued renewed supplies of phosphorus from the catchment.

In all of these processes it must also be remembered that at the scale which is important to these organisms (distances of micro- or millimetres) the environment is patchy [463, 666]. As a fish or zooplankter moves through the water excreting phosphorus, this substance becomes available to algae or bacteria in its plume but, because they may take it up so rapidly, not to those only millimetres away.

7.7.2 The nitrogen cycle in the plankton

The behaviour of nitrogen in the plankton is as complex as that of phosphorus. There are four main dissolved inorganic pools—ammonium, nitrite, nitrate and molecular N_2. The first three forms are available for uptake by most phytoplankters and may be simultaneously absorbed. Some species, particularly the Euglenophyta [462] seem able to absorb only ammonium, and often other plankters may absorb ammonium preferentially since it is energetically less costly to process in the cell. Ammonium and particularly nitrite rarely have large dissolved pools, however. Even nitrate may be barely detectable in summer and in lakes where the loading of phosphorus is large the rate of supply of inorganic nitrogen compounds may limit the growth of phytoplankton (see Chapter 6).

The turnover times of the inorganic nitrogen pools are largely unknown. Ammonia excreted from *Thermocyclops hyalinus*, the dominant zooplankter of the equatorial L.George, was measured by Ganf and Blazka [251]. They showed that the ammonia released was rapidly taken up by phytoplankton, with a turnover rate of about 1.5 day^{-1}. Parallel estimates of inorganic PO_4 turnover of about twice a day are likely to be underestimates [631]. The very small pools of inorganic nitrogen in the lake (no nitrate was detected at all) suggest that nitrogen should be cycled at least as rapidly as phosphorus and the turnover rate may be much greater than 1.5 day^{-1}.

Molecular nitrogen can be used only by nitrogen-fixing blue-green algae and by certain photosynthetic and heterotrophic bacteria. Until recently it was believed that only blue-green algae possessing differentiated cells called heterocysts could fix nitrogen. Nitrogenase, the complex of enzymes responsible, is inhibited by oxygen, and heterocysts lack that part of the photosynthetic apparatus, photosystem II, which is responsible for oxygen release [213]. Isolated heterocysts have been shown to contain active nitrogenase [773], and there is a high correlation between measured nitrogenase activity and various functions of heterocyst numbers [359]. Nitrogenase is contained in ordinary cells of some blue-green algae which do not have heterocysts [774] though it is active only under anaerobic conditions. A coccoid species *Gloeocapsa* has been shown to fix nitrogen in aerobic conditions [881] despite a complete lack of heterocysts, and this may mean that nitrogen fixation is more widespread in lakes than presently believed.

In many lakes nitrogen fixation seems, however, not to be an important source of nitrogen. Less than 1% of the total nitrogen income of L.Windermere is provided in this way [358], but production is probably limited by phosphorus supply in this lake. In lakes with high phosphorus loadings, large crops of blue-green algae may account for about half of the total nitrogen income by fixation [359].

Blue-green and other algae and bacteria secrete nitrogenous organic compounds into the water where they form part of a pool of dissolved organic nitrogen compounds about which little is known [518]. The pool is also supplied by excretion of urea by fish, and perhaps with amino acids and urea by some zooplankters [398]. Many different substances comprise the organic nitrogen pool. Walsby [837], for instance, showed at least twelve separate polypeptides to be produced by a single species, *Anabaena cylindrica*. In L.Mendota, dilute pools (<0.01 μmol) of free amino acids have been measured, with serine and alanine the most prevalent of ten acids examined. A tenfold larger pool of combined amino acids was simultaneously measured and both pools increased during decomposition of large algal populations [255].

At least part of the dissolved organic nitrogen pool, that of small molecules such as amino acids, probably turns over rapidly at rates similar to those of other simple organic compounds. Bacteria probably account for much of the uptake because algae, although capable of using amino acids [2] and other organic nitrogen compounds [52] at high concentrations, do not compete effectively with bacteria at low concentrations [351]. The vitamins cyanocobalamin (B_{12}), thiamine, and biotin, variously required by some phytoplankton species, are all dissolved organic nitrogen compounds and are rapidly used once formed. Pools of up to 8 ng l^{-1} of cyanocobalamin, 400 ng l^{-1} of thiamine and 40 ng l^{-1} of biotin were recorded in the Japanese lake Sagami by use of bioassay techniques since chemical analyses are insufficiently sensitive [603]. The vitamins seemed to be secreted by bacteria and some blue-green algae, and were taken up rapidly by planktonic diatoms.

Dissolved organic nitrogen compounds, although not solely responsible, may also act as chelators. Chelators reversibly combine with metal ions in such a way that equilibria are set up in the water between free metal ions and soluble ion-chelate complexes. As ions (such as Fe^{3+}, Mn^{3+}, Mo^{3+}) are removed from the complexes by phytoplankton, the equilibrium moves in such a way as to replace them. Such ions are readily precipitated inorganically as hydroxides or carbonates and retention of them in soluble chelator complexes ensures a steady supply for the phytoplankton. The peptides secreted by blue-green algae can act as chelators [228] and it is significant that the two most commonly used chelators in media for the laboratory culture of algae are the organic nitrogen compounds ethylene diamine tetraacetic acid (EDTA) and nitrilotriacetic acid (NTA).

7.8 SEASONAL CHANGES IN THE PLANKTON

Whilst the plankton changes hourly in the cycling of substances within the community, the community itself changes on a long-term basis with a seasonal periodicity of different species being reflected week to week or month to month. This change is imposed variously: by changes in the water caused by climate through its effects on temperature, rainfall and hence nutrient loading; by the earth's rotation with its effects of day length and light intensity, and by internal chemical changes as the community reacts to the external factors.

Almost all species which increase their population in the plankton at some time during the year are ever present in the water as small residual populations. Some species may form resting stages in the surface sediment and new ones may be brought in from time to time on water birds or by wind or floods.

There is then a great reserve of varied forms, each best fitted to exploit a particular set of conditions in the water, when its population will increase, and each unable to compete in other conditions, when its population will decline. The changing water mass throughout the year, in turn, selects the better fitted species for a particular time, and, in turn, leads to their decline. The result is a procession of overlapping, large populations against a background of small, declining populations.

If changes in weather are a major driving force in determining seasonal periodicity, the least marked periodicity must be expected in lakes at the Equator. L.George, astride the Equator in Uganda, experiences a very constant climate. Incident radiation, although irregularly intercepted by cloud, varies within a range of only $\pm 13\%$ of the mean, and the water temperature is always about 30°C. There are two dry seasons but their potential effect in determining changes in nutrient loading is offset by the Ruwenzori mountains in the catchment, whose high run-off permits a continuous inflow to the lake [831].

This constancy is reflected in the low diversity of the plankton. Over 99% of the plankton biomass is phytoplankton and of this amount 80% comprises six species of blue-green algae. Only a dozen or so other species have been recorded, compared with hundreds in more seasonally variable lakes. There is a great stability in phytoplankton biomass and species composition, and also in zooplankton biomass, which is dominated by only two copepod species, with *Thermocyclops hyalinus* comprising 80% by weight.

Lake George is, however, likely to be very unusual in the stability of its plankton communities. This property has been ascribed as general for tropical lakes but recent surveys [530, 895] have shown just as much variability in tropical lakes as exists in temperate ones. The temperate ones are better described even if imperfectly understood.

A typical pattern of temperate phytoplankton periodicity is shown in Fig. 7.13. A late winter/spring pulse of several diatom species is overlapped by one of Chrysophyta, largely *Dinobryon* species, as it declines. The onset of summer brings a wide variety of green algae, Cryptomonads, dinoflagellates and, in mid to late summer blue-green algae develop measurable populations. In autumn diatoms may grow again.

This sequence of algae is superimposed on a set of environmental changes which start, in winter, with short days and low light intensities and temperatures, but with relatively abundant nutrients. This is because run-off from the catchment is usually high, at least in late winter as the snows melt, and the incoming water replenishes the nutrient stock. As algae increase their growth in spring and summer with the lengthening warmer days, the incoming nutrient supply is diminished and nutrients may become very scarce. The water column may also become more stable as stratification sets in. Zooplankters increase their populations later than the algae as their growth is more sensitive to temperature, so that for a time grazing may be unimportant. Once it has begun it may be truncated as the newly hatched young fish start to feed in June and July. In autumn there is a general reversal of these trends as inflows increase, temperature and light intensity fall, zooplankters enter resting stages and fish cease to feed.

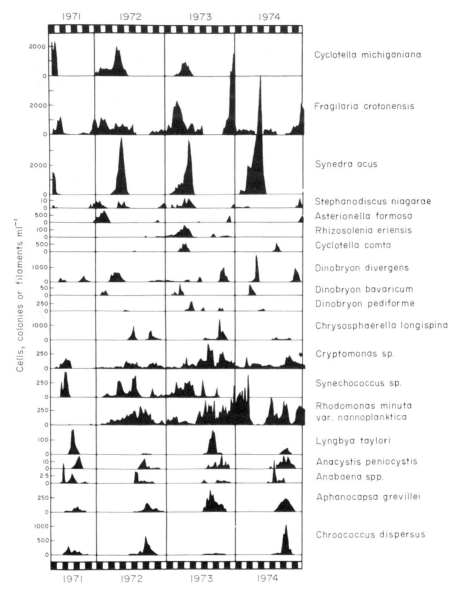

Fig. 7.13. Seasonal changes in the most abundant phytoplankton species of Gull Lake, Michigan, USA, over a period of several years. Numbers of organisms are the means of counts taken at several depths in the water column of this 30 m deep lake. The top seven organisms are diatoms, the next four are Chrysophyta. *Cryptomonas* and *Rhodomonas* are Cryptophyta, the remainder are blue-green algae. (Redrawn from Moss *et al.* [571].)

7.8.1 Mechanisms underlying algal periodicity

Some algal growth goes on in winter even under a thick ice cover. Ice is usually transparent unless it is covered by snow, which is opaque. A variety of species grow in winter and physiological investigations have shown some winter algae from polar seas to be adapted to low temperatures and low light intensities, though examples are not yet available from freshwater studies. Such species will not grow at $10°C$ or above. In temperate lakes in winter, even so, growth is not great and the levels of available key nutrients are able to build up.

As temperatures, day lengths and light intensities increase in late winter, so also do cell division rates of various diatoms whose populations reach maxima in the spring. In L. Windermere, UK, the diatom *Asterionella formosa* is the first phytoplankter to form a prominent population after the turn of the year. Lund [477] has shown that it is primarily low light intensities which minimize growth in winter of this diatom. Cells are always present in the water and will multiply in winter water if brought into more brightly lit conditions in the laboratory.

From around February onwards the *Asterionella* cells divide and exponential increase brings the population from about 1 cell ml^{-1} in January to perhaps 10 000 ml^{-1} by late spring [478, 480]. Concentrations of dissolved nitrate, phosphate and silicate, built up by the inflows in winter, all decline during this growth as the cells take them up, and in May or early June the *Asterionella* population suddenly declines. Some dissolved silicate (about 0.4–0.6 mg $SiO_3 - Si\ l^{-1}$) which is required in quantity for production of diatom cell walls, remains in the water, but the cells seem unable to take it up unless small amounts of phosphate are added [375]. On the other hand, addition of silicate without phosphate will also allow some further growth in the water at the time when the population is declining in the lake. Clearly there is a nutrient limitation but the mechanism is complex. As this point is reached the cells seem unable during division to obtain enough silicate to complete the cell walls of their daughters. Weak walls are formed and there may be invasion by bacteria. Eventually most of the population is lost to the sediments, perhaps with some regeneration of phosphorus and nitrogen from the cells as they sink. Other diatoms, faced with low silicate concentrations simply cease to grow and persist until silicate supplies are renewed [756].

The major spring diatom growth, which is a feature of many temperate lakes, does not always involve *Asterionella*, and does not necessarily end due to silicate limitation, though nutrient shortage of one kind or another is often, but not always, implicated. The genus *Melosira* is common in many fertile lakes and seems to persist in the water only when mixing is vigorous. Its walls are thick, its cells heavy and the onset of stratification, whether inverse under ice during the winter or direct in late spring, leads to sedimentation of the cells. Most phytoplankters are grazed or die in other ways once they reach the sediment, but *Melosira italica* var. *subarctica* [479] in some English Lake District lakes, survives in a quiescent state. Once mixing is vigorous enough to disturb the surface sediment, it is resuspended, expands its cell contents and divides, so long as light and nutrients are available.

Melosira species are also common in tropical lakes [795] and again their growth coincides with turbulent conditions, for example the periods when L.Victoria is mixed through the cooling effects of the seasonal trade winds [791, 792]. They may grow even in summer in temperate lakes, but usually in riverine ones where water movements produce adequate turbulence.

Nutrient shortage and reduction in turbulence are not the only factors which may bring the spring diatom growth to an end. In one Connecticut lake, the growth could be reduced or prevented in different years depending on the size of the blue-green algal population that had developed the previous year and overwintered [426]. Large populations of blue-green algae secreted substances, apparently highly specific to individual diatom species or even strains of a species, which suppressed their growth; and in lakes with small species of spring diatoms, such as *Stephanodiscus hantzschii*, there may be a phase lasting several days or weeks of growth followed by a rapid loss through grazing [449]. Increasing populations of zooplankters responding with increased birth rates to rising temperatures, graze down the diatoms and other cells before they themselves are reduced in numbers by fish predation. The fish frequently spawn in late spring and the young of the year are initially usually zooplanktivores.

The diatom spring growth is often overlapped and succeeded by chrysophyte populations, particularly those of *Dinobryon* species. It has been thought that these organisms were inhibited by even moderate concentrations of dissolved phosphate and hence had to wait until the diatoms had stripped out the stock built up during the winter. It now seems that they have very low half saturation constants for P uptake ($< 0.5 \mu$ molar P) and hence can grow in nutrient depleted water when other organisms cannot [462a].

Information on the very complex interplay of populations of the many other species that grow in the summer rather than in spring is meagre. In summer, the water is much more of a self-contained system than it is in spring when nutrients are coming in with the inflow water. The rapid cycling of phosphorus and nitrogen, and the metabolically induced changes in pH and CO_2 concentrations caused by high summer photosynthesis make interpretation difficult. Some of the green algae form summer populations because they grow slowly. Though division begins in a very sparse background population early in spring, its effects are not really shown until summer [481]. Other species, the Cryptomonads for instance, which require vitamin B_{12} and often thiamine also, may benefit from enhanced bacterial activity and secretion at the higher water temperatures. The blue-green algae of late summer may exploit the micro-aerophilic zone of the metalimnion (see earlier) because low dissolved oxygen concentrations seem to favour their growth, and if dissolved inorganic nitrogen becomes scarce then nitrogen-fixing blue-green algae will be favoured in late summer.

Low nitrogen to phosphorus ratios may favour even some non-nitrogen fixing blue-green algae [751], and other of them seem particularly adept at picking up CO_2 from very low concentrations. The increased gross photosynthesis and uptake of CO_2 in summer may force the pH value of the water to 10 or more in lakes with low alkalinities. Talling [794] has shown that *Microcystis aeruginosa* and the dinoflagellate *Ceratium hirundinella* both have abilities at absorbing CO_2 at such pH values superior to those of algae like *Asterionella formosa* and *Fragilaria crotonensis* which grow earlier in the year. Yet other late-summer species may be selected for their inedibility. As the young zooplanktivorous fish are reduced in numbers by fish-eating fish (piscivores), there may be some recovery of the cladoceran populations and as the smaller algae are removed, large ones may be placed at greater advantage. There does seem to be an association between the large blue-green alga, *Aphanizomenon flos aquae* and *Daphnia* (see above).

A general association between mixing conditions and diatoms and stratified conditions and blue-green algae is common to both tropical and temperate lakes. But because these sets of conditions are correlated with so many other factors, detailed experimentation is needed to sort out exactly what is responsible for the growth and decline of each individual species.

7.9 PRACTICAL APPLICATIONS OF PLANKTON BIOLOGY—TREATMENT OF EUTROPHICATION BY BIOMANIPULATION

This chapter has added an extra dimension to those of Chapter 6. Lakes are not just bowls of water variously fertilized with phosphorus and nitrogen and understandable purely in terms of inputs and outputs of these elements. They are ecosystems which manipulate the N and P once they have entered. This opens up the possibility of treating at least the symptoms of the problem of eutrophication (Chapter 6) by changing the structure of the open water community. For a given P concentration, there can be a range of planktonic chlorophyll *a* concentrations in a particular lake (Fig. 6.8). The trick of 'biomanipulation' is to adjust the community to give the lowest chlorophyll *a* concentrations and hence reduce the algal crops. It may also be possible to reduce the total phosphorus concentration by similar means.

That such changes are possible can be shown in simple experiments. An experiment set up in tanks on the roof of San Diego State College, California is typical [377]. Six replicate systems were set up in 30 cm deep × 2 m diameter pools. To each pool was added a 3 cm layer of sand, tap water to a depth of 20 cm and a litre of dried alfalfa pellets as a source of organic matter and nutrients. A small sample of plankton from a nearby lake and an inoculum of *Daphnia pulex* were added, and the pools left for a few weeks. Then fifty 3–5 cm fish, *Gambusia affinis* were added to each of three ponds, and observations were made on the water chemistry, plankton and benthos; some of the results are shown in Table 7.3.

The fish readily ate the larger zooplankters *Daphnia pulex* and *Chydorus sphaericus* which developed significant populations only in the fishless ponds. The less preferable, smaller rotifers increased to greater levels in the presence of fish, though they were also abundant in the absence of fish.

In the ponds with fish the reduction in zooplankton grazing allowed a very dense population of a unicellular blue-green alga, *Coccochloris peniocystis*, to develop, which markedly reduced light transmission and the growth of *Spirogyra* sp. (a filamentous green alga) on the bottom. The fish also readily ate the benthic invertebrates, either in the bottom sediments or as they emerged as adult insects, and probably took rotifers as other food supplies became short. At least, this is one explanation for the eventual decline in

Table 7.3. Effects of *Gambusia affinis* on pond ecosystems. (After Hurlbert *et al.* [377])

	Without fish	(Significance $P=0.1$, rank sum test)	With fish
Zooplankton (No. $10\,l^{-1}$)			
Daphnia pulex (2 Dec.)	92 ± 56	S	0.3 ± 0.47
Daphnia pulex (3 Feb.)	1837 ± 868	S	0
total rotifers (2 Dec.)	1134 ± 426	S	5927 ± 1911
total rotifers (3 Feb.)	32 ± 21		4.3 ± 3.2
Phytoplankton (cells $ml^{-1}\times10^{-6}$)			
Coccochloris peniocystis (10 Jan.)	0	S	116.5 ± 26
(2 Feb.)	0	S	219 ± 7.5
colonial algae (2 Feb.)	78.9 ± 55		52 ± 37
Macroscopic bottom algae (g $pool^{-1}$)			
Spirogyra sp. (6 Feb.)	312 ± 132	S	29 ± 22
Chara sp. (6 Feb.)	24 ± 7.6		31 ± 8.6
Phosphorus concentrations ($\mu g\,P\,l^{-1}$) (3 Feb.)			
inorganic phosphate-phosphorus	9.7 ± 5.0	S	0.33 ± 0.1
organic phosphate-phosphorus	18.4 ± 2.3	S	54.7 ± 31.2
particulate phosphorus	12.4 ± 3	S	271 ± 26
total phosphorus	40.5 ± 9.4	S	326.2 ± 9.8
Benthic invertebrates (No. cm^{-2} bottom sediment)			
Chironomid larvae (3 Feb.)	25 ± 3.1	S	0
Oligochaetes (*Chaetogaster* sp.) (7 Feb.)	14 ± 3.3	S	0.7 ± 0.5
insects collected as pupae or imagos at water surface between (5 Nov. and 7 Feb.)	486 ± 131	S	0
Extinction coefficients (m^{-1})			
425 nm (blue)	5.1	S	63.8
680 nm (red)	0.83	S	12.0

rotifers. Significant too was the apparently greater mobilization of phosphorus in the ponds with fish, conceivably through the exploitation by the fish of the bottom fauna and release in fish excreta of phosphate, which became available to the plankton. The presence or absence of an invertebrate-eating fish thus created two entirely different states of the ecosystem, despite similar initial physico-chemical conditions. This phenomenon can be demonstrated also in enclosures containing a community less artificial than the ones created in San Diego and set in a lake.

Lynch and Shapiro [486] for example, added increasing numbers of blue gill sunfish, *Lepomis machrochirus* to enclosures in a lake and found an increase in chlorophyll concentration and a decrease in transparency with increasing numbers of fish (Fig. 7.14). The fish ate *Daphnia* which otherwise grazed the algae. In this experiment the total phosphorus concentration was unchanged, though more of it was in the available soluble reactive form when *Daphnia* grazing was most intense.

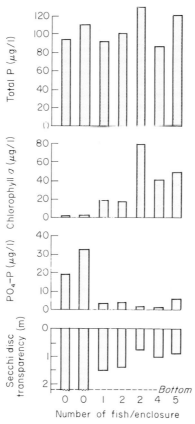

Fig. 7.14. Effects of adding bluegill sunfish to enclosures containing a plankton community. (Based on Lynch and Shapiro [486].)

Recruitment of fish from year to year is variable. It depends greatly on the water temperature just before the time of spawning and hatching and 'good years' are often associated with warm spring weather. In turn the *Daphnia* populations can vary very greatly with inverse effects on the chlorophyll *a* and total phosphorus concentrations (Fig. 7.15). Packaging of phosphorus into readily sedimented *Daphnia* faeces is one mechanism by which the *Daphnia* is able to reduce the total P concentrations.

Similar changes may follow overall trends in the fish stocks. Lake Washington was quoted in Chapter 6 as an excellent example of a lake recovering after nutrient diversion. In the 1970s, however, when nutrient diversion had been completed for several years, the lake continued to increase in clarity, and indeed became more transparent than it had ever been before. Populations of four *Daphnia* species had increased greatly but those of a major predator on *Daphnia*, a mysid shrimp, *Neomysis mercedis* [581, 582] had apparently decreased. *Neomysis* is quite a large animal—about 1 cm long—and hence very vulnerable to fish predation. Its predator, the long fin smelt, has apparently increased in numbers (Edmonson [190]), perhaps through inadvertent improvements in its spawning sites by modification of the rivers flowing into Lake Washington.

Another example of such effects is that of L.Michigan, where a number of salmonid fish (the coho salmon, *Oncorhynchus kisutch*, the chinook salmon, *O.tshawytscha*, the steelhead trout, *Salmo gairdneri*, the brown trout, *S. trutta* and the lake trout, *Salvelinus namaycush*) have all been introduced and regularly restocked to provide improved sport for anglers. These fish are

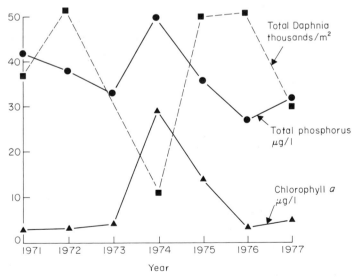

Fig. 7.15. Total phosphorus, chlorophyll *a* and *Daphnia* in Lake Harriet, Minnesota in 1971–1977. Mean surface values for the summer period are shown. (From Shapiro [735].)

piscivores, and the populations of the alewife, *Alosa pseudoharengus*, a major zooplanktivorous fish, have consequently been reduced. There have followed [713] since 1983, marked increases in cladoceran zooplankton populations and clarity of water.

There can be no doubt, therefore, that the phytoplankton crops can be adjusted by the interplay of *Daphnia* (and, of course other grazers, though *Daphnia* appears to be particularly important) and fish predation. Removal of fish, or reduction in the population of zooplanktivores could thus be an effective technique for reducing algal crops. Shapiro and Wright [737] followed changes in Round Lake, Minnesota, after it had been treated with rotenone in autumn 1980. Rotenone is a fish poison and its effects were to denude the lake of fish, whose community was dominated by zooplanktivores, *Pomoxis nigromaculatus, Lepomis macrochirus* and *Lepomis cyanellus* with relatively few predators on them. The lake was moderately eutrophic, with mean summer chlorophyll *a* concentrations of up to 17.3 μg l^{-1} and total phosphorus to 75 μg l^{-1}. Rotenone is short-lived and the lake was restocked with a much greater ratio of piscivores (large mouth bass, *Micropterus salmoides* and walleye, *Stizostedion vitreum* to zooplanktivores, than previously. Changes in the lake were followed for the next 2 years. There was an increase in transparency, in the body sizes of the zooplankton, in the proportions of *Daphnia* in the zooplankton community, and decreases in chlorophyll *a* and total phosphorus concentrations.

Deliberate biomanipulation of this kind has rarely been used though occasionally accidental circumstances have shown its value [457]. In Britain there would be resistance to removal of the initial fish stock because of the scarcity of lakes for angling. (Minnesota has a great many so liberties can be taken.) Alternatives, such as the stocking of additional piscivores, would be more acceptable but in Britain there is no native piscivore which specializes in feeding on small, open-water zooplanktivorous fish. The closest possibility is the perch-pike, *Stizostedion lucioperca*. This fish is native to continental Europe, and probably was also to Britain prior to glaciation. However, where limited introductions of this fish have been made—in rivers in East Anglia—local opinion has it that there has been a decline in other fish and that the angling has been spoiled. The declines may have been much more likely due to changes in the habitat through eutrophication and river management (Chapter 5), but this becomes irrelevant where human perception and firmly held convictions are involved.

Alternative is to provide refuges for the zooplankton and other small grazers. In shallow lakes, aquatic plants may serve to hoard large populations of Cladocera which move out at night to graze [807], or which graze on phytoplankton in water moved by wind through the plant bed. Fish seem not to enter the plant beds where they are dense except at the edges. And even if the Cladocera are unable to move back to the beds by dawn, the teeming populations available seem readily able to replace the open-water grazers the next night. Often aquatic plants have disappeared in severe cases of eutrophication (see Chapter 8), and re-establishment of them, as well as reductions in algal populations may be the aims of restoration. Provision of artificial zooplankton refuges of netting, buoyant rope or bundles of twigs may help and this technique is currently being tested [565].

Biomanipulation has great potential, but it involves a closer understanding of the ecosystem than nutrient manipulation does. This is because if it is attempted without nutrient reduction, it involves maintaining an ecosystem which may not be stable and which requires continual adjustment. Certainly its effects are less easily predictable than those of nutrient reduction. On the other hand, the latter is expensive, does not always work because of problems with sediment release of phosphorus, and is not always feasible. Biomanipulation is potentially cheap and, with a little imagination and insight, usually worth a try.

In the long run, combinations of nutrient reduction and biomanipulation will probably prove most effective. It is a pity therefore that two schools of opinion concerning what ultimately controls the algal population should be taking perhaps entrenched positions [321, 512, 648]. The 'bottom up' school places emphasis on the close relationship between nutrient supply and algal crop and regards zooplankton and further levels in the food chain as of minimal importance. The 'top-down' school sees cascading effects of alterations in the upper, predator levels as ultimately controlling the algal population. At different times and in different places both rightly and wrongly place their emphasis. Any 'school of thought' too aggressively defended tells us more of human insecurity than the operation of ecosystems.

8

The Edges and Bottoms of Lakes and their Communities

8.1 A VARIETY OF HABITATS

At the edges and bottoms of lakes the junction between water and solid imparts a heterogeneous structure. There may be rocks, sediment and plants, colonized by sessile animals and algae, together with others that move freely but stay close to the interface. In the bigger lakes these communities are probably small in effect relative to those of the open water, but the smaller and shallower a lake becomes, the more the edges and bottom assume dominance. The communities of edges and bottoms are relatively productive, perhaps first the result of access to nutrient sources in the sediments and secondly to the structural complexity of the habitats, which provides an organization in which food reserves can be fully used.

A simple framework within which to divide these habitats might divide them first into littoral and profundal. Littoral (Greek: shore) in fresh waters is taken to include the part of the lake bottom, and its overlying water, between the highest water level and the euphotic depth of algae that can colonize the bottom sediments.

The profundal (Greek: deep) includes the bottom below the euphotic zone, on which net photosynthesis is not possible. The communities which grow in this habitat depend on a supply of energy as detritus from the overlying water, or washed downslope from the littoral. In both the littoral and profundal, the organisms which are associated with the bottom are called benthic (Greek: bottom).

The profundal benthos is generally a relatively simple community of bacteria, protozoa, invertebrates, and fish. It has low diversity for the habitat is a relatively uniform fine sediment of detritus already processed by other, open water communities before it reaches the bottom. In a fertile lake, deoxygenation may confine benthic growth under the hypolimnion to the cooler parts of the year when growth is slow, and leave only anaerobic bacteria and some protozoa as active permanent residents.

The littoral supports a wider range of communities. Wave disturbance first of all determines the nature of the bottom and divides the littoral into an upper disturbed part, where erosion of particles is predominant, and a lower part below reach of the most vigorous waves, where particles may settle. The two parts are not disjunct but grade into one another along a continuum of sediment particle size from the upper coarse to the lower fine. In small lakes wave action may be very small, and fine particles may be deposited all over the littoral zone.

In big lakes with considerable wave energy, the upper shore may be of bare rocks. On the rocks will grow a community of attached algae, bacteria, protozoa and sessile invertebrates like polyzoans and the larvae of some caddis-flies. Motile scrapers like limpets and mayfly larvae will be present.

In more sheltered parts, the rocks become buried partly in gravel and sand, and there may be shores floored entirely by one or the other. Gravel is not readily colonizable. It is not so stable as either rock, being moved around by the waves, or sand and finer sediments which are only set down in less disturbed conditions. Moving gravel crushes animals and buries algae attached to it. In the quieter seasons it may be used for egg-laying by some fish, but its permanent community is small. Sand, however, though less favourable than finer sediments, does have a considerable community. The grains allow light penetration to as much as a cm or so into the bed and algae attached to them (called epipsammic) can photosynthesize. Other algae (epipelic) can move through the crevices and over the surface.

The finer sands may be colonized by aquatic plants. In infertile lakes these often include *Isoetes*, and *Littorella*, genera with short stature, excluded in fertile lakes by bigger plants, but able to root in sand. The community is completed again by invertebrates and by fish.

Finer sediments in the littoral have the most complex communities. There is a very diverse community of bacteria, protozoa and epipelic algae on mud kept bare of aquatic plants by random disturbance or, at the edges, by fluctuating water levels. To this may be added emergent, submerged and floating plants, with their associated epiphytes, invertebrates and fish. The greater productivity of this community may support herbivorous birds and mammals (e.g. coot, swan, muskrat, hippopotamus) and piscivorous birds, reptiles and mammals (e.g. bittern, heron, crocodile, otter).

Emergent plants will be most obvious in swamps at the lake edges in sheltered parts. Their biology in lakes is not fundamentally different from that already discussed in Chapter 5 for floodplain swamps. The distinction that lakes have is that communities limited to submerged plants penetrate the deeper water. There is frequently a zonation of different species determined by the changing light climate, the nature of the sediment and other factors.

There is thus an array of littoral communities, each intergrading into one another as the combinations particularly of wave action and light penetration steadily change. Figure 8.1 summarizes the main combinations, and in subsequent sections, particular attention is given to the submerged plant beds, communities of bare rock, and the profundal benthos. Finally the

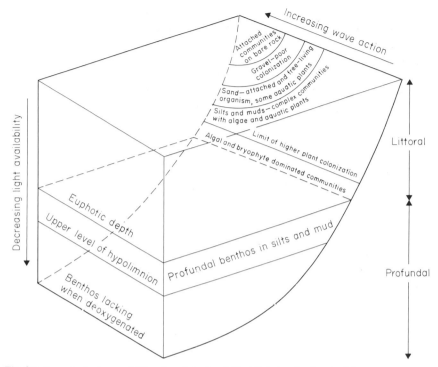

Fig. 8.1. A general scheme of littoral and profundal habitats in a freshwater lake.

mutual effects of the littoral zone and the open water on each other and of the open water communities on the profundal benthos will be discussed.

8.2 SUBMERGED PLANT COMMUNITIES IN LAKES

Below the extreme edges of a lake where exposure to waves may prevent submerged plant colonization, there is often a more or less marked zonation with depth. Vascular plants do not penetrate to beyond about 11 m and reasons for this were discussed in Chapter 6. For example, in Lake Tahoe (California–Nevada), aquatic mosses and stoneworts (*Chara*, *Nitella*) are found down to 164 m and 64 m, respectively [231] whilst vascular plants reach only 6.5 m. Light availability is likely to be an important cause of this division [762] as it is of zonation of individual vascular species with depth. Photosynthesis and problems for it in aquatic plants were discussed in Chapter 5.

The overall depth of colonization of plants (both vascular and non-vascular) is often inversely related to the extinction coefficient of the most penetrative light in a rather simple way. Spence [761] has demonstrated this for a series of Scottish lochs (Fig. 8.2). Except for cases where much inorganic turbidity (from glacial melt water, or soil erosion for instance) or organic colour (in peaty catchments) is present in the water, the depth of colonization is also inversely related to the phytoplankton standing crop and hence to the nutrient loading. The most fertile lakes, with dense phytoplankton populations, are thus likely to have rather restricted submerged plant communities, though emergent vegetation may be prolific at the margins.

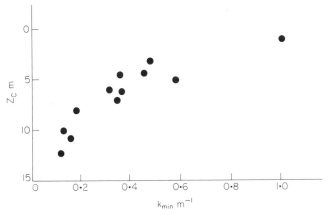

Fig. 8.2. Relationship between maximum depth of colonization, Z_c, of aquatic plants and the minimum extinction coefficient of light, k_{min} (solid symbols), in a series of lakes. (Based on data in Spence [761].)

A marked zonation of submerged plants is commonly encountered on shelving shorelines. This is at least partly controlled by the photosynthetic abilities of the species concerned [706]. Spence and Chrystal [763, 764] studied a sequence of *Potamogeton* (pond weed) species in a series of Scottish lochs. They found a distinct range for each species (Table 8.1) (the light regimes of the various lochs were not greatly dissimilar). Leaves of each species were grown at high (7.08 cal cm^{-2} hour^{-1}) light intensities under standard conditions in a glasshouse and their rates of photosynthesis

Table 8.1. Depths of colonization of some *Potamogeton* species in Scottish lochs and the extents to which their photosynthetic rates were reduced on transfer from high light intensity (7.08 cal cm^{-2} hour^{-1}) to low light intensity (3.29 cal cm^{-2} hour^{-1})

Species	Mean depth of colonization (cm)	Percentage decrease in photosynthesis
Potamogeton polygonifolius	9	42
P. filiformis	50	59
P. x zizii	120	24
P. obtusifolius	125	2
P. praelongus	190	10

measured as oxygen evolution. The leaves were then placed in dimmer light (3.29 cal cm^{-2} hour^{-1}) and their photosynthetic rates measured again. Those species which normally grew in shallow water photosynthesized in low light intensities at only a fraction (Table 8.1) of their rates at high intensity, while those normally found at depth maintained proportionately high photosynthetic rates even at low light intensities. The mechanism by which growth is maintained at low intensity was not examined, but in a macroscopic alga, *Hydrodictyon africanum*, which normally grows in low light intensities, the mechanism appears to be one of reducing respiration through the restriction of some energy-demanding syntheses [622]. One of these is the production of a photosynthetic enzyme, ribulose diphosphate carboxylase, which is a rather inefficient enzyme but forms a large proportion of the protein content of photosynthetic tissues. In restricting its production of this enzyme, *H. africanum* may reduce its own gross photosynthesis but evidently increases the difference between gross photosynthesis and respiration allowing net growth at very low light intensities.

Other mechanisms, including competition, the nature of the rooting substratum and the behaviour of seeds on dispersal also affect the depth distribution of submerged plants. Seeds of two species of *Potamogeton* in African lakes float for different lengths of time after release. This seems to favour greater depth colonization of the one which floats longest before its lacuna-filled seed coat becomes rotten and waterlogged and the seed sinks to the sediments [765].

The plant zonation is rarely regular; patchiness frequently occurs and local disturbance is probably one reason for this. Fishes may clear patches of sediment as nests for spawning. When the nests are abandoned for the year when the young have hatched the patches may be recolonized by spore or seed-propagating species (e.g. *Isoetes* and *Elatine*) which cannot compete well with rhizomatous species. Continual disturbance of the nest sites allows these species to persist, but if the nest sites are permanently abandoned by the fish, the rhizomatous species eventually invade them [93].

8.2.1 Microbial and animal communities in plant beds

On submerged leaf surfaces is a community of bacteria, fungi, algae and protozoa (Fig. 8.3) called the epiphyton or periphyton. It is difficult to study but poses some interesting problems.

As a new leaf unfolds on a water plant, bacteria suspended in the water are driven against it and some may stick and grow to form colonies. A variety of approaches to measuring the elementary properties (e.g. growth rate) of such populations may be used [21]. One approach is to measure the uptake of radioactive sulphate. Sulphur is needed for synthesis of some amino-acids needed for growth. From a knowledge of the sulphur content of the bacterial

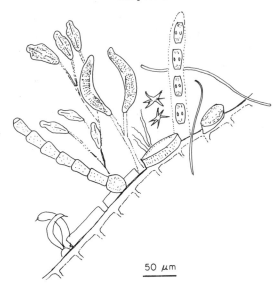

50 μm

Fig. 8.3. A typical epiphyte community.

cells and the rate of sulphur uptake, the growth rate can be inferred from the ratio of uptake per cell per unit time to content per cell.

A second approach is to measure the change in bacterial numbers on newly produced fronds [361]. The number of bacteria can be determined by staining with phenolic aniline blue in a sequence of fronds of known ages. Direct counting in this way usually gives much higher estimates (sometimes 10–20 times) of the number of cells than other methods based on growing the bacteria to form countable colonies. These methods give underestimates probably because not all cells will grow in the laboratory media. On the other hand some of the cells counted directly might be unable to divide and such direct counts may give overestimates.

Just learning the basic features of bacterial growth in a complex habitat is thus difficult. Even then, the possibility of many different species of bacteria being present and specializing on particular organic compounds must be admitted. Hossell and Baker [362] tentatively identified *Pseudomonas, Flavobacterium, Acinetobacter, Moraxella, Xanthomonas, Aeromonas, Alcaligenes, Agrobacterium, Cytophaga,* Enterobacteria and other Gram-negative and Gram-positive forms from the surfaces of *Ranunculus penicillatus*. Different plants almost certainly have different species—they certainly have different population sizes, and the populations may differ in different parts of the plant. In the submerged parts of watercress (*Nasturtium officinale*) and brooklime (*Veronica beccabunga*) there were larger populations on the undersides of the leaves than on the upper in one study [23]. The sources of the organic matter on which they feed are not properly known either. The concentrations of labile organic matter in the water moving over the plant bed are generally low, but the high concentrations of bacteria on the leaf surfaces (perhaps 10^5-10^7 cm^{-2}, or 10^8-10^{10} ml^{-1}, assuming a layer 5 μm thick to allow calculation on a volume basis for comparison with the populations (10^4-10^5 ml^{-1}) in the water, suggest a source from the plants themselves. Plants do secrete organic matter under experimental conditions (see below) and ooze it if mechanically damaged by waves or animals. Colonizing bacteria may also help themselves by causing changes in the epidermal cells of the host, for example internal swellings of the wall [690] and erosion of the cuticle [368], and may herald the eventual senescence and decomposition of the plant.

8.2.2 Epiphytic algae

Aquatic plants, like other stable submerged surfaces receiving enough light, become colonized by dense communities of photosynthetic algae as well as by bacteria and protozoa [4]. Tangled perhaps in deposited calcium carbonate (marl) and mucus secretions of the bacteria, these epiphytic communities have a substratum which is not inert. Aquatic plants secrete organic compounds [626, 858]. Such substances can readily be taken up by some algae as well as by bacteria. Allen [7] has demonstrated this with an ingenious apparatus of compartments, separated by fine but permeable cellulose ester filters and separately containing ^{14}C labelled axenic *Najas flexilis*, an aquatic plant, and representative axenic cultures of bacteria and algae. The extent to which this secretion may occur in natural situations is hard to estimate. Allen [7] has shown that ^{14}CO$_2$ supplied in polyethylene bags sealed over the flowering stem above the water surface to naturally-rooted *Scirpus subterminalis* in Lawrence L., Michigan, does appear in organic form in water close to plants whose epiphytes had been gently scraped off, or in the epiphytes themselves. In the former case, labelled organic matter could still be detected up to 3 m away from the plants after 2 hours.

Provision of such organic matter must help stimulate the growth of the epiphytic algal and bacterial community, which in turn absorbs much of the light that would otherwise be available to a submerged plant thus retarding its growth [635, 708, 711]. In a study of *Lobelia dortmanna*, epiphytes cut off 67–82% of the light available to the plants and changed the spectral nature of the light penetrating to the leaf surface. Sand-Jensen and Borum [708] calculated that the plant, presently growing at 1 m depth or above, could have penetrated to 3.5 m without its epiphytes.

It seems surprising therefore that plants should not, in the course of their evolution, have acquired chemical or physical means of restricting epiphyte colonization. Some large, filamentous algae, such as *Spirogyra* and *Zygnema*, produce very mucilaginous outer cell walls, which almost entirely prevent epiphyte colonization, but these are the exceptions. Vascular plants, mosses and liverworts, and to a lesser extent, stoneworts, are often thickly covered. Hutchinson [382] has put forward the idea that an epiphyte community may be advantageous in that it diverts the activity of invertebrate grazers from the host plant itself. The plants, however, do produce alkaloids and other substances whose taste may discourage the grazers [614].

The epiphyte community includes most phyla of algae, though diatoms are often predominant. Some are attached by stalks (e.g. *Meridion, Gomphonema*) or mucilage pads, or (*Cocconeis, Achnanthes*) by the mucilage secreted through the raphe, a long slit in the cell wall, which can also be an organelle of movement. Some species may not be directly attached but may be embedded in marl or mucilage or may crawl over the surfaces formed by the attached ones (*Nitzschia, Navicula*). As the community develops, and the weed-bed becomes denser, skeins of filamentous algae may stretch from plant to plant and leaf to leaf and in the quiet waters enclosed by this underwater forest, shoals of flagellates move. Chudbya [113] found 220 different algal species associated as epiphytes with one large algal species, the blanket weed *Cladophora glomerata* in Poland.

The community associated with plants, containing not only firmly attached, but also loosely fixed and free-living organisms is easily disturbed and thus must be sampled delicately. The complex geometry of aquatic plant surfaces—often finely dissected leaves—also confounds attempts to express the epiphyte activity per unit area of host plant. These problems have stimulated the use of artificial substrata (glass or perspex slides, plastic netting or plastic aquarium plants, placed in the water) in experiments on epiphytes. Although these substrata are much more convenient than the living plants to work with, they might not reproduce the metabolic effects of the real host. People using them have thus sought evidence to show that such metabolic effects are negligible, and that the plants are simply platforms for growth, replaceable by the artificial substrata.

A study made largely on *Potamogeton Richardsonii* in Lake Memphrema-gog, on the US–Canadian border, used plastic aquarium plants to mimic the live ones [100]. No significant differences between the two were found for the biomass of the epiphytes per unit area (measured as the chlorophyll *a* concentration of cells that could be shaken from the surfaces) or the rate of photosynthetic uptake of carbon-14 labelled CO_2 per unit area, or per unit chlorophyll *a*. This latter is one measure of the physiological status of the epiphyte cells and might be expected to be affected by the plants' activity. It suggests that the living plant might be a neutral substratum.

The activity of an enzyme, alkaline phosphatase, in the epiphyte communities was also measured, however. Alkaline phosphatase breaks down organically linked phosphate and is produced when easily available inorganic phosphate is scarce. The epiphytes on plastic plants in L. Memphremagog showed higher activity of alkaline phosphatase than those on the live plants, suggesting that the latter were providing phosphate for their epiphytes. A later study [90] in the lake, however, measured the release of phosphate from nine species of aquatic plants, which were grown in the presence of the isotope phosphorus-32 in aquaria, and then replaced in the lake. The authors use the word 'only'—others might say 'as much as'—3.4–9.0% of the P taken up by the epiphytes came from the host, and there was also some release of phosphorus to the water. Over 90% of this came from the epiphytes rather than the plant.

The Lake Memphremagog studies did not seek information on the taxonomic composition of the epiphytic algal communities. Such information can also contribute to the controversy about the inertness or otherwise of the host surfaces. Use of artificial surfaces implies the hope that the colonizing community will be similar to the natural one in composition.

Such hopes are never completely realized; in different situations quite different results may be obtained in test comparisons. Some reveal evidence of determination of the epiphyte community by the plant itself [609, 655]. Eminson and Moss [201], for example, found very different communities on four different plants growing close to one another in the same lake. Tippett [808] found considerably fewer species grown on glass slides than on adjacent plants in a small pond and maximum abundance was found at different times of the year on the different substrata. Thirdly, Fitzgerald [221] found that the nitrogen status of a host filamentous alga, *Cladophora*, directly affected the degree of epiphyte colonization. The more nitrogen it had, the more epiphytes.

Despite this evidence of an important effect of the plant on its epiphytes many comparisons have shown that communities developing on submerged glass slides are often similar, at least in their more abundant species, to those on nearby plants [382]. This particular issue will never finally be resolved in simple terms, for natural phenomena almost never have simple all-embracing explanations. The nature of the external environment, for example, may be more influential in some circumstances than others. A highly fertile water may supply so much nutrient to the epiphytes that any source from the plant becomes insignificant in determining either production [101] or the specificity of the algal community [201, 562]. In an infertile water the nature of the plant host may be far more important.

8.2.3 Invertebrates

Invertebrates, including many Protozoa [25a, 26] abound in submerged plant beds. They comprise most of the invertebrate groups found in fresh waters but simple species lists do little to explain the nature of the community because they do not tell of the distribution of the animals one to another and to the plant superstructure.

Most sampling of plant beds for invertebrates might be likened to the manoeuvring of some huge excavator over a city. Its jaws would enclose, for later sorting, the displaced office workers, ice-cream vendors, shop assistants, teachers and students, hairdressers and many others among the rubble of high rise office blocks, street furniture, department stores, schools and beauty parlours. Little can be deduced of the life of the city from this. Commonly used such samplers include nets, grabs and washings from artificial substrata.

A coarse-weave net swept through the beds misses many of the smaller (<2 mm) Crustaceans which move among the plants, while a fine net creates so much resistance that many animals can move out of the way. Any net is highly selective—animals like leeches which cling to stems and leaves tend to be underrepresented, while those that normally move around a good deal are overemphasized.

Dredges like the Birge–Ekman dredge sample an acceptably constant area of plant-bed by compressing the plants and then cutting them against the bottom between powerful spring-loaded jaws when these are released by operation of a catch. They are likely to underestimate animals which can move away during descent of the grab. Macan [492] has invented a rotary cutter comprising two concentric cylinders with teeth at their lower edges which can move against one another when the cylinders are counter-rotated by the observer. The device is gently lowered into the vegetation which it compresses and the teeth operated when it reaches the bottom. A cylinder-shaped section of vegetation is thus cut out and is retained in the cylinder by a plug of peat or sediment at the bottom and an airtight disc screwed to the top. Even with this sampler, however, there must be considerable disturbance. The mere emptying of it destroys the spatial relationships of the several plant and many animal species which are part of the under-waterscape.

Heterogeneity in plant-beds is a major problem for quantitative sampling and many replicate samples must be taken for a statistically acceptable estimate of the population of any of the organisms to be obtained. In an attempt to avoid this problem Macan and Kitching [495] have used 'artificial vegetation' made of polypropylene rope woven into a coarse mesh base weighted with flat stones, to study the weed fauna of a small lake in northern England (Hodson's Tarn). The polypropylene rope is less dense than water and floats upright, giving, with 8 cm strands, a reasonable approximation to the physical structure of the rosette-like *Littorella uniflora*, and, with 45 cm lengths, to *Carex* which fringes much of the tarn. The mats of artificial vegetation are of such a size as to be retrievable with a pond net inserted under them, and the animals can be washed out into a bucket. Although there was some bias in the composition of the communities contained in the artificial '*Carex*' compared with those in the natural vegetation, there was sufficient similarity to make this a useful technique, and it allowed demonstration of some valuable facts.

First, the density of artificial leaves determined the sizes of many animal populations, the total number of animals increasing markedly with thickness of vegetation. Secondly, the length of the 'leaves' was important. In an experiment with '*Carex*' where different lengths were used, larvae of *Leptophlebia, Cloeon* (Ephemeroptera) and *Gammarus* (Crustacea) were much less abundant when the 'leaves' were shortened from 48 to 8 cm, thus indicating a preference by these animals for the upper parts of the long leaves and hinting at the distributional complexity which other samplers destroy.

The problem experienced with epiphytic algae in using artificial substrata—of an inert versus a living platform—may also apply to the colonization of the invertebrates. Again some studies [759] have found few difficulties, whilst many ecologists will be doubly nervous of the use of artificial substrata, especially where invertebrates grazing on the periphyton are concerned.

One compromise for the sampling problem is to take a large number of small samples of the invertebrates and plants and to take them gently by enclosing small plastic boxes around the plant. A relationship between numbers of animals and amount of plant can be determined by regression analysis and used to predict the overall population from separate measurements of the biomass of the plants taken with larger, conventional grab samplers. Downing [174] found this approach to give higher estimates for the populations of the more active animals (mites, amphipods, Cladocera, copepods, ostracods and caddis larvae) than grab or net sampling.

The diets of weed-bed animals range from the periphyton to other invertebrates. Many organisms seem to use periphyton, which, because it comprises a mixture of epiphytic algae, bacteria, protozoa and detritus, must provide a rich diet for these omnivores. It seems to provide a food at least twice as rich in amino acids as underlying sediment. Chironomid larvae are encouraged to move from their winter sedimentary habitat to feed on stems coated with periphyton in spring [524]. Other periphyton feeders include mayfly larvae (*Leptophlebia vespertina* and *Cloeon dipterum*) freshwater shrimps (*Gammarus pulex*), snails and caddis-fly larvae (Trichoptera) all of which normally cling to stems and leaves.

In one Oxfordshire pond, Lodge [472] found allochthonous leaf litter to have few snails, but the aquatic plants to be associated with particular snail species. *Acroloxus lacustris*, a limpet, was most common on the smooth stems of emergent plants and the petioles of white water lilies (*Nymphaea alba*). *Planorbis vortex* was associated with aquatic grasses like *Glyceria*, and *Lymnaea peregra* with *Elodea canadensis*. For the latter two combinations, experiments showed [473] a particular preference of the snails for the epiphytes specific to the plants: filamentous green algae on *Elodea*, diatoms and associated detritus on *Glyceria*. Grazing is very active, and the turnover rate of the periphyton food may be as much as 22–45% per day [421].

Partly clinging, partly swimming and sometimes filtering detached fine material is a great variety of small Crustacea (cladocerans, cyclopoid and harpacticoid copepods, ostracods). Some larger feeders on fine particles, such as sponges (Porifera), are found attached to the plants.

Carnivores in the weed-beds may conveniently be divided into lurkers and hunters though the categories may not be distinct. The former attach themselves to plants or remain stationary among the leaves waiting for prey to pass near. *Hydra*, a small (2–3 mm) coelenterate, extends tentacles armed with stinging cells with which it is able to immobilize small prey like small Crustacea, which accidentally brush against them. Leeches attach by means of a basal sucker and, depending on the species, attack either passing invertebrates, water birds or fish, to which they may attach themselves for some time, loosening themselves from the vegetation to do so. Other lurkers include the larvae of damsel- and dragon-flies (Odonata) which are fully mobile but which also wait for their prey. They have a hinged lower mouthpart or labrum which is folded back under the head, but which can be snapped forward to seize the prey with two teeth borne at its tip. Other lurking carnivores include the larvae of the alder fly *Sialis* (Megaloptera) and a stonefly (*Nemoura cinerea*), though stoneflies, as their name suggests, in general are unusual in weed beds. Yet others include some caddis-fly larvae such as *Phryganea*, whose heavy cases, made from pieces of reed leaves or roots, hamper movement and ensure that only slow-moving prey, or eggs laid on the vegetation are taken. Members of another caddis family, the Polycentropidae, weave funnel-shaped nets among the leaves and stems and lie in wait at the apex for the prey to become entangled.

Lurking allows a predator to remain concealed, not only from its prey, but also from its own predators, such as fish, but has the disadvantage that prey organisms may very often not pass sufficiently close. Hunting reverses these features, giving greater exposure to predators but readier availability of prey. The quieter water often to be found among emergent vegetation allows those hunting insects which live on the surface tension film of water to build up their populations. These include various bugs (Hemiptera) such as the water strider (*Gerris*), and the water measurer (*Hydrometra*) which has a long head with five stylets which it uses to spear small Crustaceans swimming just below the surface. The water boatman, or backswimmer (*Notonecta*), in common with other of the surface hunters, feeds on insects which become trapped in the surface tension film upon alighting on the water. It hangs from the surface tension film by the tip of its abdomen, moves with paddle-like hind legs and seizes its prey with its front and middle legs before impaling it on its sharp, beak-like mouthparts.

The water beetles (Coleoptera), both as adults and as larvae, are voracious underwater hunters; the adults are able to replenish bubbles of air underneath their wing covers, at the water surface. With this rich supply of oxygen their activity is enhanced. (Those carnivores confined to lurking are unable to use atmospheric air and are dependent on the more restricted supply of dissolved oxygen.) There are larger predators too. Introduction of trout to a small tarn resulted in a reduction in numbers of water bugs and a confinement of some species to the thicker parts of the vegetation [493]. Two further species, however, were able to increase their numbers in the presence of fish, perhaps as a consequence of reduced competition among the smaller community of water bugs.

8.3 COMPETITION BETWEEN SUBMERGED PLANTS AND PHYTOPLANKTON

The vulnerability of submerged plants to the availability of light means that as overlying phytoplankton crops tend to increase, through eutrophication, aquatic plants may become shaded out and restricted in their distribution. In extreme cases they may disappear altogether, but it is not altogether clear that shading is always the factor responsible. Indeed the competition between plants and plankton is not a simple matter.

Infertile lakes often have aquatic plants of short stature. These include *Isoetes* and *Littorella* where catchments are of hard rock and the sediments sandy, and Charophytes and *Naias marina* in highly calcareous marl lakes in which phosphate is scarce and bound in the limey sediment. On eutrophication, greater crops of aquatic plants tend to grow at first, but not of the short-statured forms. The larger pondweeds (*Potamogeton, Myriophyllum, Hippuris, Ceratophyllum*) tend to be the major beneficiaries. Increased nutrient supply may stimulate growth of filamentous algae and epiphytes over the surfaces of the shorter plants [635], thus reducing their light supply and decreasing their competitive abilities. The taller plants may need greater nutrient supply [108] and may also have greater access to nutrients in the water than the shorter forms because of their usually greater surface area to volume ratio, though this probably does not apply to charophytes. There is a suggestion that phosphate *per se* above a low critical concentration may be toxic to some of these.

The taller plants can usually grow rapidly to the water surface, obviating the shading effects of moderate algal populations and if heavy epiphyte growths develop on their older leaves the leaves are readily sloughed off and replaced. Frequently however, despite increased nutrient loading, phytoplankton populations do not grow at first to a large extent in small lakes where aquatic plants cover large areas.

The reason for this may be to do with competitive mechanisms between the plants and phytoplankton. Fertilization of enclosures or experimental ponds containing aquatic plants has resulted in little change with no increase in phytoplankton, though sometimes some of filamentous algae [364, 568]. The plants may take up the extra nutrient as 'luxury consumption' or algal development may be prevented by grazers. The plant beds harbour a large collection of Cladocerans, some of which may move out into the open water at night to graze [807] others of which may filter water it is circulated through the beds by wind. The beds may provide dark refuges in which fish predation on the grazers is less efficient, allowing large stocks of grazers to build up (see Chapter 7). Additionally the beds at their densest may exclude large animals like fish because oxygen concentrations within them may become quite low. The mass of organic matter and reduced mixing allow stratification to establish often with high ammonia concentrations and near deoxygenation at the bottom.

Aquatic plants may also secrete inhibitors which suppress phytoplankton growth [828, 871], though through deoxygenation at the sediment surface they may also cause release of nutrients [569] and stimulate phytoplankton growth. In one case, *Myriophyllum spicatum* stimulated phytoplankton growth by an unknown mechanism [277].

However, it does appear that, once established, an aquatic plant bed has considerable buffering powers, preserving it against changing nutrient loading at least up to some critical point. Although the evidence is anecdotal, the change from plant-dominance to phytoplankton dominance which sometimes occurs during eutrophication may take place relatively rapidly. It is as if a switch had been thrown moving the ecosystem from one alternative state to another with little or no change in external conditions.

Once this change has occurred and the aquatic plants have been replaced by phytoplankton which is able then to take advantage of the abundant nutrients, a new set of stabilizing buffers comes into operation. Phytoplankton can grow earlier in the year, at lower temperatures than the plants and although circulated through the water column it on average receives more light than aquatic plants developing on the bottom. The light compensation point (gross photosynthesis = respiration) for small cells is lower than for bulky organisms as also is the CO_2 compensation point. Diffusion precludes rapid uptake into bulky tissues, so that once established the algae may compete very effectively for CO_2 [5, 488]. Uptake of CO_2, as it forces the pH to high values (9 and above) may accentuate the problem, though there may be limitation of photosynthesis of some aquatic plants at pH values as low as 7 [742].

Open water, particularly in shallow lakes, offers few refuges to large Cladoceran grazers from fish predation. Deeper lakes may have refuges in the darker depths, but where aquatic plants would otherwise be able to grow abundantly (less than a few metres) fish predation is likely to be possible throughout the water column. Once the switch to phytoplankton dominance has taken place, therefore, there is less possibility of grazer control of the algal crops in shallow water.

The problem then becomes one of what happens to operate the switch. It seems that aquatic plants can withstand severe eutrophication by operation of the buffers inherent in the community. It is not known whether they can hold out indefinitely, but experimental attempts to cause the switch to phytoplankton by fertilization alone have usually failed [364, 568]. Sometimes attempts to change phytoplankton dominance to plant dominance by reducing nutrient supplies in Lund Tubes (Chapter 6) have also failed [570].

For the forward switch, it seems that many of the buffers should operate if the plant community is intact—luxury uptake of nutrients and secretion of organic inhibitors, for example. But it may be that effects independent of eutrophication, but correlated with it, may destroy the grazer community for long enough for the phytoplankton to gain advantage. Eutrophication through intensification of agriculture is linked also with greater use of herbicides and pesticides. The Cladocera seem particularly vulnerable to very low levels of some of the latter [274, 735]. In the Norfolk Broadland, a switch to phytoplankton took place generally in the 1950s and 1960s when organochlorine pesticides were used with less control than at present, and there is some evidence of a correlation between pesticide residues and changes in the Cladocera community recorded in the underlying sediments. Even if pesticide use is restricted subsequent to the forward switch, the buffers inherent in the open water community may prevent a switch back, for efficient grazers may be unable to build up large enough populations, because of fish predation, to limit algal development.

8.3.1 Consequences of loss of aquatic plants

The littoral zone communities have important roles in lakes. Loss of aquatic plants removes these functions. Recruitment of fishes like perch (*Perca fluviatilis*) and tench (*Tinca tinca*) which usually use aquatic plants on which to lay their eggs may be limited; populations of large invertebrates—snails and dragonfly larvae for example—used by larger fish, may dwindle to the detriment of fish growth, and both herbivorous and piscivorous birds may disappear.

The herbivorous coot (*Fulica atra*) was once so abundant on Hickling Broad in Norfolk that thousands were shot annually [570]. With the loss of extensive plant beds, the population now numbers only a few hundred. Mute swans (*Cygnus olor*) have also become less numerous in many places where plants have disappeared. Their loss may be linked with the high incidence of lead poisoning recently discovered in swans [288] but this may be only the proximal reason for their deaths. The birds ingest discarded lead shot from anglers. It readily dissolves in the crop and gut, but lead shot has been a feature of river and lake edges for many years without major outbreaks of poisoning. It may be that the birds are more likely to pick up the shot as a consequence of loss of aquatic plants through eutrophication. They normally swallow small stones to aid grinding of their plant food in the crop. Eutrophication tends to be linked with increased sedimentation (from algal growth and erosion in the catchment) so that gravel beds are covered up and natural stones are less accessible. If their aquatic plant food supply is also lacking they may be driven to seek food where artificial supplies (e.g. bread thrown to them) are available. These are the areas accessible to picnicking people and to anglers. And so the chances of their ingesting shot may be increased.

Attempts to solve the problem have generally centred on a banning of lead shot and its replacement by tungsten substitutes which are less poisonous. If the interpretation above is correct, the problem is being handled by treatment of a symptom rather than of an ultimate cause.

8.4 BARE ROCKS AND SANDY LITTORAL HABITATS

The movement of waves on rocky lake shores creates a habitat superficially similar to that of rocky streams (Chapter 4). Inspection of fauna lists from such shores, however, suggests that the processing of allochthonous organic matter is not nearly so important on them as it is in streams. The major food base is the epilithic community of algae, bacteria and protozoa which covers the stones. Scrapers and predators dominate the invertebrate fauna. This is to be expected since wave action will disturb any loose organic matter arriving from elsewhere and tend to move it, under the action of gravity, down the slope of the lake shore and basin to the sediments at greater depths. A typical fauna list from a rocky lake shore might include nymphs of various Ephemeroptera (mayflies), Plecoptera (stoneflies), triclads (flatworms), gastropods and limpets, *Asellus, Gammarus* and other Crustacea, Trichoptera (caddis) larvae and others. Of course such a shore is never uniform and protected pockets, where sediment may accumulate, leaves lodge or aquatic plants grow, will have a somewhat different fauna characteristic of such conditions.

The rocky shore habitat has been most intensively examined from the viewpoint of what determines the composition of the community of invertebrates. It may best be used here for discussion of how physico-chemical factors, chance, predation and competition all combine to determine the species list and the relative proportions of organisms in it. Firstly, distribution of one group of animals, the triclads, or carnivorous flatworms, will be outlined.

8.4.1 **Distribution of triclads in the British Isles**

Flatworms occur in weed-beds as well as on rocky shores, but are certainly characteristic of the latter. There are about eleven species in Britain but four of them, *Polycelis tenuis, P. nigra, Dugesia polychroa* (then referred to as *D. lugubris*) and *Dendrocoelum lacteum*, have been very well studied by T.B. Reynoldson [674, 675] and his students. The first task was to describe the distribution of the four species in the British Isles. This was carried out with a survey of over 200 lakes through Britain which had rocky shores. Flatworms were sought by hand on the shores over a timed period necessary to collect 50–100 animals. This relative quantitative method is the best available to date, for flatworms are delicate, readily damaged, and not easily dislodged by kick sampling. The pattern of distribution found was as follows:

(i) the total number of flatworms collected per unit time increased directly with calcium and total dissolved solids concentrations in the water;
(ii) at calcium concentrations lower than 20 mg l^{-1}, *D. polychroa* and *D. lacteum* declined and were usually absent in water with less than 5 mg Ca^{2+} l^{-1};
(iii) at low calcium concentrations above 5 mg l^{-1}, *P. tenuis* was most abundant, but below 5 mg l^{-1}, if flatworms were present at all, *P. nigra* was either predominant or the sole flatworm species present.

To some extent this pattern reflects dispersal of the animals and recolonization of the lakes after the retreat of the last glaciation. Under experimental conditions, with abundant food (slices of earthworm) supplied, waters of the lowest calcium and total dissolved solids concentrations supported breeding populations of all four species indefinitely. Lakes poorest in calcium tend to be at the furthest northern reaches of the British Isles, and in northern Scotland and the offshore islands more lakes than expected had no triclads at all. Part of the triclad distribution pattern at low calcium levels may thus simply reflect a lack of time for colonization to have taken place.

Most of the pattern must be explained, however, in other terms, particularly the change from *P. tenuis* predominance to that of *Dugesia* and *Dendrocoelum* in lakes where all four species coexist. Predation on triclads by other organisms was found to be unimportant on rocky shores [150]. Of 75 potential predators tested by serological methods, only two rocky shore animals, a leech, *Erpobdella octoculata*, and a caddis larva, *Polycentropus flavomaculatus*, ate flatworms, but were shown not to eat enough to alter greatly the population size or balance of species.

Increasing the food supply led to population increases of caged flatworms far greater than those of natural populations in the same waters and it seemed that competition for food might explain part of the distribution pattern. Calcium and total dissolved solids concentrations, though not directly and causatively implicated in the distribution, are general indices of the fertility of most lakes (see Chapter 3). The production of potential prey organisms and therefore flatworms is likely to be correlated indirectly with levels of these substances as a reflection of increasing fertility of a lake. This leaves the question of the change in balance of species in lakes of increasing calcium concentration.

All four flatworms will take a range of foods, but particular and different items seem necessary for indefinite survival of each species. Serological work, examination of gut contents and choice experiments where each species was

presented with a range of foods have shown that *Dendrocoelum* prefers the
crustacean *Asellus*, which it actively hunts. Of the four species it has the best
developed sensory systems and will attach to and coil around a moving nylon
bristle the same diameter as an *Asellus* leg [44].

Dugesia polychroa eats gastropods, which are spurned by the others in
most circumstances. It was thought that the *Polycelis* species specialized on
oligochaetes, each taking worms from different oligochaete families, but
further work has shown that this is unlikely and that *P. tenuis* takes damaged
Asellus, attracted to it by oozing body fluids. *P. tenuis* is the most active of
the four, even at low temperatures, and such activity is a necessary
behavioural trait for a predator which must search for stationary prey. The
main food of *Polycelis nigra* has not yet been determined. In laboratory
experiments the 'food refuge' of each species had to be provided for indefinite
survival either in mixed or monospecific cultures, but *P. nigra* always declined
whatever food was given. In natural waters, introduction of *P. tenuis* to
locations where *P. nigra* had previously been the sole flatworm species had
led to decline (though not disappearance) of *P. nigra* in favour of *P. tenuis*
[676].

In lakes of increasing fertility the proportion and abundance of *Asellus*
and of gastropods on rocky shores often increases and this seems to explain
the coexistence of *Dugesia* and *Dendrocoelum* with *Polycelis* spp. at the higher
calcium levels.

The picture given by Reynoldson and his co-workers also goes some way
to explaining the changes in rocky shore fauna noted by Macan [491] within
L. Windermere and in a series of lakes of the English Lake District (Chapter
3). Macan listed about 36 species of macro-invertebrates on the rocky shore
of L. Windermere, of which 19 were commonly found. The total list included
mayfly and stonefly nymphs, caddis larvae, beetles, Crustacea, such as
Gammarus pulex, Crangonyx pseudogracilis and two *Asellus* species, two
gastropods, *Ancylus fluviatilis* (freshwater limpet), six flatworm species,
including all those discussed above and nine leech species. Their distribution
was not uniform among a large number of areas sampled around the lake
shore—two extreme sorts of community were found linked by a continuum
of intermediate ones. At one extreme the fauna had relatively few insects,
but molluscs, Crustaceans and flatworms were prominent. At the other,
certain stonefly nymphs (e.g. *Diura bicaudata, Nemoura avicularis*) and mayfly
nymphs (*Ecdyonurus dispar, Centroptilum luteolum, Heptagenia lateralis*) were
confined. All other species were found throughout the series of sites. The
former community type seemed to be associated with shores likely to have
been fertilized with effluent from two sewage treatment works and from
septic tank and houseboat effluent.

A rather similar pattern was recorded from a series of lakes in the same
area. Because of the varying geologies of their catchments and the greater
farming activity and human settlement in those of the naturally more fertile

areas, these lakes form a spectrum of overall lake fertility (Chapter 3) (Fig. 8.4). At the least fertile extreme, stonefly and mayfly nymphs were most abundant, while certain leeches, gastropods and flatworms reached their greatest abundance (based on a timed collection by hand and net) towards the fertile end.

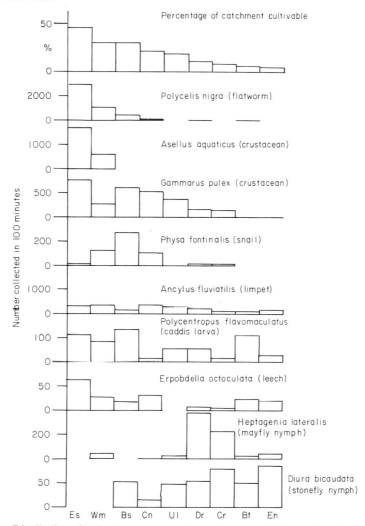

Fig. 8.4. Distribution of animals on comparable rocky shores of lakes of the English Lake District. Histograms show the number of animals of each of nine representative species collected by the same experienced observer in 100 minutes. A measure of the fertility of each lake is given by the percentage of its catchment area which is cultivable. Cultivability in this case largely reflects the availability of easily weathered rocks in the catchment. Es: Esthwaite, Wm: Windermere, Bs: Bassenthwaite, Cn: Coniston, Ul: Ullswater, Dr: Derwent Water, Cr: Crummock Water, Bt: Buttermere, En: Ennerdale Water. (Redrawn from data in Table 37 of Macan [491].)

A tentative explanation of these distributions concerns the interaction of fertility and predation [494]. As the fertility of a lake, or a station in it, increases, so does the productivity of the invertebrate community and a greater population of predators such as flatworms can be supported. There is no evidence *per se* that high fertility discriminates against nymphs of stoneflies and mayflies, and little evidence that they are major prey of the abundant flatworms. There is a possibility, however, that their eggs survive less well under crowded conditions than those of animals that are common on such shores. Mayflies and stoneflies deposit eggs which fall to the bottom and catch among the community of microorganisms on the stones. There they are vulnerable to the continual activity of the scrapers—molluscs and crustaceans—which continually work over the stones. In contrast, the eggs of molluscs are protected in lumps of jelly and those of leeches and flatworms in leathery cocoons; *Gammarus* and *Asellus* eggs are carried on the female's body until they hatch. Doubtless this is only a first level of explanation of what is really a very complex pattern, yet it illustrates some of the factors that must be considered if some pattern of predictive use is to be deduced from the mass of information on community composition that already exists in the literature.

8.4.2 Specialization in the rocky littoral

Again it should be emphasized that littoral habitats are far more complex than planktonic ones with an often fine division of niches. This can be illustrated by two examples.

The chydorids, a group of Cladocera [239, 240] includes animals specialized to the extent that a species capable of living on rock or leaf surfaces is unable to manage in sand or sediment only centimetres away. In general the chydorids are animals associated with surfaces, though many can swim as well. They move by using their first pair of trunk limbs, collect food with their second pair, and manipulate it with the third and fourth and sometimes fifth pairs so that it passes forward in the ventral food groove between the legs, to the mouth. The limbs are modified to handle particular foods: to scrape it, with the help of entangling excretions in *Alonopsis elongata*, and to sweep loosely bound material in *Disparalona* sp. The carapace is particularly important. In some species it can be clamped down on a surface so that the species with the widest carapace gape may be able to use it as a suction cup and move upside down on the underside of leaves (e.g. *Graptoleberis testudinaria*), whilst others with a flattened body and narrow gape may move more readily through soft deposits. One species is specialized as a parasite of *Hydra* and has armoured food grooves to obviate the effects of the latter's stinging cells, others feed as scavengers on dead Cladocera, or in the surface tension film. Some are very small (<0.25 mm) and have access to material in the finer crevices of rocks, whilst bigger species (> 5 mm) do not. Fryer [244] argues cogently against the combining of such disparate animals into categories for the purposes of erecting mathematical models to explain ecosystem processes. The argument is unassailable but generally ignored.

A second example concerns the fishes of L. Malawi. In L. Malawi there have evolved 'species-flocks'—groups of species which are very closely related, yet distinct. Many of them are related to the cichlid genus *Haplochromis* and similar flocks occur in L. Victoria and Tanganyika. The range of diet is not greater than elsewhere, but the specialities within the range are much narrower, presumably allowing a very efficient use of food resources. This can be illustrated by considering members of a species flock which feeds on blue-green and other algae which form a felt on rock surfaces [246]. The local African fishermen recognize, unwittingly, the close taxonomic relationships of these fish by calling all by one name, 'mbuna'.

One species, *Pseudotropheus tropheops* has small close-set teeth, like a small file, with an outer row of bicuspid tipped ones and inner rows of tricuspid teeth. It rasps the algal felt from vertical or steeply sloping rocks and bites off the rasped mass with large conical teeth at the edges of its mouth. In this way it removes even firmly attached filamentous forms. *P. zebra*, on the other hand, is larger and more mobile than *P. tropheops* and this may give some separation of niche to add to the subtle difference in diet. This species has similar outer teeth, but longer and more spaced tricuspids at the back. It opens its mouth against the rock and combs the algae so that the looser algae only are removed. A third species *P. livingstonii* has similar dentition and diet to *P. zebra*, but inhabits relatively deeper water.

A closely related genus contains *Petrotilapia tridentiger* whose teeth are long, curved and flattened at the tips into a spoon-like shape. The whole tooth is flexible. When the mouth is opened against the rock the teeth comb the looser algae, as do those of *Pseudotropheus zebra*, but *Petrotilapia tridentiger* can feed on both vertical and horizontal rock surfaces, 'standing on its head' to do so in the latter case. Lastly among this group of algae scrapers is *Labeotropheus fuelleborni* with its rigid, strong jaws strengthened in the plane of forward movement, and a mouth set on the lower side of the head. It hovers over horizontal rock surfaces then moves in to chisel the algae off with its teeth.

It would be difficult to find finer differences, based on attachment of the algae, angle of rock surface and water depth, than these, but rock-scraping fish in L. Malawi have also evolved into scrapers of other surfaces. *Genychromis mento* scrapes the body scales from a sluggish bottom-living fish, *Labeo cylindricus*. Though this is its main diet it perhaps belies its evolutionary origin by occasionally rasping epilithic algae, and points to future possibilities by biting pieces out of the fins of other fish. More specialized scale scrapers and fin biters are known also from the lake, and scale scraping is accompanied sometimes by a mimicry that allows the scraper to move unrecognized among shoals of its prey.

8.4.3 Sandy shores

Often, like gravels, sands have little fauna and flora—the continuous agitation of the sand grinds any colonizing organisms to death and washes away any fine organic particles that might act as food. The least disturbed, yet still sandy, habitats do have a characteristic set of communities, however, with microscopic algae and bacteria living freely among the sand grains and attached to them. There is also an interstitial fauna of Protozoa, nematodes, small Crustacea, oligochaetes, tardigrades (waterbears) and mites.

Epipsammic algae are sometimes of non-mobile genera (e.g. *Fragilaria* spp.) and sometimes of genera which can attach firmly but reversibly to the grains and also move freely among them. The sorts of sandy beach which have large populations of such algae (largely diatoms) are generally not continuously disturbed by water movements but mixing of the grains occurs from time to time. Diatoms attached to sand grains may thus be buried to depths of several centimetres where no light penetrates. The motile ones can then detach themselves and move back to the surface, but the permanently affixed must either die or tolerate darkness for several days, perhaps weeks. There is no evidence of much heterotrophic ability [577], but some that they can tolerate both darkness and deoxygenation for several days, while retaining their photosynthetic potential [561]. In contrast, epipelic algae (those that are always free-living and move in the surface layers of sand and other sediments) although tolerant of darkness if buried, are less resistant to anaerobiosis. They survive by rapid movement back to the surface. Deoxygenation is usual, even in sands at depths of more than a few centimetres and in finer sediments below a few millimetres.

As the sands grade into finer deposits under stiller conditions, the community of epipelic algae becomes more important as also does the supply of organic matter deposited from elsewhere. The community may be very rich in species—several hundred epipelic algae and equally large numbers of small invertebrates including the protozoa, rotifers, nematodes, oligochaetes and Crustacea. Strayer [780] found 322 species of animals in the sediments of one small N. American lake and suspected that the total was 600. The microbial community is also complex for a variety of redox conditions exist, not only with depth in the sediment, but also around decaying pieces of organic matter like leaves, animal bodies, or faeces. Such habitats have within themselves all the complexity of the lake as a whole, and although useful material for replicated experiments on ecological principles, they have been rarely used.

8.5 RELATIONSHIPS BETWEEN THE LITTORAL ZONE AND THE OPEN WATER

The littoral zone, particularly when it includes well-developed swamps and beds of submerged vegetation, stands as a link between the catchment area of a lake and its open water. Potentially it can influence the open water in three ways: as an interceptor or sink for materials which would otherwise move from the catchment to the open water; as a source of new materials to the latter which would otherwise not be supplied; and as a refuge or resource for animals which may move into it from the open water to escape predation or to feed.

Of course in any particular lake these functions might have a dominant or an insignificant effect. For a huge, deep lake, there would probably be little consequence for the open water if the littoral was covered in inert polyethylene; but for a majority of the world's lakes, which are small and relatively shallow, the influence of the littoral zone has probably been greatly underestimated.

The littoral may first of all intercept silt, remove nitrate by denitrification and act as a sink for major ions entering from the catchment. The high productivity of swamps, discussed in Chapter 5, means that organic detritus is often laid down as peat. This must occur to a greater extent in lakes than in riverine swamps where the river flow may move much of the detritus down-river. Peat contains elements other than those of carbon compounds and thus peat build-up must mean a sequestering of many elements. For most of these, which are major ions, this is of little functional importance because concentrations of them are changed only a little and they are in no sense limiting to the open water community. Burial of phosphorus, however, could sometimes be significant [812] and the removal of nitrate by denitrification must usually be important. In one lake, the swamps removed nitrate to the extent that concentrations of several milligrams in the inflow stream were reduced to less than 1 mg N in the lake itself even in winter [569]. Preservation of lakeside swamps is itself a useful device in combating eutrophication.

The littoral zone frequently acts as a source of materials to the open water. Usually these are carbon compounds resultant from the high productivity particularly of the swamps. Most of the dissolved carbon compounds, resulting from secretion or decay, that reach the open water are highly refractive yellow substances of large molecular weight. The more labile compounds are probably taken up rapidly by the large bacterial populations within the littoral itself [22]. The yellow substances, if they do little else, stain the water and influence light penetration (Chapter 6) so that a small lake heavily endowed with swamp may have brown water. Such lakes have been described as dystrophic. Small organic particles washed out of the swamp may be of greater importance in some places.

Lake Chilwa in Malawi is a good example. It is a very shallow, saline lake which dries out completely on occasions but has a clay-floored basin ringed with swamps of *Typha domingensis* [422]. The lake level rises and falls annually by a metre or so with longer-term trends towards very high or very low mean water levels. At the higher levels the water may be clear enough to allow a ring of submerged plants to front the *Typha* swamp. But at the lower, much of the time wind disturbance brings clay into suspension and this severely limits the amount of light penetrating to support planktonic or benthic photosynthesis.

In these circumstances the swamp detritus may support both zooplankton growth and that of fish feeding on the zooplankton, or that of invertebrates living in the clay bed. The zooplankton populations are aggregated close inshore to the edge of the swamp and fish catches are also greatest in this area. In the amorphous clay sediment, the chironomid, *Chironomus transvaalensis* uses pieces of detritus from the swamp to construct its burrows [505] and it too has a distribution closely following the lake's perimeter. For much of the time the central part of the open water is a depauperate clay suspension, with the only significant activity allied to the swamp ring.

There has been much discussion about the possible role of the littoral zone as a source of phosphorus to the open water. It can only be this if it is able to mobilize phosphorus from particles entering from the catchment which would not otherwise be available to the open water. Otherwise it must either be neutral, or, if it is laying down peat, a sink for phosphate. No studies have critically examined this balance, though some have shown an increase in total phosphorus in the open water when the littoral plant beds have been removed by herbicides and plant-eating fish [745]. The experiments have never been carried out long enough, however, to establish what the new 'stable state' would be if the littoral plants were kept from redeveloping over a period longer than 2 or 3 years.

Aquatic plants certainly do mobilize phosphorus from sediments, either through uptake from sediment and subsequent decay in the water column or through destruction of the oxidized microzone (Chapter 6) at the surface of the sediment within their beds. In temperate lakes the senescence and decay of the plants may release large quantities of phosphate to the water [31]. However, this occurs at the end of the summer and in autumn, when the released amounts will be diluted by incoming water and washed out, or re-absorbed by the re-oxidized sediment surfaces. Laboratory experiments are usually the basis on which phosphate release rates are measured. Calculations of the potential contribution of this loading to the lake are largely misleading if season and hydrology are not taken into account. In tropical lakes where growth occurs year around, a steady leak of phosphate from the sediment through the plants might be significant, but, in many tropical lakes, phosphorus is not a particularly scarce nutrient, compared with nitrogen.

The littoral zone may have its greatest influence as a refuge. Plant detritus or the plants themselves are the substrata upon which many species of fish lay their eggs. Species like pike (*Esox lucius*) lurk in the edges of plant beds for their prey. Older, larger fish take larger and larger invertebrates as food and these are to be found in the littoral zone, whilst smaller invertebrates like Cladocera may find refuge there from open water fish predation. An example has been given above of how reduction of the algal population in the open water may depend on the use of plant refuges by the zooplankters. The swamps of Lake Chilwa, lying closer to the incoming streams also act as refuges for fish during the periods when the central part of the lake dries down completely. One species, a catfish, *Clarias mossambius*, can bury itself in the damp mud of the swamp and, air-breathing, await the reflooding of the hard, cracked clay basin which much of the lake becomes.

Intuition would suggest that where they are extensive relative to the open water, littoral zones, with their considerable complexity and diversity, might have many overt and subtle influences on the open water. However, the complexity of these habitats is such that good information is scarce and presently only their potential importance can be emphasized [92, 856].

8.6 THE PROFUNDAL BENTHOS

The profundal benthos is much less structured than most of the littoral zone communities. To a casual observer its existence is suggested only by the clouds of midges (chironomids) which emerge from the bottom through the water column to swarm and mate around the lake. A simple comparison between littoral and profundal benthic invertebrate communities recorded at the peaks of their development in the same lake illustrates just how less diverse is that of the profundal [47, 411].

At 2 m water depth in L. Esrom, in Denmark, the littoral plants and their underlying sediment provide habitat for at least 40 species including oligochaete worms, caddis-fly and dipteran larvae, gastropod and bivalve molluscs. A comparable sampling of sediment from under 20 m of water, in the dark hypolimnion, revealed only five species: an oligochaete worm, three dipteran larvae, and a bivalve mollusc. Exhaustive search of both habitats would reveal more, rarer species, but would not alter the picture greatly. Ciliate protozoa, however, were undoubtedly numerous [214].

The profundal benthos, living in a structurally less complex habitat, has reduced diversity, but has not necessarily a low productivity for its food supply may be copious. To some extent the deoxygenation to which the profundal benthos may be subjected may have limited its variety, but even in infertile lakes with oxygenated hypolimnia the profundal benthos is not species-rich compared with adjacent littoral communities. Because the hypolimnia of deep tropical lakes are warm and hence have low oxygen levels even when initially saturated with oxygen, they rapidly deoxygenate and it is common for there to be a complete absence of profundal multicellular invertebrates in such lakes.

Many profundal communities are dominated by chironomid larvae (Diptera) and oligochaete worms, with sometimes some small bivalve mollusc species. Many attempts have been made to classify lakes on the basis of their benthic invertebrates [67] and although these have been successful only for limited regions or purposes, a general trend in benthic fauna related to lake fertility can be seen. Well-oxygenated hypolimnia have a variety of larvae of *Chironomus, Tanytarsus* and other chironomid species, *Pisidium* spp. (bivalve molluscs), some Crustacea, such as the amphipod *Pontoporeia* and sometimes insect larvae like *Sialis* (alder-fly), but relatively few oligochaetes. At the other extreme of a sediment covered by anaerobic water for part of the year, the chironomid fauna is reduced firstly to one or two detritivorous species, for example *Chironomus anthracinus*, and about the same number of species of larval predators. These often include *Chaoborus*, the translucent 'phantom-larva' which migrates to the epilimnion to hunt zooplankton at some stages of its life history. There is usually also a much greater biomass of oligochaete worms such as *Tubifex* and *Ilyodrilus*. Frequently the benthic animals of such lakes are pink or red in colour from contained haemoglobins, though the role that these might play in respiration or survival under deoxygenated conditions is not clear. The profundal benthos of such a fertile lake, L. Esrom, has been studied in some detail [411, 412, 413] and will serve to illustrate something of the biology of these organisms.

8.6.1 Biology of selected benthic invertebrates

Chironomus anthracinus
Chironomus anthracinus in L. Esrom (Fig. 8.5) starts life as an egg mass deposited between sunset and darkness in May on the lake surface. Mated female gnats dip their abdomens into the surface water film. The egg masses are about 2×2.5 mm in size when dry but swell a hundredfold when wetted, and are deposited in the lee of beech woodlands on the western shore of the lake. These create the calm conditions necessary for the flight of the gnat, but water currents subsequently distribute the eggs throughout the lake and they sink to the bottom.

By June the eggs have hatched into instars less than 2 mm long. This is the first of four instar stages which precede pupation and emergence of the adult some 23 months later from populations in the profundal. Growth is rapid in June and the second instar (3–4 mm long) has hatched into the third in July. The previous growth coincides with a ready supply of food falling from the epilimnion where a major spring population of diatoms has formed (see Chapter 7). The first instar is transparent and probably moves along beneath the sediment surface, swallowing sediment as it moves. The second

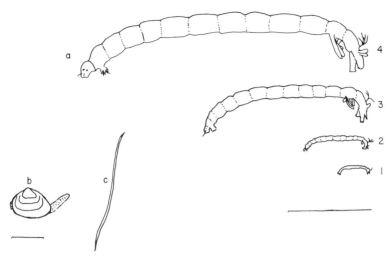

Fig. 8.5. Common detritivores in L. Esrom. Scale lines represent 5 mm. (a) *Chironomus anthracinus*, instars 1–4, (b) *Pisidium*, (c) *Ilyodrilus*. ((a) after Jonasson [411].)

instar has heavier musculature and has acquired some haemoglobin. It builds a tube, open at the surface and tapering at the base. In this it lives, pumping a current of water past its mouth by undulations of its body. The larvae may also emerge partly from the tube and gather sediment encircling it; this leads to exposure of patches of the underlying dark, deoxygenated sediment until the oxidized micro-zone can reform.

Lake Esrom is fertile and the dissolved oxygen concentrations fall rapidly in its hypolimnion. By July the lower few metres are almost deoxygenated and at around 1 mg O_2 l^{-1} growth of the *Chironomus* larvae stops (Fig. 8.6). This may be because the limited respiration rates then possible produce sufficient energy only for maintenance, and partly because the supply of sedimenting food has changed in quality or amount. In a lengthened journey, delayed in the metalimnion (see Chapter 6), decomposition of labile, energy-rich compounds has been going on to the advantage of the zooplankton and bacterioplankton, but to the detriment of the benthos. On the other hand, if growth of *Chironomus* stops, so does predation on it by bottom-feeding fish, which cannot tolerate the deoxygenation and move to shallower waters. The third instar larvae build a tube, lined with salivary secretions, which projects a little way above the sediment surface, and feed on the deposits around the tube. They spread a net of salivary threads over the mud, to which particles stick, then drag the net down into the tube to eat it.

Most littoral animals cannot survive the low or even zero oxygen concentrations that *C. anthracinus* and the related *C. plumosus* can (Walshe [838]). The respiratory rate of most littoral snails, for example, falls

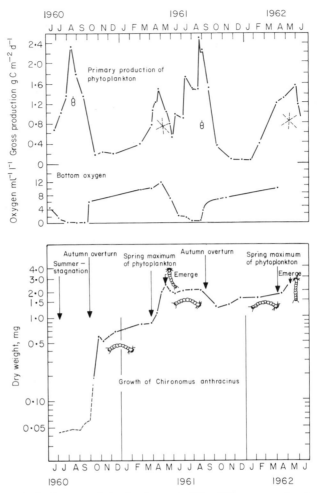

Fig. 8.6. Events in L. Esrom and the development of the *Chironomus anthracinus* population. (Redrawn from Jonasson [413].)

progressively with oxygen content of the water, whilst that of the *Chironomus* species remains high and constant over a wide range of oxygen concentrations. The haemoglobin they contain can store enough oxygen to maintain full activity for only a few minutes, so that it is unlikely that it helps the animals during prolonged deoxygenation. Animals beneath the sediment surface are probably at lesser risk from predators than those at the surface, however, and the oxygen stored by the haemoglobin may minimize the time that the animal has to spend exposed at the surface, without loss of activity.

After the overturn of stratification in L. Esrom in September, growth of *Chironomus* resumes and the third instar larvae moult and change into fourth instar larvae. These are bigger, over 1.5 cm, heavily muscled and bright red in colour. They produce tubes of sediment particles glued with salivary secretion which have 'chimneys' projecting 1–2 cm above the mud surface. The role of the projection is unknown, though it flattens easily and may prevent swamping of the larva with mud during the autumn and winter when the sediment surface may be disturbed as the water column mixes. Swamping would scarcely seem to be a problem for a burrower in mud, however. The height of the chimney, on the other hand, may be sufficient to reach water only a few millimetres from the sediment surface which is much better oxygenated than that immediately in contact with the sediment.

Growth of the fourth instar larvae depends on the supply of phytoplankton and detritus reaching the bottom in autumn, but is soon reduced as the shortening days diminish plankton production. Growth resumes the following spring, but, because of the pause enforced by deoxygenation during the previous summer, most of the larvae are not mature enough to pupate and form adults. They must therefore remain as fourth instars to achieve further growth in the spring and autumn of the second year. Pupation follows in the succeeding spring when the pupae emerge from the tubes and float to the water surface. This happens in the evening; within about 35 seconds of reaching the surface the pupal skin splits and the adult emerges, turning from the pupal red to black. The adults (imagos) rest on the surface until sunrise when the rising warmth quickens their metabolism sufficiently for them to fly away and later to mate and lay eggs on the water. The adult life of the midge is very short—only a few days, compared with the 23 months spent as a larva in the profundal.

At lesser water depths, where there is no summer deoxygenation, *C. anthracinus* is able to grow sufficiently in one year to emerge the next. The shallow water populations also have higher respiratory rates at any given oxygen concentration, which probably also accelerate development of the larvae. A few of the deep water population do manage to emerge after only 12 months, in the spring following their birth, but they do not contribute any young to the population. Their eggs are mostly eaten as they reach the sediment by the large population of remaining fourth instar larvae.

Ilyodrilus hammoniensis and Pisidium casertanum

Ilyodrilus is an oligochaete worm (Fig. 8.5c), a centimetre or so long when mature. Its development in the profundal of L. Esrom is completed relatively rapidly (within about 2 months) so that a succession of generations may be found in the sediment. It hatches from egg cocoons, each containing 2–13 embryos and populations of 25 000 worms m^{-2} build up in the summer of some years. The worms burrow through the subsurface sediment, eating each day 4–6 times their own weight of sediment of which they expel most, probably having digested bacteria from it. They are thus very important sediment processors, causing mixing in the layer in which they live and maintaining the bacterial population in a fast growing state by continual grazing. In the alternate years when *C. anthracinus* exists as a population of fourth instar larvae, the oligochaete population is low, but it increases in the intervening years when these smaller animals seem to compete more effectively with the younger chironomid larvae.

The tiny pea-shell cockle *Pisidum casertanum* (Fig. 8.5b) also lives below the mud surface, in burrows parallel to it. Through the openings of the burrows it draws a current of water to maintain its oxygen supply, except of course in summer, when, like the other profundal animals, it is inactive and does not grow. It feeds by sucking in through a tube, or siphon, a current of watery sediment. This is passed over the gills where steadily beating flagella direct a stream of particles to the stomach. The life history of an individual probably lasts more than a year and the young are retained, until fully developed, within the shell of the hermaphrodite adults.

Carnivorous benthos—Chaoborus and Procladius
As in all ecosystems the abundance of carnivores in the profundal benthos of L. Esrom is small (about 10%) compared with that of the detritivores. There are two main species, one of which feeds mainly on the plankton and only partly on the sediment community, *Chaoborus flavicans* (syn. *C. alpinus*), and the other, *Procladius pectinatus* also a Dipteran larva, about which relatively little is known. Both species ultimately emerge as flies for brief adult lives in which they lay eggs.

Chaoborus, the phantom-midge, is so called because of its transparent body, punctuated only by its dark eyes and apparently black (an optical effect) air sacs at the hind end. These possibly allow it to regulate its buoyancy when it moves between the epilimnion and the sediment. Eggs are laid in late summer by the adult midges and first instar larvae appear in the plankton in September. Zooplankters, on which *Chaoborus* mainly feeds after seizing them with its prehensile antennae, are reasonably abundant in autumn and the *Chaoborus* larvae quickly pass into their second and third instars. These spend some time in the plankton, particularly during overcast weather when their visibility to prey is least, and part of their time in the sediment. Fourth instar larvae are generally produced the following spring, and these migrate nightly from the sediment to the surface water, where zooplankters are again abundant. During the winter *Chaoborus* spends most of the time in the sediment. Its food then is unknown but it may feed on *Ilyodrilus*. Pupation and emergence are in July.

Procladius is a chironomid species, known to be carnivorous and feeding probably on the smallest *Chironomus* larvae and *Ilyodrilus* juveniles. Its maturer larvae emerge in the spring, while less developed ones may migrate to the shallows, for none can be found in the profundal mud in summer. Their respiration rate is severely reduced as oxygen concentrations decline.

8.6.2 **What the sediment-living invertebrates really eat**

Organic sediment is, to a large extent, the remains from feeding by a range of terrestrial and aquatic organisms. As a result, it might not be expected to constitute the most nutritious food, and indeed the organic matter by itself seems unable to support the growth of benthic invertebrates. Studies by McLachlan *et al.* [511] have uncovered in the invertebrate community of a bog lake in Northumberland a story paralleling that of the feeding of shredders in streams (see Chapter 4).

Blaxter Lough is a shallow basin set in the peat of an extensive blanket bog. Erosion by waves of the peat at the windward edge provides a ready source of organic matter to the bottom of the lake, but a major detritivore, *Chironomus lugubris*, is conspicuously distributed at the side of the lake basin opposite to the source of peat particles. The eroded peat is washed by water movement across the lake bottom to the leeward shore and as it moves it is broken down into smaller particles. However, although *C. lugubris* does have some selectivity for the size of particles it can eat, this does not seem to be the reason why it does not colonize the area of the lake where the peat is freshly supplied. Suitably sized particles sieved from eroding or *in situ* peat would not support growth. On the other hand if fresh peat was allowed to become colonized by microorganisms over a few days, it would support growth of *Chironomus*, whether in the laboratory or in chambers placed in any area of the lake. The natural distribution of the Chironomids in the lake seems, therefore, to reflect the distance travelled by suitably fine peat particles in order for them to become colonized by palatable bacteria and fungi. The microorganisms absorb nitrogen compounds from the water and by the time they become palatable the peat particles have increased in calorific content per unit weight by only 23%, but have doubled in protein content.

The peat ingested by *Chironomus lugubris* contains a rather greater bacterial than fungal biomass but the balance changes as the microorganisms are digested in the gut. The voided faecal pellets are relatively large, coherent and dominated by fungi. *C. lugubris* is not coprophagic, that is, it does not eat its own faeces and then digest the new generation of microorganisms which has grown during the interim, as, for example, *Gammarus* (see Chapter 4) seems to do. The faeces are too big for it to eat and *Chironomus* seems to depend on a supply of small, colonized peat particles. Its faeces form the food source of a small cladoceran, *Chydorus sphaericus*. This animal is able to rasp material from the faecal pellets, presumably digesting most of the microorganisms, and producing fine faeces of its own. These are small enough to be ingestible again by *Chironomus*, once recolonized by bacteria. A rather neat reciprocal relationship therefore seems to exist between the two animals, resulting, with the essential help of microorganisms, in the ultimate breakdown of the peat, an initially rather poor food source. It seems highly likely that similar mechanisms and principles might apply to other sorts of sediments.

8.7 INFLUENCE OF THE OPEN WATER COMMUNITY ON THE PROFUNDAL BENTHOS

In moderate to large-sized lakes, the profundal benthos receives most of its food from the overlying water. (The littoral may contribute the bulk in smaller lakes.) Early attempts at classification of lakes by the nature of their benthos assumed that the productivity of the open water was a major determinant of the profundal community. But the relationship is a complex one. Studies by Johnson and Brinkhurst in the Bay of Quinte, on the northern shore of L. Ontario, will illustrate the point [408, 409, 410].

The Bay of Quinte is long, narrow, and winding. It is shallow (about 5 m deep) at its inner end, where several towns enrich it with sewage effluent, and opens out, over 100 km distant, into the 30 m deep waters of inshore L. Ontario, where depth and dilution have reduced the fertility of the water. At four stations (Big Bay, Glenora, Conway and Lake) along this gradient the amount of sedimenting material and its fate as it was processed by the bacteria, benthic invertebrates, and fish, was followed. The results emphasize not only the quantity of sedimenting material as important in determining the productivity of the benthic animals, but also the nature of it.

In the inner bay, Big Bay station had a benthic community dominated by chironomid species, while the third station, Conway, had a rich association of bivalve molluscs (*Sphaerium*), oligochaete worms, chironomids and crustaceans. The second station, Glenora, had a community transitional between those of Big Bay and Conway. In L. Ontario, at the fourth station, there was a diverse community of *Sphaerium* spp., Crustacea, and many other species. In general, the diversity of the community increased towards the main lake, along with the gradient of decreasing fertility and a gradient of summer bottom-water temperatures, which were above 22°C at Big Bay, but around 10°C in L. Ontario.

The first task was to determine the rate of sedimentation, the rate of supply of materials to the bottom communities. The sedimenting seston was collected in 20 cm diameter funnels fitted into bottles to retain the sediment and suspended about 1.5 m above the bottom. The traps were emptied weekly, as far as possible, and the inorganic and organic parts of the sediment measured. Of course seston is colonized by bacteria while it is still suspended and these bacteria continually decompose it, even in the sediment traps. The rate of this decomposition was found to be, for example, about 5% d^{-1} between 9 and 14°C and appropriate corrections were applied to give the true rate for the four stations. As might be expected the sedimentation rates decreased from the inner bay to its mouth, though not steadily, for Big Bay had a much greater rate of sedimentation of organic matter than the others.

The overall rate of processing of this material as it reached the bottom was measured as the community respiration, that is the sum of bacterial, invertebrate and fish respiration. The sum of the first two components was measured as the rate of oxygen uptake from water overlying sediment in small cores of the sediment and its community, obtained by a special sampler. The sediment and its overlying water were sealed from the air by a layer of oil and incubated in the laboratory at the temperature at which they had been obtained. Dissolved oxygen was measured chemically in samples taken at the beginning and end of an incubation of several hours. The respiration rates of fish feeding on the benthic community could not be directly obtained. They were estimated from studies elsewhere as about 50% of the total invertebrate production (see below), and this was added to the measured sediment core respiration to give total community respiration. One can see from Table 8.2 that although community respiration decreased from Big Bay to L. Ontario, it did not do so to the same extent as sedimentation rate. This point will be returned to later.

Table 8.2. Mean sedimentation rates and community respiration rates at four stations in the Bay of Quinte

Station	Organic matter sedimented g m^{-2} day^{-1}	Community respiration g O$_2$ consumed m^{-2} day^{-1}	Mean temp. °C
Big Bay	3.01	0.35	17.1
Conway	0.71	0.25	11.8
Glenora	0.29	0.22	10.8
L. Ontario	0.28	0.15	9.1

It was then necessary to separate the activity of the benthic microorganisms from that of the macro-invertebrates (those larger than about 1 mm and therefore not the protozoa and organisms like nematodes which had to be included with the microorganisms). This was done by measuring the respiration rates of invertebrates separately. Representative animals of the most abundant species were placed in small jars with a substratum of sand, almost free of microorganisms, and their rates of oxygen uptake measured at a variety of temperatures. By numerous experiments, Johnson and Brinkhurst were able to derive equations relating respiration rate to the two major factors affecting it: size (as dry weight) of animal and temperature. They did this for all of the major species, and, for instance, the relationship for *Tubifex tubifex*, one of the oligochaete worms, was:

$$\log_{10} R = 2.6 + 0.046T - 0.27 \log_{10} W,$$

where R is respiration rate in μg O$_2$ mg^{-1} ash-free dry weight d^{-1}, T is temperature in °C and W is ash-free dry weight (i.e. the organic content) of the animal in milligrams.

This meant that from routine samplings in which temperature was measured, and animals were counted and weighed, respiration rates for the whole macro-invertebrate community could be determined. It was also possible to find net production, the increase in amount of animal tissue per unit time. This was done by keeping representative animals in small mud cores, freed of other animals by previous heating or freezing, and by measuring by how much their weight increased over several days or weeks. This method has the disadvantage that growth rates may be altered in the absence of competition with other species, but at least gives some indication of the rate. Growth rates were expressed as percentage increases in weight of animal per unit time, and allowed extrapolation to the natural community using the relationship:

$$\text{Production} = \text{growth rate} \times \text{biomass.}$$

Production of animals which formed distinct cohorts in their life histories was determined by a variant of the Allen curve technique (see Chapter 4).

Assimilation rates, approximately the sums of respiration rates and net productivities, were then also calculable and these are given in Table 8.3. Biomass of the macro-invertebrates increased towards the outer lake—this contrasts with sedimentation rate and community respiration—and assimilation, production and respiration all reached peaks at Glenora, the second station, and were generally similar or lower at Big Bay and the outer two stations. The high sedimentation rate at Big Bay, therefore, did not support comparably high invertebrate production. The turnover rate of the community, measured by the yearly production to biomass ratio did, however, follow inversely the gradient outwards, probably reflecting the decrease in mean temperature.

Table 8.3. Productivity of the benthic macro-invertebrate community in the Bay of Quinte. All rates in kcal m^{-2} year^{-1}, and biomass in kcal m^{-2}

Station	Biomass (B)	Assimilation	Production (P)	Respiration	P/B
Big Bay	5.45	108.7	74.3	34.3	13.6
Glenora	29.9	368	233	136	7.8
Conway	25.6	142	65.8	75.8	2.6
L. Ontario	38.0	165	51	115	1.3

From all of these data, the diagrams in Fig. 8.7 could be constructed. They show the flow of energy through the sediment community (bacteria, invertebrates and fish) at the four stations, and all quantities have been converted to cal m^{-2} day^{-1} for rates, and to kcal m^{-2} (in brackets) for the standing biomass of the various components. IM is the incoming organic matter, the amount of sedimentation. Community respiration (R_{com}) degrades

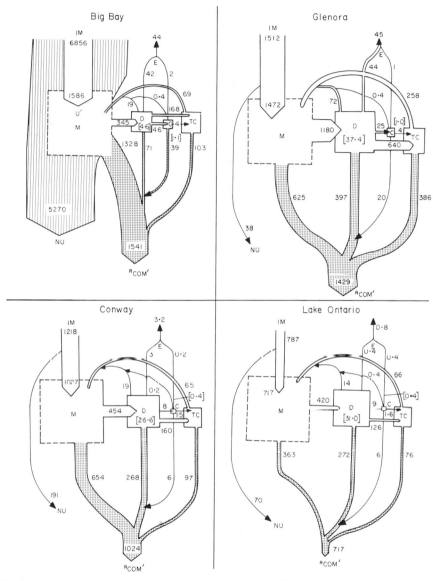

Fig. 8.7. Mean rates of energy flow at four stations in the Bay of Quinte. Boxes represent standing crops and stocks (kcal m^{-2}) and pipes the rates of flow in cal m^{-2} d^{-1}. IM: incoming organic matter; M: microorganisms; U: amount utilized; NU: not utilized and stored in permanent sediment; E: emerging insects; D: detritivores; C: carnivores; TC: top carnivores (fish); R$_{com}$: community respiration. Stippled pipes show respiratory losses. (Redrawn from Johnson and Brinkhurst [410].)

less than a quarter of it, and animal production less than a twentieth at Big Bay, so that much of it is not utilized (NU) and forms the permanent sediment. Apparently it is not used because it is refractory and difficult even for bacteria to degrade. Its low quality probably reflects partly the chemical nature of the cell walls of blue-green algae which are abundant in the phytoplankton there, but also the large import of more fibrous organic matter left after terrestrial decomposition and washed into the Bay.

The amount of unutilized matter is small at the other three stations and most of the sedimenting seston was degraded by microorganisms (M) or used in animal production. This is reflected in the low organic content of the sediment at the L. Ontario station, 3–4%, compared with 32% in the sedimenting material. The organic content of sediment has, in the past, been used as an indicator of potential benthic animal production in lakes. These studies indicate that it represents only the net result of several processes of accumulation and degradation. These processes are amenable to a general treatment which may have implications wider than those for the Bay of Quinte.

Several measures of the efficiency of energy use may be deduced for the benthic community. The proportion of incoming energy used by the whole community (microorganisms, invertebrates and fish) is:

$$a = \frac{R_{\text{com}} + E}{IM} = \frac{U}{IM}$$

where E is the energy lost in the emergence of adult insects which fly away, and U is the energy used, i.e. not stored in permanent sediment.

The proportion of usable energy channelled through the macro-invertebrates is:

$$b = \frac{R_{\text{D}} + R_{\text{C}} + E}{R_{\text{com}} + E}$$

where R is the respiration of detritivores, D, and carnivores, C. Thirdly, E_{gc} the net growth efficiency of the invertebrates, is defined as:

$$E_{gc} = \frac{A_{\text{D}} - (R_{\text{D}} + R_{\text{C}})}{A_{\text{D}}} = \frac{P_{\text{D}+\text{C}}}{A_{\text{D}}}$$

where A is assimilation and P is production of both detritivores and carnivores.

Table 8.4 gives these quotients for the four Bay of Quinte stations. They show clearly the lower percentage utilization of incoming matter at Big Bay ($a = 0.23$), but also the fact that the microorganisms at Big Bay take a greater proportion of the utilized material (($1 - b$) = 90%) than they do at the other stations (61–73%). This may be interpreted in terms of the quality of the sedimenting seston. If it is difficult to degrade, a greater investment of its

Table 8.4. Efficiency of energy use in the benthic communities of the Bay of Quinte

Station	a	b	E_{gc}
Big Bay	0.23	0.1	0.68
Glenora	0.97	0.31	0.64
Conway	0.84	0.27	0.40
L. Ontario	0.91	0.39	0.34

contained energy must be made by bacterial activity to convert it to a form (bacterial cells) usable by the invertebrates.

The product of *a* and *b* gives the proportion of incoming energy usable by the invertebrates, and is very low at Big Bay, 2.3%, and 23–25% at the other stations. A general equation for utilization by the animals is:

$$U = a \cdot b \cdot IM$$

a probably decreases with increased allochthonous import of material (usually of a low quality, refractory nature) from the catchment, while *b* reflects the cost of processing the material and is low when allochthonous matter, or tough-walled algae, particularly blue-green algae, are present.

The quantity E_{gc} can now be incorporated into the model. Production of macro-invertebrates bears some relationship, *c*, to utilization of organic matter by the macro-invertebrates:

$$P = c(U)$$

and

$$U = R_{D+C} + E$$

E, the insect emergence, is generally only a small proportion of the total energy flow and may be neglected so that:

$$P = c(R_{D+C})$$

Because

$$E_{gc} = \frac{P_{D+C}}{A_D} = \frac{P_{D+C}}{P_{D+C} + R_{D+C}}$$

$$R_{D+C} = P_{D+C}\frac{(I - E_{gc})}{E_{gc}}$$

Also, because

$$P = c(R_{D+C})$$

$$c = P/R_{D+C} = \frac{E_{gc}}{(I - E_{gc})}$$

and because

$$U = a \cdot b \cdot IM$$

$$P = a \cdot b \cdot c \cdot IM$$

This is now a relationship which describes how production of invertebrates is related to the supply of incoming organic matter. The two quantities would be directly dependent only if a, b, and c were constants, which they are not— they all decrease as IM increases; a and b decrease for the reasons stated above, c does so because as the import of organic matter increases, the rate of deoxygenation in the surface sediment and the water just above it also increases. Animals must then use more energy in obtaining oxygen (in body movements to keep water circulating over the animal's surface or in production of haemoglobin). Or, like *Chironomus anthracinus*, they may have periods when they lie quiescent, respiring, albeit at a low rate, but probably not feeding. This means that a smaller proportion of their energy supply is available for growth. It might be predicted that c should also be low at low levels of import because of the extra activity then necessary in seeking food. The relationship between benthic invertebrate production and import of organic matter to the community may then have a maximum at intermediate import levels flanked by minima due to food scarcity on one side, and to low food quality and to environmental changes due to deoxygenation on the other.

9

Fish Production and Fisheries in Lakes

For many people, fish and fisheries are the reasons for study of freshwater ecosystems. In the developed world recreational fisheries may be much prized, but in parts of the developing world the supply of inland fish may be the most important protein source. Freshwater fish ironically may be crucial to health not only in diet but also through the several worm parasites (Chapter 5) to which fish play intermediate host, particularly in the Far East.

Fish yields—the crops taken by man—are broadly related to fish production in a lake, which in turn is related to the overall productivity of the system (Chapter 6). The yield is always less than the total production for some fish are unsuitable for sustained fisheries, and others may be difficult to catch or unacceptable as food for sociological reasons. Fish that have not yet reproduced must also not form part of the yield.

Maintenance of a fishery without its collapse through overfishing, when more fish are taken than are replaced by recruitment and growth, ideally requires three things. First, there should be a knowledge of the biology of the available fish species; secondly, a wise choice of those whose populations will be able to stand up to fishing; and thirdly, a close monitoring of catch statistics and imposition of regulations on the type and amount of fishing that can be done (fishing effort) to protect the stocks. The first part of this chapter will deal with these issues.

These requirements cannot always be achieved, particularly in the developing world. Expertise may not be available and simple collection of statistics from a large number of fishing villages widely spaced around a large lake may be impossible. Cruder methods which involve deciding on the potential overall yield of all the species collectively and regulating fishing up to that yield, may be more practicable, if less certain. Methods for doing this will be considered in the second part of the chapter.

Population and economic pressures now demand increased fish yields. Often these have involved fishing for inappropriate species or introduction of supposedly high-yielding species. The problems that these have caused, and the better solution of pond culture to the need for increase in fish yield, are considered in the third part.

9.1 SOME BASIC FISH BIOLOGY

When fish are considered species by species, very large differences are found. These differences have great implications for the suitability or otherwise of a species for a fishery. The trout (*Salmo trutta*), the Nile perch (*Lates niloticus*), a 'tilapia' (*Sarotherodon niloticus*), the walleye, *Stizostedion vitreum*, and the grass carp, (*Ctenopharyngodon idella*) (Fig. 9.1) have been chosen to illustrate the variety among fish because of the range of their reproductive biology, diets and zoogeography.

Trout are carnivores, eating largely benthic invertebrates; they are indigenous to Europe, N. Africa and W. Asia, but have been introduced to suitable waters elsewhere for their sporting qualities. Nile perch are voracious piscivores in the R. Nile and its associated, or once associated, great Lakes Albert and Turkana. They are valued food fish in some places, being large (specimens weighing over 45 kg are common) and have also been introduced

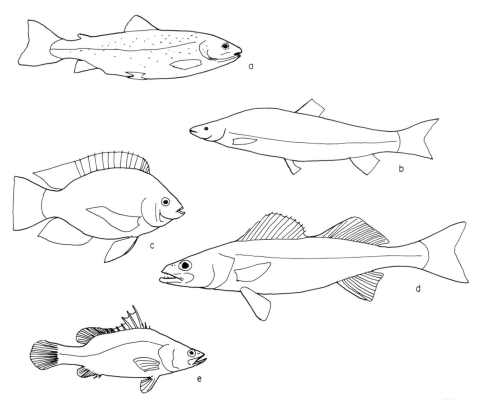

Fig 9.1. Five fish species of contrasted biology (see text). (a) Brown trout, *Salmo trutta*; (b) Chinese grass carp, *Ctenopharyngodon idella*; (c) 'tilapia', *Sarotherodon niloticus*; (d) pike-perch, *Stizostedion lucioperca* (the walleye is a similar though stockier fish); (e) Nile perch, *Lates niloticus*.

to other African lakes (see later). The tilapia in question feed on fine bottom detritus and on phytoplankton and are also native to the Nile watershed but have been widely introduced to other African lakes and rivers. 'Tilapia' is a common name applied to cichlid fish of the genera *Tilapia*, *Sarotherodon* and *Oreochromis*. Walleye are N. American piscivores which feed on zooplankton when young. Some comparison will be made between walleye and their close European relatives, the pike-perch, *Stizostedion lucioperca* (Marshall [521a]). Lastly, the grass carp is an avid feeder on submerged and even emergent aquatic plants. It is endemic to the R. Amur and parts of eastern Asia but has been widely introduced to other areas of Eurasia for control of nuisance aquatic plants and pond culture.

9.1.1 Eggs

All fish hatch from eggs, mostly released into the water for their development. *Sarotherodon* species are exceptions; once the eggs are externally fertilized the female may gather them into her mouth for brooding. Under such close protection, relatively little investment in numbers of eggs produced is required. *S. niloticus* certainly protects the young fish in this way when predators approach, but may merely guard them as eggs in a shallow depression scraped on the sandy bottom by sweeping movements of its tail. Trout also carefully excavate a nest or redd in gravel, much as already described for salmon (Chapter 4) [235].

The Nile perch and grass carp take little or no care of their eggs and must produce very large numbers of them to ensure ultimate survival of enough young to maintain the population. The eggs of both species are planktonic and probably slightly denser than water. The difference is minimized, however, by incorporation of oil globules in the former and a water-filled cavity between the egg membranes in the latter. While *S. niloticus*, which guards its eggs in the nest, may produce only a few thousand, the Nile perch releases several millions to the open water.

Walleye also take no care of their eggs. They are scattered onto a large area of gravel in well-oxygenated water having been fertilized during release from the female. The walleye's European relative, the pike-perch, is similar in morphology, but quite different in spawning behaviour. The male pike-perch excavates a nest on a muddy organic bottom in such a way that the roots or rhizomes of aquatic plants are exposed. The eggs are sticky and adhere to the exposed roots. The female guards them and fans water over them with her tail—this may increase the rate of survival in a habitat which is generally unsaturated with oxygen.

Many fish in temperate regions lay eggs which adhere to stones (e.g. minnows, *Phoxinus phoxinus*, barbel, *Barbus barbus* and gudgeon, *Gobio gobio*) or to weeds (e.g. roach, *Rutilus rutilus*, carp, *Cyprinus carpio*, bream, *Abramis brama*) in lowland rivers and shallow lakes, and often these eggs are unguarded. With limited resources available for egg production there has been a tendency for large numbers of small eggs to be produced by those species which do not protect their eggs by either hiding or guarding them, and for smaller numbers of larger eggs to be produced by those that do. In the latter case the invididual probability of survival is doubly enhanced, for a larger egg contains more yolk for sustenance of the fry. On the other hand, the spawning requirements of the 'protective' type may be more stringent and less easily available and the larger egg, with its smaller surface area: volume ratio may require greater external concentrations of oxygen to enable fast enough diffusion into it. The general trend is not always followed, however.

The eggs of the pike-perch, which are guarded, are smaller (0.8–1.5 mm diameter) than those of the walleye (1.4–2.1 mm diameter), which are not. Furthermore, the walleye produces only 30 000–65 000 eggs per kg of body weight, compared with 110 000–260 000 in the pike-perch. This reversal of the expected trend is probably related to the contrasted habitats in which these fish live. The walleye inhabits well-oxygenated waters, the pike-perch is successful in stagnant or slow-moving productive waters with low oxygen concentrations and high turbidity, although under experimental conditions adult walleye can tolerate lower oxygen concentrations than pike-perch. Nonetheless the smaller size of pike-perch eggs must enhance gas exchange in a nesting habitat which is within highly organic sediments, and the large numbers of eggs produced, even though guarded, may be a necessary ploy in a fertile habitat usually characterized by cyprinid fish which prey on eggs to a greater extent than the species coexisting with walleyes.

9.1.2 **Feeding**

The fry which hatch from the eggs do so after varying periods of development, dependent on temperature, among other factors. Trout spawn in autumn, walleye between March and June, the tropical Nile perch and tilapia probably at most times of the year. The grass carp spawns in rising flood waters in spring when increased turbidity may camouflage the eggs somewhat against predators. The times at which hatching occurs, however, have become adjusted through natural selection to the periods in which food is available for the fry. The yolk carried in the egg is only large enough to keep the fry for a very short period and any increase in the amount of yolk produced per egg would mean a corresponding decrease in the number of eggs produced. On balance, the maximum number of fry are hatched in the spring and summer in temperate regions, when suitable food for the fry is most plentiful.

Small fish can eat only small portions; large fish can eat much larger items, and the investment of energy in finding large items is proportionately less than that needed to find the same amount of food in smaller units. The diets of fish thus often change markedly as the fish increase in size. Trout alevins dart up from their shelter in the gravel to take small chironomid larvae and Crustacea in the weeks after their yolk is used up. When they have grown to about 4 cm they station themselves in the water column and space themselves about 8 cm apart, defending their 'water-space' against neighbours by aggressive darts at intruders. They then feed on invertebrates moving past them. As a trout ages it will also actively hunt food, particularly if the prey moves along the bottom, and may take fish fry and larger fish when it is bigger than 30 cm in length. Trout have relatively wide mouths and many backwardly directed teeth which efficiently hold prey once it is grasped. Other fish species feeding in mid-water, but on smaller items like zooplankton, have a much narrower mouth, protrusible into a lengthened tube with which they suck up prey, like a vacuum cleaner. The tropical elephant snout fish (Mormyridae) are good examples.

The adult Nile perch is a voracious feeder even on other piscivorous, but smaller, fish like the tiger fish (*Hydrocynus vittatus*). When it is younger, however, it takes invertebrates. The fry, 0.3–1.35 cm in length, feed on planktonic Cladocera in L. Chad and inhabit shallow, weedy areas [357]. At about 20 cm they begin to take larger invertebrates—a bottom-living prawn (*Macrobrachium niloticum*) and snails—and some small fish. As they grow they take more and more, larger and larger fish; they themselves grow to 2 metres in length. The Nile perch is a very active pursuer of prey. It has a large head, a widely gaping mouth and serried ranks of backwardly directed teeth. The dorsal and anal fins are situated well back (see Fig. 9.1), giving the fish a powerful tail thrust which, as in the trout, allows bursts of high speed enabling capture of prey, which may itself dart rapidly away.

Newly-hatched walleye begin to feed on plankton—large diatoms, rotifers, nauplii of crustaceans. They form schools which may ensure a greater survival rate of the fry, which are only 6–9 mm long, than if they moved individually. Progressively, they eat larger zooplankters and such open water insect larvae as those of *Chaoborus* (see Chapter 8). Finally they become piscivorous, seizing their prey then manoeuvring it until it can be swallowed head first. This is probably because some favoured prey species, such as yellow perch (*Perca flavescens*), have spiny pectoral fins which would lodge in the throat if the prey was swallowed tail first. Walleye have rather elongate gill rakers (the strips of bone which protect the delicate gills from damage by large particles pulled in with the respiratory water current) and it is on these that spiny fins could catch. The pike-perch does not have such elongate rakers and is able to swallow at least some of its prey tail first. From the point of view of the prey, spiny fins have some advantage, for in the time taken for a fish, seized usually at the tail, to be manoeuvred into a head-first position for swallowing, there is a greater possibility of escape. Spines also may be used by a fish to avoid capture if they can be jammed into rock crevices.

The grass carp becomes predominantly vegetarian at lengths above about 30 mm. As fry, it eats rotifers and crustaceans, occasional chironomid larvae and perhaps some filamentous algae. Between 17 and 18 mm in size it takes more chironomids and fewer of the smaller zooplankters are eaten, and by 27 mm, higher plant food becomes prominent in the diet. Although thereafter it unavoidably takes invertebrates associated with water plants, the grass carp is well adapted to a plant diet. This is unusual, for relatively few fish eat water plants.

Like other Cyprinidae the grass carp has a toothless mouth, but it has strong and specialized projections of the pharyngeal bones, which line the region behind the mouth, just before the entrance to the oesophagus. These pharyngeal teeth lie in sockets in the pharyngeal bones. The row on the upper bones has two small teeth on either side, and on the lower bones are four or five teeth with serrated cutting surfaces on each side. In older fish these become flattened, with both cutting and rasping surfaces. This change is related to the fact that young fish eat only the softer, submerged plants, whereas older ones can tackle more lignified emergent plants as well. The teeth tear and rasp the plant food into particles 1–3 mm in diameter. Only the cells which are rasped and ruptured are digested and half of the food passes out, undigested, as faeces [347]. The pH of the gut secretion is quite high: 7.4–8.5 in the anterior part, around 6.7–6.8 in the rectum, and such values are not particularly noteworthy. They provide a relative measure for comparison with those of *Sarotherodon niloticus* which, perhaps because of its extremely low gut pH, between 1.4 and 1.9, is one of the few fish yet examined that can digest even the blue-green algae [550].

Fry of this tilapia may feed on insect larvae—chironomids and *Chaoborus*—but the fish soon turn to an algal diet. In different lakes this may be epiphytic or periphytic, for example in L. Volta (Ghana) *S. nilotica* eats the weft of algae loosely hanging from submerged dead trees in this man-made lake (see Chapter 10), and bottom detritus is also taken. In L. George, Uganda, which has a very dense blue-green algal population, much of which rests on the bottom but is circulated into the water column by regular diurnal mixing, *S. niloticus* takes the blue-green algae, but whether in the water column or from the bottom is not clear.

Feeding makes little use of the teeth or jaws. Food is sucked in with the respiratory current and entangled with mucus secreted by glands in the mouth, and then it is passed back into the pharynx. The food is thus not filtered out, as it is in many other plankton feeders, by fine projections on the gill rakers. The first lot of food taken during the day is not well digested, but as the stomach secretion falls below pH 2 digestion takes place and 70–80% assimilation of the blue-green algae *Anabaena* and *Microcystis* has been noted in laboratory experiments [551]. Assimilation from natural phytoplankton populations in L. George was lower, about 43%, but this is still considerable in view of the long-held belief that blue-green algae are indigestible as live cells. *Sarotherodon niloticus* shows some selectivity in its food. A comparison between the percentage representation of different foods in the gut (*r*) compared with that in the external environment (*p*) may be used to calculate an electivity index $r - p(r + p)^{-1}$ [395]. In L. George, *S. niloticus* showed a positive selection for *Microcystis*, *Lyngbya* (blue-green algae) and *Melosira* (a diatom) but discriminated against the smaller species, *Anabaenopsis* (blue-green alga) and *Synedra* (diatom).

9.1.3 Breeding

The changes associated with the onset and completion of spawning are major events in the life of a fish. The gametes alone may constitute a quarter of the body weight and the energy demands in producing them and in accomplishing the act of spawning may be very great. A 'spent' fish (one that has just spawned) is weak and more vulnerable to predation, and the extensive migrations which some fish undertake before spawning may have exhausted them so much that bacterial and fungal infections are common. Nonetheless, most fish spawn in several successive years after reaching maturity.

Breeding occurs at different ages in different fish, presumably at a time which represents for each a compromise between several factors. If left too late there is a high chance of failure to breed through early death. If attempted too early the fish may be too small and unable to cope with the energy demands. Trout first breed when they are 3 or 4 years old, walleye from 2 or 3 years and *S. niloticus* at only a few months. This reflects to some extent high growth rates resulting from the higher water temperatures and lack of seasonal food scarcity in most tropical fresh waters. In the main part of L. Albert (Uganda, Zaire), *S. niloticus* reaches a size of 50 cm and breeds when it has attained 28 cm. In the Bukuku lagoon, a part of the lake now completely isolated by a sandbar and forming a 600 m triangle, the fish breeds at 10 cm and achieves only 17 cm at most. This appears to be a phenotypic effect, not a genetic difference and seems to be a response to the extreme environment of the lagoon, which is very saline and must dry out almost completely from time to time. A fish able to mature earlier and therefore produce more frequent generations may thus have a greater chance of survival than one whose breeding cycle is longer than the survival of its environment.

Breeding rituals are common in fish. They serve to preserve adaptive differences between closely related species and may need specific environmental conditions. Examples are provided by the European three-spined stickleback (*Gasterosteus aculeatus*) and the African *Haplochromis burtomi*, a cichlid fish. Sticklebacks are silvery-brown, small (5–8 cm) and eat small invertebrates. Sometimes they winter in estuaries. In spring, the males leave the mixed shoals and take on breeding colours: bright blue and red underparts and a translucent appearance to the scales on the back. Each male chooses a small territory on the substratum, a sandy or silted bottom, and defends it against other males. Within the territory, a depression is excavated by the male sucking up the sand or mud and expelling it some distance away. Strands of aquatic plants or filamentous algae are collected and formed, with the help of a secretion from the kidney, into a tunnel-shaped nest about 5 cm long and broad, which lies in the depression.

At this stage the male becomes responsive to swollen, gravid (bearing ripe eggs) females, though not to thin, spent ones. If a suitable female enters the territory, the male first creeps through his nest and then performs a 'courtship' dance in which he zigzags towards the female, then turns towards his nest. This may not attract the female, though she may swim nearby, adopting a characteristic posture with the head up. The male may then dart at the female with his spines raised and usually she will then be led down to the nest. By inserting his snout the male points at the nest entrance, then backs off and swims on his side with his back towards the female. She appears to inspect the nest opening, and then may retreat to the water surface. The whole courtship process may be repeated several times before the female enters the nest, with something of a struggle, for the opening is narrow. Eventually her head is well in and her tail sticks out of the entrance. The male then puts his snout against the base of her tail and quivers violently. The tail begins to rise as the male continues quivering and when it is raised high the female releases a stream of 50–100 eggs into the nest. When the last egg is laid, the female, now much thinner, rushes out of the nest as the male bursts in and releases a cloud of spermatozoa quickly over the eggs. Thereafter the female is chased away, the male prods the fertilized eggs deep into the nest, adds sand to reinforce and camouflage it and wafts water over the eggs with his fins. He remains guarding the fry until about 10 days after hatching when his breeding colours have also faded.

Spawning behaviour in *Haplochromis burtomi* in L. Tanganyika (Fig. 9.2) is equally complicated. Little is known of the preliminaries of courtship but at its culmination the female lays a very small batch of quite large eggs on the lake bed while the male courts attendance. When she has laid the eggs the female quickly turns and scoops the unfertilized eggs into her mouth. Then the male sweeps past the female, displaying his anal fin as he does so. On it are light, circular markings which, against the darker fin background, resemble eggs. The female is apparently deceived by these and moves to pick up what she thinks are eggs. In doing so she sucks in water from near the male's genital aperture from which he has just released sperm and the eggs are fertilized in the female's mouth. The process is repeated several times until her mouth is full of fertilized eggs where they are protected until hatching.

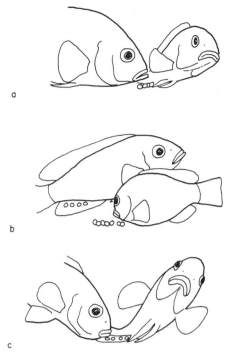

Fig. 9.2. Spawning in *Haplochromis burtomi,* a fish species of L. Tanganyika. (a) The female has laid a batch of eggs; (b) she turns and takes them into her mouth before they are fertilized; (c) the male sweeps past the female, displaying the 'egg-dummies' on his anal fin. In attempting to collect the dummies the female takes in spermatozoa released by the male and fertilization of the eggs takes place in her mouth (After Fryer and Iles [246], based on a film made by G. H. Wickler.)

9.2 CHOICE OF FISH FOR A FISHERY

For a sustained fishery annual mortality due to natural causes and fishing combined should be no greater than annual growth of the stock plus recruitment of new fishes to it. A high-yielding fishery is thus one which uses fishes of prolific growth and recruitment matched by high natural mortality. Ideally the natural mortality is replaced by removal in the fishery. A second requirement is that the fishery methods should not damage the habitat and hence interfere with spawning and recruitment.

Where only subsistence fisheries are concerned, with individuals or very small groups using relatively inefficient methods like spears or hooks and lines, almost any fish species will sustain a limited amount of hunting. Where more people, using intensive methods such as large nets, in a commercial fishery are concerned, it is clear that the most suitable fish are those of the open water (pelagic fish), followed by the bottom-living fish (demersal fish) of the profundal. Quite unsuitable usually are bottom-living fish of the littoral zone.

Pelagic fish live in an unstructured habitat where, like the zooplankton (Chapter 7) they are continually at risk from their own predators. Many eggs are usually produced to drift in the water and mortality at all stages is high. The fish tend to spawn early and grow fast. The 'structure' of the open water is not damaged by fishing methods. In contrast, the littoral habitat can be severely damaged by trawled nets, which as well as disrupting nesting sites (e.g. of sticklebacks or *Haplochromis burtomi*) may also despoil the habitat of the rather more specific food sources on which littoral fish (e.g. the rock-dwelling fish of L. Malawi, Chapter 8) feed. Egg-guarding and hence low egg production and mortality mean low recruitment rates, and the cover provided by a well structured habitat limits natural mortality rates. Demersal fish form an intermediate case with a much less vulnerable habitat than the littoral, but one which nonetheless can be destroyed by too frequent disturbance.

The key to choice of suitable fish is thus that it should be adapted to a relatively unstable habitat in which its risks of natural death are high. Whole lakes, littoral zone included, may sometimes fall into this category if they are subject to such disturbances as frequent drying out. Lake Chilwa (Chapter 8), in Malawi, is one such. It has limited fish diversity (Chapter 2), with species which are omnivorous and unfussy about their conditions for breeding. They are capable of living in a variety of habitats, and of growing fast and reproducing quickly when they recolonize the lake from the rivers after a drying phase (Morgan and Kalk [548]).

Fishes such as these have been naturally selected for r (high growth rate) as opposed to K (stable population) characteristics [245, 476]. These terms come from the logistic growth equation:

$$\frac{\mathrm{d}N}{\mathrm{d}t} = rN \cdot \frac{K - N}{K}$$

where N is number, t is time, r is the intrinsic rate of natural increase and K is the 'carrying capacity' of the environment. r-selected fish tend to invest much energy in growth and reproduction. Examples are the five fish compared in 9.1. K-selected ones make efficient use of the resources to which they are closely adapted. Examples are the littoral fish of L. Malawi discussed in Chapters 2 and 8. This brings us to a consideration of fish growth and reproduction for the management of a sustained intensive fishery. Ideally this should include estimates of the production of the fish before fishing starts.

9.3 MEASUREMENT OF FISH PRODUCTION

Production is estimated as the increase in weight per unit time and usually per unit area. Due allowance must be given to production lost to predators and disease, as must the large proportion of the body weight that the gametes comprise, particularly in females. The first problem is to estimate the absolute population size per unit area of waterway.

The most widely used method of sampling fish in planning a fishery is to use a net or trap which removes a sample of the population. These fish are marked (by dyes or fin clipping) or tagged in some way to make them identifiable, then released again. After a period to allow them (it is assumed) to mix randomly with the rest of the population, this is again sampled and the number of marked and unmarked fish counted. The ratio of recaptured marked fish (n_{rm}) to the total number recovered in the second sampling (N') is supposed to be equal to the ratio of the number of fish originally marked (n_m) to the total population (N):

$$N = \frac{n_m \cdot N'}{n_{rm}}$$

N may be placed on an areal basis from the area of the lake. In a big lake the population may move only over a restricted area which must be determined by marking fish and then sampling over a wide area to discover the extent to which the marked fish have moved. Practical problems involved are in ensuring that the samples are not selective of particular fish sizes, and that the marking technique does not alter fish behaviour or increase the chance of death. These ideals are probably never attained.

For inshore fish, beach seine nets may be useful. These involve the laying out of a small meshed net in a wide arc with the end of the arc at the shoreline of a suitable shelving beach. The top of the net has floats to keep it at the surface and the bottom of the net is weighted to hold it at the bottom. After setting, the net is hauled in from the shore. Seines may also be set in the open water, but must have a rope threaded into the bottom so that as the net is drawn in, it is closed at the bottom into a 'purse'. This requires a mechanical winch on the boat used. Fish escape around the seine nets but this may be minimized by an elongated sock of netting bulging out at the centre and called a cod end. The fish tend to move into this as the net is hauled in. Trawling is also a relatively non-selective method for bottom-living fish. The trawl is a bag of netting, kept open either by its resistance to movement through the water or by a wooden beam across its entrance, and pulled by a sufficiently powerful boat.

Gill nets are highly selective. The net is floated passively in the water at a desired depth and fish moving against it may move through it and pass out the other side, or they may be too big to penetrate it at all. Particular size groups, however, move part through it until the widest part of their body becomes jammed and their gill covers catch on the netting as they try to move back. Such nets, dependent on their mesh size and to some extent the slackness with which the net is set, are very selective for particular size ranges. A set (or fleet) of different mesh sizes may be used for experimental fishing, but most fish are damaged on recovery.

9.3.1 **Growth measurement**

By use of a marked sample the fortunes of a fish population can be followed throughout life. The individuals of most temperate fish species spawn over a limited period of the year and the newly spawned generation or cohort can be treated as a unit recognizable in successive years because of the annual or other periodic rings laid down in the skin scales or in the otoliths (the ear bones). When a population is sampled it is held in an aerated tank while each individual, or a random sample if there are very many, is weighed and its length measured. A length/weight graph may be used to avoid the need to weigh as well as measure in future samplings. The fish are usually lightly anaesthetized to calm them during these measurements. Scales may be removed from a part of the body, often high on the back just below the dorsal fin, where their rings are known to be clear. If this is done carefully with blunt forceps the fish should suffer little damage, though its chances of becoming infected by fungi, protozoa or bacteria subsequently may be increased. Removal of otoliths requires dissection and is fatal.

The scales may be used simply to age the population and identify separately the different cohorts (often called year classes) or they may be used on an old fish to estimate the growth in previous years, for the distance between the rings is believed to be proportional to the growth in the year the ring was laid down. Rings, however, may be reabsorbed during periods of starvation.

For a given cohort an Allen curve (see Chapter 4) may be plotted relating the mean weight of fish (on the vertical axis) to the numbers remaining in the cohort for a sequence of sampling occasions. At first there are very many small fry, but as individual weights increase the numbers decline. The area under the curve as it approaches the abscissa, when the last survivor dies, gives the total production. Production in a given year can be found by determining the area under the curve between points representing the times in question. The curve is roughly hyperbolic, but has irregularities—numbers decline in winter in temperate lakes though the increase in weight of the survivors may be negligible. It may even be negative as fat stores laid down the previous summer are used up. A similar decrease in weight occurs during spawning when the weight converted to gametes can be determined from the area under the curve as it reverses direction during the spawning period. By adding up the production of each of the several year classes present in a given year, and by doing this for each species, a measure of the total fish production may be obtained.

9.4 COMMERCIAL FISHERIES

Fisheries do not differ from any other commercial venture in that the aim of their management must be to maximize the yield of harvestable fish. The particular problem of fishing, however, is to sustain the yield from year to year by harvesting no more than the equivalent of the annual increment of fishes becoming large enough to be worth catching plus the annual growth of those already fishable. It is relatively easy for fishermen when not subject to control to remove more than this annual recruitment and growth, in which case the fishery will eventually become extinct.

Figure 9.3 shows the increase in biomass of a newly-hatched cohort of fishes. The rate of increase is at first high but declines as the maximum potential biomass, represented by the curve's asymptote, is reached. Simultaneously some members of the cohort die and the rate of mortality is at first high, but declines as the fishes become older, bigger and less vulnerable to predators. The net effect of growth and death is to create the biomass curve; growth exceeds death in the early part of the cohort's existence, but eventually mortality predominates. There is a point at which the biomass of the cohort is maximal. The aim of a fishery is to catch that portion of the

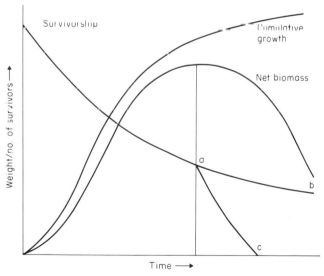

Fig. 9.3. Ideal exploitation of fish stock. The net biomass represents the balance between cumulative growth and death, and at some time reaches a peak. Provided spawning has by then occurred the peak represents the time from which most efficient removal of fish can take place. The natural survivorship curve, ab, is replaced by the steeper fishing mortality curve, ac, as fishing removes biomass before it can be further diverted to disease organisms, parasites or other predators. (Based on Fryer and Iles [246].)

biomass which, after the maximum has been attained, would be lost to natural mortality. The more rapidly the fish can be removed after the maximum the better. The intention is to replace the natural survival curve ab with the fishing mortality curve ac, the area between the curves representing the yield. It is important that fishing does not remove fishes of an age at which growth exceeds natural mortality, for this will reduce the potential yield and may also remove fish that have not spawned. This may reduce the potential recruitment to the next season's fishery and is called overfishing. It is also important that fishing mortality should only ever replace natural mortality, and never exceed it. Fish that can safely be removed are thus mature and should not be allowed to grow to their potential maximum size, by which time most of the biomass of the cohort will have been lost by natural mortality.

The year to year changes in fish population can be represented by the Russell equation [698]:

$$P_2 = P_1 + (R+G) - (F+M)$$

where P_1 and P_2 are the fish stocks (biomass) in 2 successive years, R is the annual recruitment of mature fishes to the fishery, G is the growth made by those already fishable but not yet removed, F is the annual mortality due to fishing and M the natural annual mortality. In a well-run fishery, ideally $P_1 = P_2$ and $M = 0$, but this is never achieved, for natural environmental fluctuations will alter R and G in ways that a fisheries manager cannot predict in time for him to regulate the amount of fishing (F). P should, however, fluctuate around a mean without showing a general tendency to increase (underfishing) or decrease (overfishing). The year to year changes in a fishery can be monitored and fishing methods regulated through devices such as controls on net mesh size, number of nets allowed, and season for fishing, to keep F at the desired level.

F has been referred to in the past as the maximum sustainable yield, but the concept is outmoded not only for the reasons given above that it will be affected by natural environmental changes and is not a constant, but also for economic reasons. Figure 9.4 shows the relationship between yield of a fishery and fishing effort. As the latter increases the yield reaches a peak (the maximum yield) above which the yield is reduced as overfishing interferes with recruitment or growth by removing spawning fish or fish not yet growing at their maximum rates. The yield curve is also a curve describing the value of the catch. As fishing effort increases, so proportionately does cost (the straight line graph in Fig. 9.4). Overfishing eventually leads to cost being greater than value and some fishermen go out of business. This is the usual state of an unregulated fishery.

At an earlier stage, however, the greatest profit (value minus cost) is reached at a lower yield, the maximum economic yield (MEY), than the supposed maximum sustainable biological yield (MSY). A well run fishery,

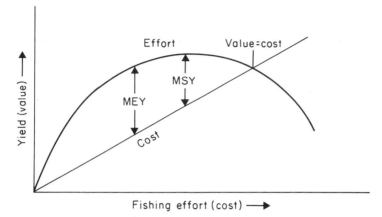

Fig. 9.4. Relationship between yield of a fishery (also its commercial value) and fishing effort (the curved graph) and costs of a fishery and fishing effort (straight line graph). MEY is maximum economic yield, MSY is maximum sustainable biological yield.

in the interests of both fishermen and fish will thus attempt to keep yields around this point.

Fisheries can be managed through sets of equations, or models. Much sophistication has been achieved in the development of such models for marine fisheries [137, 138, 139, 305, 640] where very good data can be obtained from fish landings. Most commercial freshwater fisheries, however, are subtropical or tropical and involve large numbers of fishermen operating individually or in small groups around the indented margins of large lakes. Catch statistics are then not easy to obtain. A fishery for *Oreochromis (Tilapia) esculentus* around North Buvuma island in L. Victoria, however, is notable in that landings are made over a restricted area where fishery scientists managed to account for changes in the population [257, 258]. The fishery has now been over-exploited for political and commercial reasons, and thus provides a doubly interesting example.

9.5 THE NORTH BUVUMA ISLAND FISHERY

An essential of fishery management is that the fished population should be recognizable and discrete. In a small lake the entire population of a given species may provide recruits, but in a large lake there may be several different stocks or sub-populations, separated by geographical barriers, such as deep water, for inshore species, or stretches of swamp. Such sub-populations may be distinguished on the basis of slightly differing blood proteins, detected serologically, by minor morphological differences, or if neither of these exist, by careful observation of the movements of tagged fish. The absolute size of such a 'unit stock' may be determined by the mark, release, recapture method outlined above but an absolute measure of population size, as opposed to a relative one, is not usually necessary for fisheries management. The catch obtained (numbers or weight landed) per unit fishing effort (for example, hours of fishing, numbers of nets set per night, total length of nets set, or numbers of trawls made), called the stock density, is such a relative measure.

The aim of management is to predict future populations of the fish and the effects on them of different fishing intensities. This is best done separately for each age cohort, and depends on recognition of separate cohorts. In temperate regions annual scale or other rings enable this, and length of fish is usually well correlated with age. For tropical fishes like the Buvuma *Oreochromis* the scale rings may be irregular or laid down at 6-monthly spawning intervals. For practical management of the *Oreochromis* fishery it was most convenient to consider the fish in successive length classes rather than age classes.

Consider a cohort of fish of a given length range, for example 23.0–23.9 cm, designated as b. N_b is the number of such fish and they die at the rate M_b from natural causes and F_b from the fishery. It takes a time t_b for a fish to grow through the length range b, so that the number of fish surviving into the next group, $(b+1)$ is:

$$N_{b+1} = \frac{N_b}{\exp{(F_b + M_b)t_b}}$$

This represents an instantaneous abundance, for F and M are instantaneous mortality rates. A measure of the average abundance N_b is given by integration of the above equations for each length category in the range considered:

$$\bar{N}_b = \frac{N_b e}{F_b + M_b}\left(1 - \frac{1}{\exp{(F_b + M_b)t_b}}\right)$$

The problem now becomes one of estimating t, M, and F so that future values of N_b can be predicted for successive length groups until the cohort is completely removed. N_b is given by catch per unit effort after a correction has been applied for any differences in selectivity of different sorts of fishing gear used.

9.5.1 Estimation of t_b, F_b and M_b for the Buvuma *Oreochromis* fishery

t_b can be calculated from measurement of the rate of growth of the fish. Fish grow proportionately more slowly the bigger they become and this growth is often mathematically expressed by the von Bertalannfy equation:

$$L_{b+1} = L_\infty - \frac{(L_\infty - L_b)}{\exp(Kt_b)}$$

where L is length, and L_∞ the asymptotic maximum length attained by the fish as its growth rates drops in old age. K is a constant defining the rate of deceleration of growth. Fish in the length groups b, $(b+1)$ etc., must be aged and L_∞ established from a plot of L against age. K can be calculated from rearrangement of the equation, expressed graphically and an estimate of t_b can then be obtained.

Fishing mortality is related to fishing effort and a graph can be plotted of total mortality for a length or age class against effort (Fig. 9.5) from several years' data. Total mortality is given by:

$$N_{b+1} = \frac{N_b}{\exp(Z)}$$

where Z is the coefficient of total mortality and N_{b+1} and N_b are the numbers (given by catch per unit effort) of the cohort at successive times. The intercept

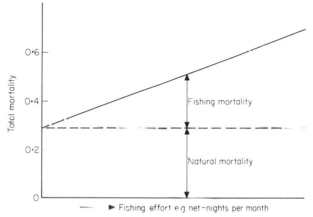

Fig. 9.5. The determination of natural mortality from the relationship between total mortality and fishing effort.

on the total mortality axis gives the mortality in the absence of fishing, M, and the curve gives the fishing mortality, F, for any given degree of effort. A conservative assumption is that M does not decline as F increases. Fishing mortality can also be inferred by tagging fish and finding the proportions of tags recovered by the fishery. The fishing mortality in the *Oreochromis esculentus* fishery can be regulated very closely by the use of different sizes of mesh in the gill nets that are used.

The original equation for N_b, above, can now be solved for all length groups (approximately age cohorts) in the fishery and the following variables calculated for future periods of the fishery: biomass (the product of abundance and average weight of the length group as a total for all length groups); numerical catch (the average abundance multiplied by the relevant fishing mortality in each length group); catch in weight (the product of numerical catch and mean weight as a total for each length group).

Table 9.1 gives values for M_b and F_b for the L. Victoria *Oreochromis esculentus* fishery. It can be seen that 5 in mesh (14.2 cm) does not catch fish below 26.9 cm length but is very effective in killing the 31.0–31.9 cm category whereas 4 in mesh (11.4 cm) kills fish optimally at length 26.0–26.9 cm and not at all above 31 cm.

Table 9.1. Mortality and net selectivity in the Buvuma Island *Oreochromis* fishery. (From Garrod [257])

Length (cm) (*b*)	Total mortality per year (Z_b)	Natural mortality per year (M_b)	Fishing mortality per year (F_b)	% of catch in a particular length category for nets of particular mesh size		
				4 in	4.5 in	5 in
23.0–23.9	0.275	0.01	0.265	9.9		
24.0–24.9	0.307	0.01	0.297	38.6		
25.0–25.9	0.325	0.01	0.315	82.5	4.4	
26.0–26.9	0.372	0.01	0.362	98.9	19.1	
27.0–27.9	0.396	0.016	0.380	65.5	51.6	1.0
28.0–28.9	0.465	0.024	0.441	24.0	89.5	11.0
29.0–29.9	0.517	0.122	0.395	4.9	98.0	32.0
30.0–30.9	0.625	0.266	0.359	0.5	82.6	73.5
31.0–31.9	0.761	0.536	0.225		46.9	99.0
32.0–32.9	0.929	0.790	0.139		25.5	91.5
33.0–33.9	1.314	1.018	0.296		10.2	63.5
34.0–34.9	1.996	1.124	0.872		4.1	34.0
35.0–35.9	4.954	1.138	3.816			17.5

Figure 9.6 shows the effects of various levels of fishing effort (which is proportional to fishing mortality) on the steady state biomass of the stock, the weight of catch, and the catch per unit effort. The effects are more extreme at the smaller mesh sizes because they crop fish whose growth rate is still higher than the natural mortality.

There would thus be advantages in using only 5 in mesh nets, for these keep the population biomass high, thus favouring the maintenance of spawning and recruitment and keeping the fish mortality high, between 0.5 and 1.0. Decreasing the mesh size further does not increase the catch, but decreases the stock slightly and the catch per unit effort markedly. This means an uneconomic use of labour and gear. For the fishery in question, $F = 1.0$ is equivalent to the setting of 46 000 net-nights per month or about 1500 nets per night. Each net is a standard length.

For various reasons the fishery has not been rationally exploited as it might have been with the help of Garrod's model. Mesh sizes much smaller than 5 in have been used and events have taken the following course (Fig. 9.7).

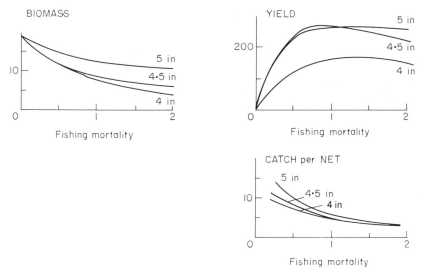

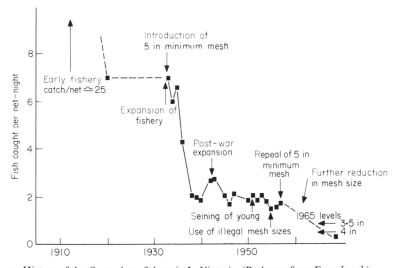

Fig. 9.6. Predictions of the fishery model for the Buvuma Island (Kavirondo Gulf) *Oreochromis* fishery for different levels of fishing mortality (related to fishing effort) and different mesh sizes of gill net. (After Garrod [257].)

Fig. 9.7. History of the *O. esculenta* fishery in L. Victoria. (Redrawn from Fryer [241].)

Between 1953 and 1956 the fishery was fairly stable, with $F = 0.3$ and 153 000 net-nights fished per month. The model shows that a catch per net of 1.44 fish was obtained, which was marginally profitable for the fishermen. It was not the maximum yield that could be sustained biologically, but the amount of fishing needed to attain this is often uneconomical. In 1955 and

1956, for unknown reasons (perhaps poor recruitment) the catch per net fell. Legislation in force forbade nets of less than 5 in mesh and had been in force since the 1930s. Illegal $4\frac{1}{2}$ in nets were used and the yield temporarily went up because a section of the population previously unexploited was now being fished. The minimum 5 in net legislation was repealed.

Under the new conditions the model would predict a catch of 1.8 fish per net-night and catches of around 2.0 per net-night actually recorded provided part of the validation of the model. The increased catch provided incentive to the fishery, which led to more nets and a renewed decrease in catch per net to 1.36. This was smaller than that previously, but with a 25% reduction in biomass of the stock. It is believed that this reduced the rate of recruitment to marginal levels.

In recent years both 4 in and $3\frac{1}{2}$ in nets have been used! These certainly remove some fish before they have begun to breed and must mean that the yield of the fishery can now only be temporarily sustained at very low level. The catch per net was 0.35 in 1968. More recent statistics are not available but in any case the fishery has been disrupted by introduction of the Nile perch to the lake, a matter which is discussed later.

9.6 APPROXIMATE METHODS FOR YIELD ASSESSMENT

The single species approach to managing a fishery, described above, despite its limitations, offers the best method in principle. It has worked well in the regulation of marine fish stocks by the developed nations though it has been less used in fresh waters. It is in the developing world, however, that inland fisheries are of the greatest importance and here the diverse fish faunas and catches, lack of enough trained fishery biologists and more diffuse nature of the fisheries, generally preclude such detailed methods. Events also may overtake the models as has happened in the *Oreochromis* fishery described above.

Frequently a *laisser-faire* attitude with no regulation at all leads to great variations in yield with failures at some times, and the policies of governments may be towards maximum short-term exploitation to solve immediate problems with little planning for the future. In these areas some approximate method of predicting a safe yield of groups of fishes is better than no regulation at all (Ryder [700]).

The basis of such methods is first to predict the likely total stock of fish, then to determine an annual yield of this from simple empirical equations. One way of predicting the yield directly is from the relationship between morphoedaphic index (Chapter 6) and yield determined from a series of lakes in the area and then applied to further lakes. A second way is to use the particle size hypothesis of Sheldon *et al.* [739].

This hypothesis claims that approximately equal concentrations of biomass occur in a food chain for different particle sizes, segregated by order of magnitude. In the equatorial Pacific Ocean Sheldon *et al.* found about equal concentrations of zooplankton (10^3 μm), small fish and squid (micronekton) (10^4 μm) and tuna fish (10^5 μm).

Thus if estimates are available of the biomass of one particle size group which is easily determinable, for example the phytoplankton or zooplankton, that of the group of exploitable large fish can be inferred. Then this value (B) can be inserted in the empirically determined yield equation of Gulland [306]: $Y = kMB$, where Y = yield, k is an empirically determined constant (about 0.4) and M is an estimate of the average natural mortality per year of the group of fish concerned. For Lake Nasser in Egypt, B was determined as 10^4 tonnes [202] from zooplankton estimates. Data were available for M for a variety of fish (Table 9.2) from the ages of the fish caught. To obtain the

Table 9.2. Determination of long-term potential yield of L. Nasser by use of the particle size hypothesis. (After Ryder and Henderson [701])

Species	Mortality coefficient (M') (year^{-1})	Proportion of total biomass (B')	$M'.B'$
Sarotherodon niloticus	0.25	0.56	0.14
Alestes sp.	0.4	0.18	0.072
Labeo sp.	0.25	0.09	0.023
Lates niloticus	0.1	0.07	0.007
Bagrus bayad	0.15	0.04	0.006
Clariidae	0.3	0.01	0.003
Schilbeidae	0.6	0.01	0.006
Other	0.5	0.04	0.020
		$\Sigma M'B' = 0.28$	

product MB, the individual mortality coefficients (M') were multiplied by the proportion that each fish constituted of the total biomass (B') and then $\Sigma M'B'$ was multiplied by 10^4, to give a final value for Y of about 1100 tonnes [701]. This value was similar to that obtained by other techniques and suggests that the particle size hypothesis might be more widely applied. The power of this general approach seems either corroborated or rendered highly suspect by its ability to predict the stock of the Loch Ness monster [715, 738] an animal which may not even exist [555].

9.7 CHANGES IN FISHERIES

Management of any fishery is vulnerable to the fact that fish growth and recruitment are subject to natural environmental change particularly of climate. More often, however, problems are caused by changes wrought by

human activities. These may be indirect, having an ultimate effect on the
fishery, or they may be deliberate, but unsuitable manipulations of the fishery
itself. An example of the interaction of both kinds is that of the N. American
Great Lakes, and of the latter, some of the E. African Great Lakes.

9.7.1 **The North American Great Lakes**

The North American Laurentian Great Lakes stretch almost half-way across
the continent, and are among the largest lakes in the world. They drain a
huge area, originally of conifer forest to the north and deciduous forest to the
south, and were once probably all well oxygenated, clearwater lakes with
maximum total phosphorus concentrations probably less than $2 \mu g \, P \, l^{-1}$,
that is towards the lower end of the fertility continuum (see Chapter 6). The
waterway they provided from the Atlantic Coast to what is now the mid-west
and plains region of the USA and Canada was a main route by which the
continent was explored by Europeans in the seventeenth and eighteenth
centuries. The discovery of minerals and cultivable land led to great increases
in population. For example, the catchment of L. Erie, which in 1750 was a
largely unexploited wilderness supporting perhaps 100 000 people, now
contains some 12 million people with their associated industry and agriculture.
L. Superior has changed the least in 200 years—its catchment still contains
much intact forest—and changes have been greatest in the shallow L. Erie,
and intermediate in the other three Great Lakes, Michigan, Huron and
Ontario. The most apparent change in the twentieth century has been
eutrophication, though other changes have been equally significant and
began in the eighteenth and nineteenth centuries. These changes are reflected
in the commercial fish catches of the waterway [112, 749, 750].

Total commercial fish yields in the American Great Lakes over the past
century have remained relatively constant in Lakes Superior, Michigan and
Erie, but have declined in Lakes Huron and Ontario. The data mask the fact,
however, that great declines in the fisheries for prized salmonid, coregonid
and other fish have been compensated for by catches from a much less
diverse fish community supported by increased fertility and production in
the water. The reasons for the changes, in the order in which they became
significant were: intensive selective fishing, modification of the tributary
rivers, invasion or introduction of marine species, and lastly, eutrophication.

The changes began in L. Ontario, the lowest basin on the waterway and
the earliest to be settled by Europeans. Progressively the changes have spread
upstream so that the state of the upper Great Lakes fisheries in 1970 was
approximately that of L. Ontario in 1900. Atlantic salmon, *Salmo salar*, were
only ever present in L. Ontario in the Great Lakes system, for their upstream
spawning migration from the sea was blocked by the Niagara Falls. Fishing
for salmon began in the 1700s, was reduced by 1880, and had ceased
altogether by 1900. The fishing was intensive, but the species survived it for

many decades, and seems to have disappeared because of changes in the streams flowing into the lake where it moved to spawn. In the nineteenth century, forests were cleared over much of the catchment, dams were built on the streams so as to operate water-power for the saw-mills, and much waterlogged sawdust floored the streams themselves.

Clear felling of the trees has two effects on drainage streams, other than chemical ones (see Chapter 3). It reduces their flow in summer because more water evaporates than previously, and it increases their temperature, because the streams are no longer shaded. Repeated attempts to re-establish an Atlantic salmon fishery in L. Ontario have failed because the spawning stream waters are now too warm and the flows insufficient to maintain the cool, oxygenated water and gravel bottom which the salmon require.

In the upper lakes and in L. Ontario, after the salmon declined, whitefish (*Coregonus clupeiformis*) and lake trout (*Salvelinus namaycush*) were both heavily overfished, and fisheries for them had declined in L. Erie by 1940 and L. Huron by the 1950s. A trout fishery is now maintained by annual stocking. Gradually, however, as the trout and coregonine fisheries became less profitable, lake herring, or cisco (*Leucichthys artedi*) and deep-water ciscoes (*Leucichthys* spp.) were progressively fished out. Currently, the lake fisheries depend on percids and other fish which have been favoured by different changes taking place in the lake. Overfishing for the salmonid and coregonid species was probably not the only cause of their decline.

The sturgeon (*Acipenser fulvescens*), although a valuable commercial fish, was deliberately removed because of the damage it did to nets. By 1890–1910 it had almost disappeared, partly from overfishing, as its valuable by-products (gelatin, isinglass, a bladder extract used in clarifying beverages and sizing textiles, and caviar) were prized, and partly from ruination of its spawning habitat (see below). It is a particularly vulnerable fish, with a low growth rate and late sexual maturity. Even following a ban on commercial fishing, it is now common only in parts of L. Huron.

The changes in tributary streams which so affected salmon reproduction in L. Ontario became widespread elsewhere in the late nineteenth and early twentieth centuries. Most of the commercially exploited fish were those of shallow water, which entered streams to spawn. Although the physical environment for spawning may not have changed too seriously for many of the species, congregation of sturgeon, coregonids, and percids in the water below mill dams provided easy fishing with seines, dipnets, and even spears. In this way two agents of change may have combined to produce declines. In the case of the sturgeon, drainage of swamps and marshes associated with the headwaters removed a favoured breeding habitat.

Between 1860 and 1880, two marine species, the sea lamprey (*Petromyzon marinus*) and the alewife (*Alosa pseudoharengus*) entered L. Ontario. They may have come up the St. Laurence River, which opens out at the northern edge of their ranges, or via the canal built in the early 1880s between the

Hudson River and L. Ontario. The Hudson River enters the sea at New
York. The lamprey is not a fish, but a cyclostome, which is parasitic on fish.
It feeds by rasping fish flesh with a tongue after it has attached by a sucker
which forms its jawless mouth. Both species could have entered L. Ontario
at any time in the previous centuries, but if they did, they were unable to
establish significant populations. Possibly the community changes caused by
fishing and stream modification provided suitable niches for them. The Erie
and Welland canals, both of which, for navigation purposes, bypass the
Niagara Falls, removed an otherwise impassable barrier to migration of
these species upstream to the upper lakes. The lamprey reached L. Erie by
1921, L. Huron and L. Michigan in the early 1930s and L. Superior in 1946.
The alewife generally lagged behind, reaching Erie and Huron in 1931–1933,
Michigan in 1949, and Superior in 1953.

Both immigrants like deep water for parts of their life cycles, and are
neither abundant nor problematic in the relatively shallow L. Erie. In the
other lakes they have caused major changes. The pattern appears to have
been one of parasitism by the lamprey, firstly of the larger deeper water
carnivores, such as lake trout, burbot (*Lota lota*), and deep water ciscoes,
then of smaller species, until the lamprey population itself declined.
Reduction of the large piscivores then apparently allowed increase of the
alewife, predation on which was reduced, while the alewife feeds aggressively
on large zooplankters and benthic Crustacea, such as *Pontoporeia*. These are
also the main food sources of the young piscivores. whose populations cannot
then recover from the lamprey depredations. Alewives may even increase
their competitive advantages by also feeding on the young of the large
piscivores, and have been among the commonest Great Lakes fish.

With all of these influences it is difficult to separate out the effects of
progressive eutrophication this century. The mechanisms by which certain
species disappear on eutrophication are not fully known. They may involve
loss of gravelly spawning habitat as increased sedimentation covers it with
organic deposits which easily become too deoxygenated for salmonid eggs to
survive. The burbot lives and spawns in the deepest parts of well-oxygenated
lakes, from where hypolimnial deoxygenation may force it at an early stage.
Extensions of marginal plant-beds, in providing cover, spawning habitat and
abundant invertebrate food may favour some species unable previously to
compete successfully. Because of the complexity of fish biology, however, it
is impossible yet to state quantitatively at what stage of enrichment major
changes in fish community will occur.

Conditions leading to summer total phosphorus or chlorophyll *a*
concentrations in the epilimnion of the order of 20–30 µg l^{-1} have led, in L.
Erie and in similar large lakes such as the European Bodensee, to
predominance of percid and cyprinid fish. Extreme eutrophication, which
may result in total phosphorus concentrations ten times as high, leads to little
further change in the fish communities, compared with that which is

associated with increases from less than 10 ug l^{-1} to more than about 20 μg l^{-1}.

This account of the N. American Great Lakes has been much simplified. Such large lakes are not uniform and the diversity of their biota and the complexity of its interactions are still very great. Attempts are being made to reverse the changes that have occurred by use of phosphate-stripping (see Chapter 6) to limit eutrophication, and restriction (ultimately it is hoped a complete ban) of detergent phosphate use, by restocking of some fish and use of lampricides. But the very long flushing times of the water masses of the larger lakes delay the effects of remedial action. Long-term planning and foresight are thus essential. The fact that four of the lakes straddle two countries, the USA and Canada, adds administrative complications, which are, apparently, being overcome. The sheer volumes of the water masses concerned have fortunately buffered the changes, which are still going on. The introduction of salmonid game fish in L. Michigan (Chapter 7) is apparently reducing the alewife population and may have yet unforeseen effects on other fish.

9.7.2 The East African Great Lakes

Tropical Africa has undergone many geological and climatic changes. Yet none of them has been so devastating and widespread for the freshwater fauna as the glaciations which removed completely the freshwater habitat from much of the temperate land surface. African fish communities have thus had a long period of development in which there has been an extensive series of fresh waters, and speciation has occurred to a high degree (Chapter 2). The Great Lake basins have resulted in a variety of ways (Chapter 6) and in them the fisheries have developed from the technically crude but the biologically sophisticated to the commercially advanced but ecologically unwise.

Initially the subsistence fisheries depended on one of five main methods [397]: addition of natural plant poisons to the water; spears and harpoons; hooks and line; non-return basket traps; and baskets scooped through the shallows. Sometimes these methods have been very cleverly used [875].

On L. Albert, the Banyoro tribe collected grass or brushwood and tied it into bundles which they lowered to the lake-bed in 6–10 m of water attached to a line buoyed by pieces of ambatch wood. Ambatch (*Aeschynomene profundis*) is a leguminous swamp tree with a light corky wood used also for making canoes. Overnight the bundles became colonized by fish, mainly the cichlid *Haplochromis* species, presumably taking cover from predators. The bundles were hauled up and the *Haplochromis* baited alive on small hooks on lines with which the larger tiger fish (*Hydrocynus vittatus*) were caught. In turn, the tiger fish were used to bait large barbed hooks to catch Nile perch, the ultimate quarry.

Baskets, woven from papyrus or other reed-like stems or pliant tree branches were widely used. Women of the Jaluo on N.E. L. Victoria moved into shallow water in groups of seven or eight, each with a small basket on her head and a much larger, wide-mouthed basket to hand. The women converged in a circle and simultaneously swept the large baskets through the encircled water, scooping out the fish and depositing them in the head baskets. Pelicans feeding in the same area apparently employ similar methods—driving fish into a small area which they surround before scooping with their mouths.

Such methods are now rarely to be seen around the E. African Great Lakes, for the demands of increasing population have brought about an expansion of artisanal fisheries, largely based in shallow water and catching the readily filleted and tasty tilapias. The keys to development of these fisheries have been the introduction of gill nets and of sailing boats, probably originally by the Arab traders in the nineteenth century. Nets were at first of twine and then flax or cotton, but rotted easily. Rayon, laboriously picked from the linings of old car tyres, provided a more durable material, eventually to be replaced by custom-made monofilament nylon nets. These gill net fisheries have been extremely successful. They employ cheap gear and a large number of people, and exploit a group of fish which has many of the characteristics of the ideal fishery fish [245] in the offshore open water down to about 20 or 30 m. Artisanal fisheries seem ideally suited to the local conditions.

However, commercial ambitions, resulting initially from various groups of European colonists, have led to an expansion of mechanized fishing, Greek settlers brought in the skills of Mediterranean purse-seining to exploit the fish at the centres of the large lakes, where the larger boats were less vulnerable to storms than canoes and sailing dhows. Several species of sardine-like fish (Clupeidae), particularly *Limnothrissa miodon* and *Stolothrissa tanganyikae* are successfully fished in this way in L. Tanganyika. The sardines eat zooplankters, moving to the surface to do so. Predators (*Lates* spp.) of the sardines move with them and are also caught. Lights are used to attract the fish towards the boats which use large nets closed and drawn in by power winches. Such a fishery is very efficient but also liable to overfishing; catches of four species of the predators were halved over 7 years, though those of the sardines have, as yet, increased.

The main problem with mechanized fisheries, however, is that because catches are large, big boats, large-scale docking and cold stores onshore, roads and fuel are all needed. This means investment of capital and also considerable unemployment of fishermen for a large boat run by a crew of six can catch as much fish as a fleet of over a hundred, three-man canoes. Because two different groups of fish are usually concerned, however—the tilapias inshore and the sardines offshore, it is possible for both fisheries to coexist.

Mechanized fishing by power trawls, however, is probably much more damaging than purse seining in several ways [120]. It is inevitably destructive of the littoral zone and takes species which are unsuitable for a fishery as readily as those that are. It was introduced to Lake Malawi in 1968 and took as many as 160 species, 80% of them small haplochromines. Many of these are endemic and highly specialized [237] and have a very restricted distribution, perhaps as little as 3000 m^2 of lake bottom in the cases of some species. Significantly 20% of them disappeared from the catches between 1971 and 1974 and there is a real danger that many may become extinct. If trawling is allowed to continue, extensive conservation areas will need to be set aside if this fish community is to retain its diversity. However, the sustainable yield cannot remain more than modest and a better policy would be to abandon trawling and to support the artisanal fisheries very strongly.

If mechanized fishing poses severe problems, a possibly worse aspect of the moves to intensify fishing in the African lakes is that of introduction of alien species. Fish introductions on a world scale have been extensive in the last 40 years, with perhaps 160 species being transferred, *inter alia* among 120 different countries [27]. Fishery officers have sometimes had the misapprehensions of those agriculturalists who believe that natural selection can be improved upon. The introduction of the Nile perch to Lake Victoria [10] has been a disaster [29] which was foreseen at the time (Fryer [238]) by scientists who were simply ignored.

The Nile perch was once a member of the L. Victoria fish fauna, but this was in the geological past. Its bones are fossilized in sediments some hundreds of thousands of years old. It has been absent from the lake for a long interim in which evolution has produced a remarkable flock of highly specialist small fishes of the genus *Haplochromis*. The argument for introduction of the Nile perch was that it would feed on the small fish ('trash fish') which were not marketable, and would package their flesh into its own very large and easily catchable body.

After an introduction probably around 1957 (the details were not fully recorded) the Nile perch took a little time to expand its population and was not a problem in 1971 [397]. Since then, however, it has become distributed throughout the lake. It has demonstrated its omnivorousness and voraciousness by eating not only the small haplochromines but also other commercially valuable fish, like the tilapias, and has apparently reduced the fish stocks so much that it is now forced to take prawns (*Caridina*) as a major food.

Catching Nile perch requires large boats so that some of the artisanal fishermen on tilapia have been put out of business and it is much less valued at the lakeside (1 shilling per kilo) than the tilapia (30 shillings per kilo). Frozen fillets, however, find a good market in the large towns. As a top predator its yield must inevitably become lower than that which its prey previously provided. Furthermore it is an oily fish which, for rural storage and distribution in an equatorial climate, needs to be smoked, which requires fuel (the less oily tilapias can simply be sun-dried). There is evidence of consequent deforestation, with its attendant erosion and other problems in the area. Clearly the Nile perch population itself cannot be sustained at its present high level, but what the situation will be once some sort of equilibrium has been established is difficult to predict.

With this example and others like that of *Cichla ocellaris* in Lake Gatun (Panama Canal zone) [884] it seems odd that introductions to the African lakes are still being mooted. (*Cichla ocellaris* was introduced to ponds as a sport fish but escaped to the main lake in floods [884]. One of its consequences, apart from a major simplification of the food web, has been an increase in malaria carrying mosquitoes, whose larvae were eaten by one of the fish that *Cichla* has almost eliminated.)

The problem is that some introductions have proved useful, though these have been usually to man-made lakes (see Chapter 10) where a previously riverine fish fauna did not have fish capable of exploiting the open water plankton. *Limnothrissa miodon*, for example, has been introduced to L. Kariba (formed in the 1960s) without problem, and even to the older, but still comparatively recent, L. Kivu, formed perhaps 20 000 years ago. Almost half the British fish fauna is also introduced and there have been few problems. The British fauna was severely depleted by glaciation and an early isolation of the islands by the English Channel prevented much recolonization from Eurasia.

It is a quite different matter to propose introducing the Lake Tanganyika clupeids to Lake Malawi, however, as has been suggested [818]. L. Malawi has its own open water sardine, *Engraulicypris sardella*, which some fishery biologists have claimed to be 'inefficient' because *Chaoborus* larvae, a potential prey of the fish, persist in the lake but are absent from L. Tanganyika. The alleged inefficiency may be grossly misplaced [185].

Introductions, once made, are usually irreversible. A programme of, it is hoped, destructive overfishing has been suggested for Nile perch in L. Victoria, coupled with captive propagation of its threatened prey [677]. It is suggested that the lake could be restocked with the endemic *Haplochromis* species once the Nile perch has been reduced to low numbers. However, quite apart from the theoretical problem of maintaining the endemic species in a habitat for which they have not been rigorously naturally selected, the Nile perch may rise again. The problem is now an indefinite one.

We are left, however, with a quandary. There is a problem of high population, short of protein, in Africa, as elsewhere in the developing world. In their attempts to meet this problem, fishery officers and government officials, however misguided, have doubtless felt that they have been doing the right thing in encouraging mechanized fishing or making introductions. Yet most of the wild fish populations of such ancient lakes are unsuitable for much increased exploitation. A better solution perhaps is an expansion of pond culture [346].

9.8 FISH CULTURE

Pond culture of fish has been carried out for a long time. The Egyptians were culturing tilapia in 200 BC and the Chinese have cultured fish from at least 500 BC [60]. Common carp were retained in ponds in mediaeval Europe to provide winter protein. More recently there has been interest in the culture of highly prized marine and game fish in the developed world [784]. Trout and salmon farms have increased markedly in number in the UK in the last 10 years (see Chapter 4) but produce a luxury product which can be sold for a high price to offset the high costs of production. Such farms depend on feeding high protein food (generally of fish meal made from small marine fish) and are net users of protein and energy. They provide no solutions to food problems though they may be very important as sources of employment in areas such as Western Scotland.

Greater fundamental significance attaches to fish farming in the developing world. Here there are two trends: one towards high technology culture of tilapia (656), which could also become a luxury industry serving rich sub-tropical countries like the Gulf States, and the other towards low technology culture on a village basis [406]. The latter is far more important for it not only helps to preserve the wild fisheries, but also provides protein to those who are really short of it.

Tilapia are ideal fish for both sorts of culture. They taste good, have no fine intramuscular bones, breed early and easily, thrive on cheap plant and algal food and hence produce high yields. They are tolerant of wide temperature and salinity ranges, are relatively free from parasites and diseases and hybridize readily. This latter helps breeding programmes.

At one stage, a rather unsuitable tilapia, *Oreochromis mossambicus* was widely distributed from its African range for culture in the Far East. It has proved a problem for it breeds so prolifically that large populations of small, black, rather unattractive fish are produced, which compete with native fish. Other tilapias, however, if managed well, are very suitable for culture with only moderate technological help.

These species include *Oreochromis andersonii, O. macrochir, O. niloticus* and *O. aureus*. They also breed frequently and large populations of stunted fish may result. This can, however, be controlled by stocking predators like *Channa striata* with them, and by culturing first generation hybrids which are almost entirely male and do not breed further. Males also grow faster than females. Crosses between female *O. nilotica* and male *O. aureus* produce more than 85% males and if *O. urolepis hornorum* is used as the male parent, 100% males can be almost guaranteed in the F1. Treatment of the fry with androgenic steroids can also be used to turn all the fish into males but this begins to introduce a high-technology and potentially costly element into the operation.

Tilapia are not the only suitable freshwater fish for culture. Cyprinid fish

Table 9.3. Culture of fish in Indian ponds. Details are given for one year's management at a fish farm at Killa, Cuttack. (After Jhingran [405])

	Pond number			
	1	2	3	4
Pond area (ha)	0.13	0.16	0.4	0.13
Pond depth (m)	1–2	1–2	1–2	1–2
Species stocked in ratio used				
catla	1	3	2.5	2
rohu	5	4	5	6
mrigal	1	2.5	2.5	2.5
silver carp	4	6	5	5
grass carp	3	1.5	2	2
common carp	3	2.5	2.5	2.5
gourami	3	—	—	0.3
kalbasu	—	0.5	0.5	—
Stocking density (fingerlings ha^{-1})	4450	6250	5000	5075
Initial weight stocked (kg ha^{-1})	204	282	11	241
Fertilization				
inorganic (N:P:K = 6:8:4) (kg ha^{-1} year^{-1})	113	1380	975	1725
organic (cow dung) (kg ha^{-1} year^{-1})	25 000	21 500	21 500	25 000
Artificial feeds (kg ha^{-1} year^{-1}).				
Mustard oilcake and rice bran (1:1)	0	3000	3000	2300
Aquatic plants (*Hydrilla, Naias, Ceratophyllum, Azolla*)	14 400	11 700	—	—
Spirodela	—	—	14 660	—
Lemna	—	—	—	14 000
Gross weight of fish harvested (kg ha^{-1})	2234	5040	3575	4200
Survival (%)	73	87	95	87
Economics (rupees ha^{-1} year^{-1}) Costs				
pond preparation	100	100	100	100
fingerlings	450	600	250	500
fertilizers	310	680	625	920
artificial feed	—	1200	625	625
labour	625	625	1175	980
net depreciation	500	500	500	500
Total cost	1985	3705	3275	3655
Value of catch	8400	11 500	13 400	15 800
Profit	6415	7795	10 125	12 145

are used in India and China, often in mixtures which allow a very full exploitation of the available food supply. For example, the major carps, catla (*Catla catla*), rohu (*Labeo rohita*), and mrigal (*Cirrhina mrigala*) will take zooplankton and other invertebrates, the white bighead or silver carp *Hypophthalamichthys molitrix* eats phytoplankton, the grass carp (*Ctenopharyngodon idella*) eats plants, and the Chinese black roach (*Mylopharyngodon piceus*) and European common carp (*Cyprinus carpio*) will eat benthic invertebrates. Stocked in combination in ponds which can be cheaply fertilized with cow dung, grass cuttings, by-products of agriculture such as rice bran, or even cheap artificial fertilizer, high yields can be obtained (Table 9.3.). Even raw human sewage can be used if diluted and the ponds stocked with air-breathing fish like catfish and murrel originating from deoxygenated swamp habitats (Chapter 5). Fish from such ponds must be well cooked because of the presence of various trematode parasites and tapeworms for which the fish form an intermediate host. Pigs may be penned over ditches leading to such fish ponds, which they fertilize and ducks may also be used to provide continuous fertilization [346]. Rice fields, after the rice has been harvested and the stubble left, can also be used for culture though in recent years the use of pesticides to improve rice yields has prevented many fields from being used.

In general, low-technology culture of several species together is very valuable and poses few problems, beyond those of parasite transmission. Some of the Indian major carps will not spawn easily but can be induced to do so by a simple injection of ground-up pituitary gland preserved in alcohol from a wild fish which was just about to spawn. There is a vast number of small ponds and ditches in the developing world that can be used—those created for irrigation, stock watering, water chestnut cultivation and flood control for example. Yields, particularly where the ponds are fertilized with village wastes, are much higher than those of wild fisheries. The latter might produce, in the Tropics, up to a few hundred kg ha^{-1}; intensive pond culture can realize as much per m^2, though this requires expertise and heavy feeding. On a village scale yields 10–100 times those of wild fisheries would not be unreasonable, however. Such village polyculture represents a sensible and helpful approach to the problems of the poor, whilst complementing the wise management of lakes and their fisheries.

10

The Birth, Development and Passing of Lakes

10.1 INTRODUCTION

Like people, lakes have a distinct start—some of the ways in which their basins are formed were discussed in Chapter 6; they also have a lifetime of development and change. Eventually they may fill in and are lost to the landscape. This chapter is about the life history of lakes; it seeks any common pattern that might accompany their development.

There are several problems in studying lake history. Most natural lakes were formed at least several thousand years ago, and though occasionally a new lake will be formed by a landslide, or the retreat of a polar glacier, these events may not be readily amenable to study. Secondly, the life spans of lakes are much greater than those of freshwater ecologists, so that direct observations of development are not possible. Thirdly, there is the problem that any fundamental pattern in the development of lakes is likely to be obscured by the severe human interference of the past century.

On the other hand, in the last 30 years we have created a large number of new lakes by the damming of rivers for hydroelectric power generation and the storage of water for domestic, industrial and irrigation use. This continues a tradition beginning at least 2000 years ago with the ancient dams at Anuradhapura in Sri Lanka and Angkor Watt in Cambodia, though the number and size of such dams has recently increased quite remarkably. An amount of water equivalent to a third of that present in the Earth's atmosphere at any one time is now stored in reservoirs, some of which are so large as to be claimed the only man-made structures visible from space. Often the early stages of formation of these lakes have been studied and this gives us some information about the infancy of lake development.

The course of later change over periods up to many millenia can be followed indirectly through a study of the sediment deposits in the lake. These are often laid down in an orderly chronological sequence and contain a wealth of information both chemical and biological. From them can be learnt the details of changes in both lakes and catchment so that to some extent the effects of natural and human-induced changes can be teased out; and when the lake ceases to exist as a body of water, its sediments remain to give details of its final stages.

'I nere is an important applied element to these studies also. Creation of new lakes is usually heralded as an improvement to the lot of mankind [444, 624]. For the vast majority of people affected one way or another by them this may in retrospect be seen as the propaganda of a powerful dam-building industry. The lakes may create more problems—of social disruption and disease in the tropics—than they solve.

Applied aspects of palaeolimnology (the study of lake history from sediments) are also important for they may give an objective picture of the changes made to lakes by eutrophication and acidification. Used appropriately, palaeolimnological data can be used to influence decisions on how the environment should be managed.

10.2 MAN-MADE LAKES

In the 1960s and 1970s, several very large lakes were created in Africa (Fig. 10.1), and much information resulted from the early stages of these [506]. The building of big dams has largely ceased there in the 1980s, but in China, Brazil and South East Asia equally large schemes are currently being enacted or are proposed [102, 103, 104].

The water rising behind a newly-closed dam is often turbid from suspended sediment in the river and from erosion of the newly-flooded soils at the water's edge. Production is then low, for the turbidity prevents much phytoplankton growth. Gradually, however, as the new lake becomes larger the volume to circumference ratio increases and the products of erosion are much diluted. Progressively, also, the incoming river silt is deposited in deltas at the river mouths as the current flow is decreased by the water mass. In L. Kariba (Zimbabwe–Zambia) there was no early turbid phase, for river silt is deposited in vast swamps just upriver from the lake basin, but in other lakes it has been prominent.

Flooding of the river valley vegetation kills it and it starts to rot underwater. This has two consequences. First, there may be deoxygenation of the bottom water with production of H_2S. In L. Volta in Ghana (Fig. 10.1) the whole of the water was deoxygenated for a time as the flooded softwood forests began to decay. Secondly, the decomposition and the waterlogging of previously terrestrial soils releases nutrients such that the conductivity of the water may rise, and there is a considerable internal loading of nitrogen and phosphorus compounds. The rise in L. Kariba was from 26 mg l^{-1} total dissolved solids in the river to 67 mg l^{-1} in the early lake. This in turn stimulates increased growth of phytoplankton, aquatic plants and fish. L. Kainji (Nigeria) which has a very short water retention period, and hence is frequently flushed out, consequently did not undergo these stage, nor did L. Nasser-Nubia, where 80% of the basin was of rock and desert sand.

In L. Kariba, however, the most spectacular example of this burst of

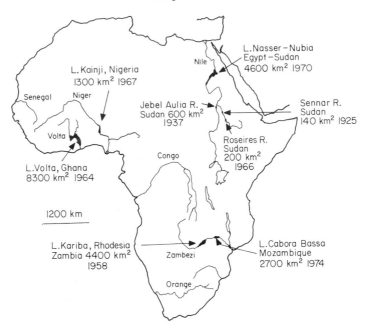

Fig. 10.1. Some major man-made lakes in Africa, with their areas and dates at which the lakes began to form. (Based on Lowe-McConnell [474].)

production was the rapid colonization, in the first few years after the dam was closed in late 1958, of a floating fern, *Salvinia molesta* (Fig. 10.2), then called *S. auriculata*. *Salvinia molesta* is a robust plant, up to 30 cm long with overlapping leaves densely covered with hairs. These prevent wetting and waterlogging of the plant, which also has air spaces which give it buoyancy. *Salvinia* spp. normally can reproduce sexually—sporocarps are produced on trailing underwater stems—but *S. molesta* vigorously produces stolons and can rapidly build up a dense mat on the water surface. This may itself support a 'sudd' vegetation of other species, like *Scirpus cubensis*, rooted in it. Subsequent investigation has shown that *S. molesta* sporocarps are mostly empty and sterile. It seems to be a pentaploid hybrid of two related South American species, *S. biloba* and *S. auriculata*, perhaps of horticultural origin [542].

   S. molesta was very much a problem plant in the new L. Kariba in the early 1960s for it impeded navigation and the use of fishing nets. It also created anaerobic conditions in the shallow margins of the lake which were otherwise likely to be most productive of fish for the shoreline villages. *Salvinia* spp. do not form monospecific stands in South America, where they are endemic, and it is a mystery how the hybrid species has reached East and Central Africa and also Ceylon and Indonesia, where it does form such stands. Probably it was introduced by botanists, aquarists or water gardeners,

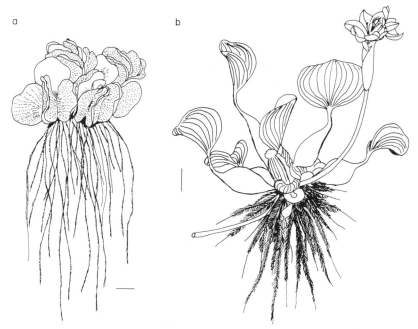

Fig. 10.2. Two floating aquatic plants which have caused severe problems in the early stages of new reservoirs. (a) *Salvinia molesta* (Scale bar = 1 cm); (b) *Eichhornia crassipes* (water hyacinth) (Scale bar = 3 cm).

as were other noxious water plants like the water hyacinth, *Eichhornia crassipes* (Fig. 10.2). Removed from contact with their endemic competitors and grazers, such plants have frequently become problems demanding expensive control. *Salvinia* was known in the R. Zambezi at Kazungula above the Victoria Falls in 1949 but the river flow presumably prevented build up of large populations until the lake started to form downstream. By 1962 *Salvinia* covered a quarter of the (4400 km^2) lake.

Around 1962 considerable concern was felt about the *Salvinia* problem in L. Kariba, for it seemed that it might permanently hinder navigation and fishing on the lake, and might eventually also cause problems in the turbines of the power station built underground next to the dam. At the Gebel Auliya dam on the R. Nile *Eichhornia crassipes*, a very bulky, beautifully-flowered, floating angiosperm has continually to be controlled to prevent its spread into irrigation ditches below the dam. It would rapidly block the water flow in them. Masses of the S. American *Eichhornia*, originally introduced as a business gift to Louisiana, now clog canals in most of the southern States of America and cost millions of dollars per year in control by herbicides and other means, to which an entire scientific journal is devoted!

Fortunately the growth of *Salvinia* in L. Kariba declined in the mid-1960s

as the supply of nutrients, provided by the rotting of flooded vegetation, was
flushed out of the lake. The water replacement time of Kariba is about 4
years. *Salvinia* growth now seems limited by the availability of nitrate and
the plant covers less than 10% of the lake surface, mainly in sheltered inlets.
It seems that wave action, in the middle of the lake, would in any case have
prevented its covering the whole surface. One view is that it now provides an
extra habitat for wildlife, particularly wading birds which can wander over
it.

Fears that serious floating weed problems would permanently beset other
new lakes have not been realized, though early periods of high biological
production have been characteristic of most. The reason for the Kariba
problem lay in the presence of an exotic species able to exploit the new
conditions in the absence of any competitors. Because aquatic plants
frequently reproduce very rapidly by vegetative means, problems caused by
introductions may be very severe in general. Apart from the water hyacinth,
the spiked water milfoil, *Myriophyllum spicatum*, a native of Europe, has
caused much clogging of ditches and littoral zones in the USA, the pond
weeds *Elodea* and *Egregia* have caused similar problems in the reverse
direction. Currently Great Britain is facing a small, but probably now
growing problem from the introduction, long ago, of the Australian swamp
stonecrop, *Crassula helmsii* [620].

As the internal nutrient loading declines in the new lakes, the hypolimnia
progressively became more oxygenated; the lakes have settled to a less fertile
state, and the initially highly productive fisheries have also declined.

10.2.1 Fisheries in new tropical lakes

The stages of filling of a man-made lake create an extended period of rising
river level for the original river fish fauna (Chapter 5), and not surprisingly
most of the river species flourish initially. As the lake level stabilizes some of
them disappear from the main waterbody but may persist near river mouths
if they need flowing water for spawning. In L. Volta the mormyrids (elephant
snout fish), which were important river species, are examples [634]. In
contrast, species which feed on submerged plants, detritus and periphyton,
such a *Sarotherodon galilaeus, Tilapia zillii* and *Oreochromis niloticus*, have
much increased their proportion of the total fish population in the new L.
Volta. Some niches in the lake may be unoccupied because food sources,
scarce in the original river, may have no appropriate species to exploit them.
No predator on open water zooplankton was present in L. Kariba until two
such fish species *Limnothrissa miodon* and *Stolothrissa tanganyicae* were
introduced from L. Tanganyika in 1965. This might be seen as an unwise
move (see Chapter 9) especially in view of the initial problems caused by the
unintended introduction of *Salvinia molesta* but just as the latter now forms

no problem and increases the diversity of habitat in the lake, the fish introductions seem not to have been retrograde in this case [520]. The fish are eminently suitable for a commercial fishery and were introduced to a relatively depauperate local fish fauna (28 species). The original riverine habitat was subject to severe draw downs which undoubtedly favoured rather generalist fish. As far as is known none of these have been greatly affected by the sardine introductions and most grow better than they did in the river.

As well as the provision of suitable spawning habitat during the initial filling stage, a second factor has contributed to the early high fish production of new tropical lakes. This is the high production of invertebrates in the littoral zone where many of the original trees may remain, dead but standing, for several years. The submerged branches became covered with abundant periphyton, stimulated by the high nutrient levels, which in turn supported grazing invertebrates. Notable among these in L. Kariba and L. Volta was a mayfly, *Povilla adusta*. Its larvae burrowed into the bark and rotting wood of the softwood trees flooded in the southern part of L. Volta, and took advantage of holes bored by beetles (*Xyloborus torquatus*) in the hardwood trees around L. Kariba. The beetles attacked the drowned trees when they were exposed to the air during the annual drawdown (to accommodate the annual flood) of the lake level once the basin was full. *Povilla* normally is found in dead papyrus stems and rotting submerged branches in African lakes.

About a fifth of the future basin of L. Kariba was cleared mechanically of trees and bushes before inundation, and the debris burned. This was to create areas free of snags for the nets of a future fishing industry. At L. Volta such clearance was not done for economic reasons and because it was thought that the softwood vegetation would soon rot down. Artisan fishermen in L. Kariba have found it most profitable to fish, with appropriate small-scale methods, in the areas where submerged trees still remain because fish production is apparently higher there. The submerged trees are an ephemeral habitat, however, which must eventually disappear under the action of wood borers, bacteria and water movement. The production they foster represents one of the last phases of the initial productivity surge of the new lakes, which gives way, after a few years, to a less productive, relatively stable ecosystem.

The experiments unwittingly carried out in the creation of a new lake were of great importance in the development of understanding of limnology. According to long-held ideas on lake fertility, the initially fertile water should have remained highly productive and gradually increased in fertility, as nutrients from the catchment area were presumed to accumulate in the basin. The fact that these lakes became less fertile as an initial supply of nutrients was washed out or fixed in sediments is a major piece of evidence that lakes do not maintain their fertility unless an external loading of nutrients is continually applied (Chapter 6).

10.2.2 **Effects downstream of the new lake**

Once a new lake has filled, the average river flow below the dam may only be a little less (because of evaporation over the expanded water surface of the lake) than it was before. If water is removed from the lake for irrigation, the flow may be less at some times but greater at others—in the dry season when the irrigation network drains to the river. The seasonal pattern of flow is, in most cases, considerably changed and this may cause great changes in the lower river ecosystem.

First, the silt load carried by the river is deposited in the lake itself, and the turbidity of the outflowing water is much decreased. In the R. Nile this silt provided an annual fertilization of delta lands and also a detrital food source which supported a valuable inshore sardine fishery in the Mediterranean Sea off the Nile delta. The fishery has declined greatly since the closure of the Aswan High Dam which impounds L. Nasser-Nubia, and agriculture on the delta now requires increased use of artificial fertilizers [704]. In the R. Niger, below the Kainji Dam, fisheries seem to have been at least temporarily affected because traditional fishing methods depended on the fish being unable to see the nets used in the turbid waters. The regulated river flow no longer drains some of the downriver swamps at times essential for rice cultivation in them [389].

Below the Volta dam a valuable shell fishery could have been completely eliminated if the dam had been located only 10–15 km further downstream [456]. The clam, *Egeria radiata*, is collected from submerged sand banks free of weed in the river and sun-dried for sale. The adult clams thrive in fresh water, but the spawn and juveniles (veligers) require water with about $1\%_0$ salinity. Spawning occurs in the dry season when, before impoundment of the river, sufficient salt water moved in from the coast to give a salinity of $1\%_0$ about 30 km from the river mouth. During filling of L. Volta the river flow was reduced and the $1\%_0$ level moved upriver to about 50 km from the sea. This caused some problems in that the collecting industry was displaced from its usual markets but these were not serious. With impoundment the dry season flows have increased so that the $1\%_0$ level is now only 10 km from the coast, but had the dam been placed much lower downstream this point would have been displaced into an area of the estuary unlikely to have provided suitable habitat for the clam. The significance of this example is that such effects were not foreseen when the dam was planned and it is only by good fortune that a problem was avoided.

Fish migration is another feature of the original river system which may be interfered with by a dam, and has been most studied in temperate rivers. Anadromous fish, like salmon, must be 'helped' over the dam by a pass or lift (Chapter 4).

10.2.3 New tropical lakes and human populations

Modern limnological studies are incomplete if they ignore the effects of man on lakes and rivers. The creation of new lakes in the Tropics has led to some new problems and intensified some pre-existing ones. The new ones are largely outside the scope of the book for they are sociological and psychological. River valleys in the tropics are relatively densely populated and the new lakes have displaced as many as 50 000 people (Kariba), 42 000 (Kainji), 80 000 (Volta) and 120 000 (Nasser–Nubia). Although the moving of the villages was planned well in advance, the modification of a culture closely geared to the seasonal flooding and shrinking of the river and the fishing and farming opportunities presented by this was no easy matter. Chapter 1 provides an example of this. During the period after the move there was evidence of increased mortality, not all of it attributable to infectious disease. It is, however, with water-borne diseases in the Tropics (particularly schistosomiasis (Chapter 5) and malaria) that limnological study can aid public health measures. Water-borne diseases were, of course, features of the rivers before inundation. But the increased shoreline of the lakes and particularly the networks of irrigation channels which are associated with some, for example, L. Nasser–Nubia, have exacerbated locally what is a widespread and serious problem.

10.2.4 Man-made tropical lakes, the balance of pros and cons

The larger man-made lake projects have attracted much criticism and scepticism that their overall costs would exceed the benefits they would confer. Much of this criticism has come from western observers, living in countries which have long enjoyed the benefits of abundant power and well-watered agriculture, and much of it, based on the early 'problem' period of formation of the new lake, has not been borne out once the lake has filled and reached some sort of equilibrium. Other criticism has been well founded, particularly that concerned with the problems of people displaced from their traditional homelands by the flooding of huge areas [282, 283]. Chapter 1 gave an allied example of a general insensitivity on the part of Governments to real human needs. Dams are usually justified on cost-benefit analyses which balance the costs of building against the commercial benefits. Properly constructed analyses should include also the social, environmental and health costs. Frequently these are 'paid' by the local rural populations whilst the benefits accrue to multinational companies gaining cheap electricity in return for capital invested in the dam and to urban elites in the country concerned [322, 598].

Without doubt the Volta scheme could provide abundant cheap electrical power for Ghana for many years. Its effects have been felt only in a minor way, however, for the loans of about £100 million (US $170 million) necessary to build the scheme are yet being paid off, and the non-Ghanaian Kaiser Aluminium, a company which financed the aluminium smelter and which uses much of the generated power, will enjoy a pre-agreed low cost for its power for the rest of this century. Nonetheless, Tema, where the smelter is situated, has established some new industry to supplement the national income and to reduce dependence on imported goods. It was expected that 20 000 tons of fish would be harvested per year from the lake once the initial production surge had passed, but it seems that this was pessimistic for catches are currently running at 40 000 tons years^{-1} after the peak of 60 000 year^{-1} in the late 1960s. This has compensated for the reduced imports of beef protein from countries to the north affected by the Sahelian drought [289].

Looked at in the long-term, the costs of the Volta scheme should decrease as the industrial benefits flourish. For the moment the huge lake has disrupted land communications, and water transport has yet to replace them. A planned irrigation scheme has, for lack of finance, not yet been constructed, but cultivation of the wet mud flats left on drawdown of the water level allows cropping of maize, tomatoes, cow peas, sweet potatoes and others at times when lack of rain prevents growth elsewhere in the country. Schistosomiasis has certainly increased, and it is difficult to justify yet a case that the dam has been beneficial to the people of Ghana as a whole [296, 322].

The Aswan High Dam in Egypt has provoked the greatest controversy and its advantages and disadvantages are on a large scale [455, 704]. Again it has provided a constant supply of energy which will allow industrial development, but also it supplies an extra 19×10^9 m^3 of irrigation water for regular and numerous crops in a country almost exclusively dependent on agriculture but only 3% cultivable. The Aswan scheme has increased this by a factor of 1.6. Early fears that the lake would result in a net loss of water by seepage and evaporation have not been borne out.

The drawbacks of Aswan are more serious, however, than those of the Volta which may largely be those of delays in being able to take advantage of the opportunities offered. Egypt is arid and the evaporation of irrigation water is leading to increased soil salinity, which could threaten yields. The lack of freshwater flow to the Nile delta is also leading to encroachment of sea water further inland and soil salination. Lake Nasser produces 10 000–13 000 tons of fish annually (Chapter 9), but the detritus carried to the sea by the unimpeded Nile supported a sardine fishery of 15 000 tons year^{-1} and probably other fisheries as well. The sardine fishing has declined completely, and the L. Nasser fish must be transported long distances to the centres of population which are near the coast. The fishery balance is certainly a negative one. The problems of erosion of the Nile delta and the need for artificial (and expensive) fertilizers to replace the once free silt have already been mentioned.

Overall there is no simple answer to whether or not the building of big dams and creation of large lakes is a good thing. There is now a considerable body of opinion that most have not been the assets they were predicted to be. The facts that the power benefits and the food produced by irrigation usually go elsewhere than to the local population are frequently cited; the insensitive displacement, and, it is alleged, destruction of rural people and cultures [105] cannot be regarded as a tolerable snag to the fuelling of international industrial markets but only as an outrage. Nonetheless, despite the clearly documented problems of the past, schemes for damming of even greater scope are being prepared. The Grande Carajos scheme is intended to industrialize one-sixth of Brazil, with parts of it flooding up to a third of a million hectares of Indian land, displacing 50 000 people in 34 tribes. And there are proposals to dam the Yangtze River in China to generate 40% of the country's electricity. Some 2 to 3 million people will have to be moved. Spokesmen for the Governments concerned are usually highly enthusiastic for such schemes; whether they have a right to be must be assessed in each.

10.3 THE DEVELOPMENT OF LAKE ECOSYSTEMS

On a famous British radio programme of the 1950s, a bucolic character when asked any question, would slowly drawl that the answer lies 'in the soil'. Translated into an aquatic context—the sediment—he would not have been far wrong if questioned about lake development.

Because they are at the lowest points of their catchments lakes become filled inevitably with particles eroded from the land as well as produced and sedimented in the water. The rate at which they are filled depends on the locality but is often of the order of a few millimetres per year. In water deeper than a metre or so the deposit laid down is a brownish or greyish lake mud rich in planktonic and littoral detritus. In very shallow water it is often a more structured, fibrous peat derived from emergent plants with their more lignified supporting tissues. Build up of a peat deposit from the edges may result in encroachment of vegetation on the previously open water and eventual conversion of the lake to a swamp and even land vegetation. This process is not inevitable, however. In a very large lake, wave action may continually erode the peat deposit of a marginal swamp so that no encroachment occurs. Photographs of the edges of Scottish lochs (lakes) taken over half a century [760] have shown no encroachment, except where the swamp was located on a river delta where allochthonous sediment deposition may have been greater than the rate of erosion by waves. Material eroded from edges, on the other hand, may be deposited in deeper water as lake mud and contributes to the filling process in this way.

At first sight lake mud is somewhat unprepossessing stuff: an apparently amorphous, dark-coloured, often evil smelling, watery goo. With proper analysis, though, it contains far more information about the conditions in the lake under which it was laid down than we can hope to interpret. It contains a record of changes both natural and man-made in the catchment area as well as the lake. It may even harbour a record of global events such as changes in climate and magnetic field.

Ideally, lake mud has been laid down chronologically from the time of origin of the basin. In large lakes with steeply descending shores some slumping may upset the chronological sequence, but with experience of a particular lake this can be recognized and interpreted. From most lakes a continuous sequence of sediment can be obtained using one of a variety of corers designed to take an undisturbed column of the sediment from top to bottom of the deposit. The pathway to discovery of a lake's development is then one of obtaining the sediment core, dating it, analysing it by study of suitable chemical and biological fossils, and then interpretation of the data [48].

The simplest corer comprises a piece of plastic drainpipe, sliced longitudinally in half, then taped together again with waterproof tape. If the water and mud are both shallow, as they might be in lakes formed only a few centuries ago, this can simply be pushed into the deposit until it penetrates the basin material and then pulled up. Success depends on the basin material and sediment being of a consistency such that they stick in the tube when it is pulled up. The overlying water can be siphoned or poured off and the pipe opened to reveal the core. For deeper water or deeper sediments more sophisticated devices are needed. The Mackereth [498] corer (Fig. 10.3) has proved a valuable instrument in recent years, though several other sorts of corer exist.

10.3.1 Dating the sediment

Samples of mud from a corer can be dated by a number of methods, mostly using radioactive isotopes. These usually involve measurement of concentrations of an isotope thought to be produced at a constant rate on the earth and of a derivative formed at a known rate (λ) from its decay. The time (t) during which an amount Ao has decayed to A ($Ao - A$ being measured as the amount of decay product formed or by the concentration of the original isotope left, assuming a constant and known initial concentration) is given by:

$$A = Ao \, e^{-\lambda t}$$

and its derivative, $t = \lambda^{-1} \log e \dfrac{Ao}{A}$.

Choice of isotope depends on the estimated age of the sediment to be dated. The older the sediment, the longer must the half-life (the time taken for half of a given initial amount to decay) be to ensure that it can still be detected. For sediments up to about 30 000 years old, the ^{14}C method can be used and for those up to about 150 years a method using ^{210}Pb is useful.

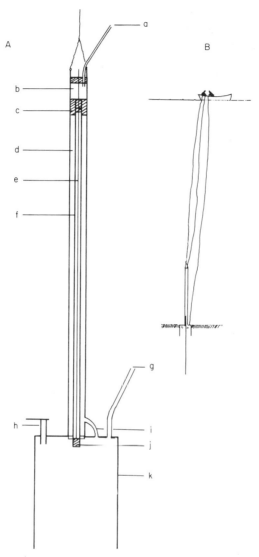

Fig. 10.3. (A) Detail of a Mackereth corer and (B) the corer in operation. The corer is lowered to the sediment and air is sucked out of the anchor chamber (k) through outlet (g). Compressed air is then injected through (a) into chamber (b). This pushes piston (c) down through tube (d) and pushes the corer tube (f) past a fixed piston (j) into the sediment. As (f) fills with sediment, air is displaced from inside it through a fine tube (e). When (c) passes the junction of tube (i) with tube (d), the compressed air is diverted into the anchor chamber (k), forcing it from the bottom of the lake and buoying the whole apparatus back to the surface. A valve (h) releases some of the air in (k) so as to control this ascent. Back in the laboratory the sediment core can be extruded with a piston. (Based on Mackereth [498].)

10.3.2 **Radiometric techniques**

The ^{14}C method depends on the steady formation of radioactive ^{14}C by bombardment of nitrogen molecules by neutrons at the top of the atmosphere. The neutrons are part of the 'cosmic ray' flux from the sun.

$$^{14}N_7 + {}^1n_0 \rightarrow {}^{14}C_6 + {}^1p_1$$
$$\text{(neutron)} \qquad \text{(proton)}$$

^{14}C atoms are produced at the rate of about 10^2 cm^{-2} of the earth's surface min^{-1}, giving a ratio of ^{14}C to the stable ^{12}C of about 10^{-12}. It is assumed that carbon isotopes are taken up, ultimately through photosynthesis, into living organisms in this ratio and that ^{14}C then decays with a half-life of 5700 years. The ^{14}C:^{12}C ratio is measured in material to be dated and compared with the initial ratio as follows:

$$\left(\frac{^{14}C}{^{12}C}\right) \text{current} = \left(\frac{^{14}C}{^{12}C}\right) \text{initial } e^{-\lambda t}$$

For ^{14}C it takes, on average, 8200 years for each atom to decay so that:

$$t = 8200 \log_e \left(\frac{10^{-12}}{^{14}C/^{12}C}\right) \text{current} = 8200 \log e \left(\frac{^{12}C}{^{14}C}\right) \text{current}\cdot 10^{-12}$$

where t is the age of the material being analysed.

The ^{14}C method is not usable for recent (<200 year) sediments, since the widespread burning of fossil fuels with very low $^{14}C/^{12}C$ ratios has upset the previously assumed constant ratio of 10^{-12} in the atmosphere.

^{210}Pb is present in the atmosphere ultimately as a result of naturally occurring ^{238}uranium in the earth's crust. One of the products of uranium decay is ^{226}radium which decays to the rare gas ^{222}radon. This diffuses into the atmosphere where it decays by a series of short-lived daughters to ^{210}Pb. It is washed out of the atmosphere in rain, where its average concentration is around 2 pCi l^{-1}, within a few weeks of its formation, reaches the lake sediments and decays with a half-life of 22.26 years to ^{210}Bi. Measurement of the atmospherically derived ^{210}Pb can then give a useful method of dating recent sediments. A correction has to be applied for the ^{210}Pb derived directly from ^{226}Ra decay in the sediment minerals because the initial concentration of this varies from place to place.

10.3.3 **Non-radiometric methods**

Radiometry is, in general, time consuming and therefore expensive. Other methods that are cheaper, if less versatile, are also available. The earth's magnetic field has changed in both its horizontal and vertical components (declination and inclination) in past times and a record kept in London since before 1600 AD shows a change in declination through some 30° since then.

Some minerals, laid down in lake sediments, become magnetized in the direction of the earth's field at the time they are deposited and some of this magnetization, the remanent, persists. The direction of magnetization can be measured in carefully prepared core slices, kept orientated relative to a fixed line on the corer tube, with a sensitive magnetometer and then plotted against depth in the sediment column. An oscillation in recent sediments from east to west and back has been discovered and compared well with the records kept at the London Observatory. For older sediments comparison with ^{14}C dates shows an oscillation in declination with a period of about 2800 years, with the west and east peaks being contemporary for different lakes. The pattern of remanent magnetization may therefore form a time-scale for other lakes where ^{14}C dating at frequent positions in the core proves too expensive.

Under anoxic hypolimnia, in deep lakes, disturbance of the sediment by water currents and burrowing animals may be insufficient to obscure fine variations in deposition between spring and winter and late summer each year. In these cases light and dark layers, or paired varves, may be detectable. The light member of the pair often contains carbonates deposited as a result of intense late summer photosynthesis. The pairs can be shown to be annual by detection of typical spring and summer sedimented tree pollen in the appropriate members of each pair and dating carried out by counting of the pairs. Such visual varving is unusual, however. Varving detectable by more sophisticated methods may be commoner. X-ray photography of thin longitudinal sections of cores from L. Washington has shown prominent dense bands which are not themselves annual, but their finer striations may be. Similar examination with a stereo-scan electron microscope has shown fine bands of diatom fossils with pairs of layers characterized by spring and autumn species. However, even if these prove to be usable for dating, this method will still be a relatively expensive one for routine use.

10.4 SOURCES OF INFORMATION IN SEDIMENTS

10.4.1 Chemistry

Both inorganic and organic particles are washed into a lake from its catchment and these accumulate in the sediments. Inorganic precipitates produced within the lake are also sedimented and of course the organic matter accumulating from the plankton and the littoral zone also contains elements other than carbon, hydrogen and oxygen. Analysis of sediment for particular elements can thus give information on changes and erosion of soils of the catchment, as well as of production within the lake.

For example, iodine is a scarce element in soils, but is present, from seawater spray, in rainfall, and is strongly adsorbed onto clay and humic colloids which are washed into lakes. Its concentrations therefore seem to reflect the amount of erosion from the soils of the catchment and may indicate soil disturbance through climatic change or human activity, such as deforestation and agriculture [629].

Much information can also be determined from phosphorus determinations on sediments. Because it is a relatively insoluble element, the phosphorus content seems to reflect the rate of loading on the lake, particularly when the rate of sedimentation is taken into account. It is sometimes possible to detect changes in sewage treatment practice (diversion, expansion of the works) from the phosphorus content of the sediments of lakes into which effluent is discharged [612, 778].

Sediments are often highly organic; gas chromatography and mass spectrometry among other techniques are capable of revealing thousands of separate organic compounds in them. Many are derivatives of compounds originally deposited which have undergone slow change or diagenesis, and cannot yet be associated with particular groups of organisms. The compounds as yet identified include hydrocarbons, fatty acids, amino acids, sugars, alcohols, ketones, steroids, and plant pigments such as chlorophyll derivatives and carotenoids.

The organic matter of sediments comprises partly the refractory materials washed in from the catchment (litter, 'humic acids') and autochthonously produced substances. There has been some question as to which source contributes most. In the English Lake District Mackereth [500] noted that the total sediment depth laid down during the period since the lake basins were carved out by glaciers was broadly similar in lakes such as Wastwater and Ennerdale, which have igneous catchments and are very infertile, and in lakes like Esthwaite and Windermere, which lie in more fertile catchments containing some softer rocks. He concluded therefore that the bulk of the sediment was catchment-derived and that organic matter produced in the lake (greatest in Esthwaite, least in Wastwater and Ennerdale) was fully oxidized in the plankton or by the benthos and did not contribute to the permanent sediment (see Chapter 8). This hypothesis seems to hold for these lakes except in the last few decades when several lakes other then Wastwater and Ennerdale have been fertilized particularly by sewage effluent. In these recent sediments organic analysis for carboxylic (fatty) acids has shown a different array of n-alkanoic acids from that in older sediments and in the recent sediments of the unfertilized lakes [123]. Acids with 16, 22, 24 and 26 carbon atoms predominate in the former array, but the C_{16} fraction is absent in the latter. C_{22}–C_{32} acids are abundant in soils but C_{16} acids appear characteristic of algae. The evidence thus indicates a greater preservation of autochthonous material in fertile lakes. Probably this is associated with more severe deoxygenation of the sediment surface in the fertile lakes. Analysis of the carbon skeletons of sedimented hydrocarbons also appears promising, for alkanes between C_{23} and C_{33} appear characteristic of higher plants, C_{31}

is predominant in acid peat and C_{27} and C_{29} in base-rich forest soils [124]. Changes in the vegetation of the catchment may thus be detected in sediments and the information used to complement or further interpret studies of the pollen in the sediment (see later).

The greater preservative properties of deoxygenated sediment surfaces are also reflected in the photosynthetic pigment derivatives preserved. In dead cells most pigments are decomposed by oxidation so it is not surprising that a greater pigment content, and a greater diversity of pigments is preserved in sediments laid down under fertile conditions [712]. The contribution of catchment area vegetation to the sediment pigment content is usually small as the litter has been long exposed to the air before it reaches the lake bed. Extraction of chlorophyll derivatives and carotenoids with methanol or acetone and subsequent measurement by spectrophotometry thus helps confirm phases of increased fertility, but if the pigments are separated chromatographically, and separately measured, even more valuable information may be obtained.

The carotenoids present in different groups of algae are often highly specific and have been used to help classify the varied algal phyla. Detection of particular pigments in sediments may then indicate changes in abundance of particular algal groups in the history of a lake. To date, this technique has been applied to the pigments myxoxanthophyll and oscillaxanthin [79] which occur only in the Cyanophyta (blue-green algae). In the sediments of L. Washington (USA) sediment cores taken in 1967 showed maxima of oscillaxanthin derived from the large populations of *Oscillatoria rubescens* and *O. agardhii* which had appeared in the phytoplankton in the 1950s as the lake became heavily fertilized with sewage effluent [304]. The effluent was diverted completely from the lake by 1967 and progressively the *Oscillatoria* populations decreased. In new cores taken in 1972 it was found that the oscillaxanthin maximum was still present, but now buried under 5.45 cm of new sediments which contained little oscillaxanthin [303]. The pigment thus preserved a record of the blue-green algal growth which characterized the highly fertile phase which the lake had passed through. A deeper oscillaxanthin maximum was also detected and seems to record a phase when raw sewage was discharged to the lake for a few years.

10.4.2 Fossils

A wealth of algal and plant remains is preserved in sediments—the heterocysts and sometimes whole filaments of blue-green algae, colonies of the green algae *Pediastrum* and *Botryococcus*, the silica walls of diatoms and the silicified scales and cysts of Chrysophyta, pollen from higher plants in the lake and its catchment area (and perhaps further afield), and lignified cells such as sclereids, fibres and xylem vessels from aquatic plants. From animals there are the tests of certain amoebae, sponge spicules, Bryozoan statoblasts, parts of the exoskeletons of Cladocera and other small Crustacea, mollusc shells, head capsules of chironomid larvae, mite exuviae and even an occasional fish scale.

Because of the closer relationship that algae and plants, rather than animals, have with the physico-chemical environment, it is generally easier to interpret the meaning of fossils of algae, particularly diatoms, and of pollen. The former give information mostly about the lake itself, the latter about its catchment. The occurrence of particular animals is a function not only of the physico-chemical environment, but also to a large extent of competitive and predatory relationships (see Chapters 7 and 8). Nonetheless, animal remains are useful and may even throw light on the past operation of some of the predatory interrelationships.

10.4.3 Diatom remains

Diatom cell walls are finely patterned, with dots and lines which characterize particular genera and species. They were studied for a long time by microscopists before they came to the interest of limnologists and consequently a large body of information on the habitats in which particular species occur has accumulated. Considerable reconstruction of past environments is possible from an analysis of the diatoms in sediments. Permanent microscope slides of them can easily be prepared from preparations of diatom walls (frustules) made by heating the sediment with an oxidizing agent such as nitric acid and resuspension of the residual diatoms in distilled water. Diatom walls are of two overlapping halves held together by bands and there are four main groups distinguished by their wall patterning (Fig. 10.4).

With experience, an observer can deduce much about the balance of planktonic (Chapter 7), epipelic (sediment living) and attached communities (Chapter 8) and hence about changes, for example, in water depth and the abundance of plant beds. Certain genera, such as *Eunotia* and *Frustulia*, are characteristic of generally infertile water and others, such as *Melosira granulata*, of very fertile conditions. There are distinct marine species and distinct freshwater ones and some indicators of brackish conditions. In African lakes, *Navicula elkab* and *Nitzschia frustulum* characterize saline inland waters and may be used in sediment analyses to recognize drops in lake level and periods of endorheicity [678].

Diatoms have also been much used recently to trace changes in the pH of waters believed to have been acidified, and to help distinguish between rival theories of the cause of the acidification. Careful taxonomy is necessary and study of contemporary lakes suggests that some species are associated with very high (alkaliphilic) or very low (acidophilic) pH. Others are associated with neutrality (circumneutral) whilst intermediates (acidobiontic and alkalibiontic) are also found. An index which compresses information on the characteristics of all the species from a lake can be calculated. For example

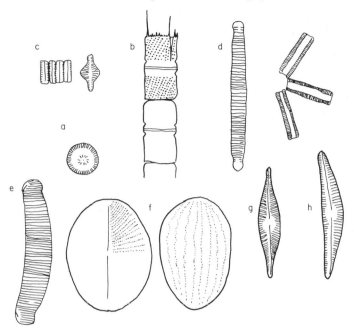

Fig. 10.4. Some representative diatoms. Species are shown after acid cleaning to remove organic matter and reveal the characteristic pattern of their silica walls. (a),(b) Centrales which are usually planktonic: (a) *Cyclotella*, (b) *Melosira*, seen in side (girdle) view. (c),(d) Araphidineae: (c) *Fragilaria*, (d) *Diatoma*—both valve (top) and girdle views are shown; (e) Raphidioidinae, *Eunotia*; these latter groups are usually attached to surfaces, though some may be planktonic. (f) Monoraphidineae, *Cocconeis* seen from each of its two valve surfaces, one of which bears a raphe, the dark line, the other of which does not; raphes are used for movement or attachment. This group is usually attached to plants or rocks. (g),(h) Biraphidineae: (g) *Navicula* (h) *Cymbella*; this group is most common moving over sediments or other surfaces and has raphes on each half of the cell.

index B of Renberg and Hellberg [667] is based on counts of the species present and calculations of the percentages of each group of the total counted:

$$\text{Index B} = \frac{\% \text{ Circumneutral} + 5 \times \% \text{ acidophilic} + 40 \times \% \text{ acidobiontic}}{\% \text{ Circumneutral} + 3.5 \times \% \text{ alkaliphilic} + 108 \times \% \text{ alkalibiontic}}$$

The index can be related to pH of lake water by regression of the measured pH against counts made from the diatoms in the contemporary sediment [223]. For example in a series of 33 lakes in southern Scotland:

$$\text{pH(log units)} = 6.3 - 0.86 \log(\text{Index B}) \ (r^2 = 0.82, \ P < 0.001)$$

Changes in value of pH can then be calculated from measures of the index obtained from a sequence of sediments (Fig. 10.5).

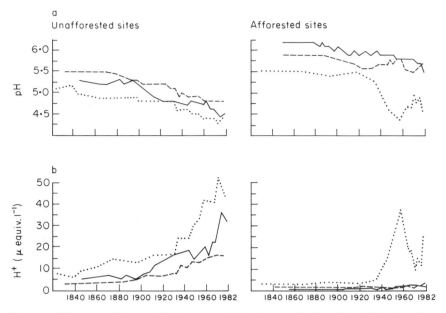

Fig. 10.5. The recent history of lake-water acidity changes in three Scottish lochs, where catchments have not been forested (left hand side), expressed as (a) pH and (b) H⁺ ion concentration, and for three for which part or all of the catchment has been planted with conifers (right hand side). The data (based on reconstructions of pH from the sediment diatom communities) suggest that factors other than afforestation have been responsible for the recent acidification, where it has occurred. (Based on Flower *et al.* [225].)

10.4.4 Pollen

The use of pollen preserved in peat and lake sediments has a distinguished history and has formed the basis for recognition that climate has changed greatly over periods of only thousands of years. The walls of many pollen grains are of particularly refractory waxes ('sporopollenin') which resist the most vigorous oxidizing and corrosive agents. Grains may be concentrated and separated from sediment samples by digestion with hydrofluoric acid. Genera, and often species, are easily recognizable in most groups, though grass pollens for example are not easily separable to species. However, major changes in catchment vegetation can be detected. Agricultural activity is recorded by the appearance of the pollen of characteristic annual herbs which accompany cultivation and evidence of deforestation may be found by a complementary decrease in tree pollen.

10.4.5 General problems of interpretation of evidence from sediment cores

Vallentyne [823] has pointed out that 'no anatomist or physiologist in his right mind would ever base a study of the life history of an organism on the analysis of its accumulated faeces. This is, however, precisely the position of a palaeolimnologist with respect to the developmental history of a lake. Sediments are lacustrine faeces, the residue remaining after lake metabolism'. Clearly, therefore, interpretation of lake development is fraught with difficulties and it is wise to examine some of them before considering interpretations from particular lakes.

Sediment is not necessarily deposited uniformly. Higher rates of deposition may be measured near inflows and in the centres of basins which are conical in shape. Water movements and gravity tend to move fine sediment from the edges to the deepest parts in a process now called sediment focusing. Cores from several places may thus be necessary to obtain a full picture, though even in quite large lakes it is remarkable how similar the picture obtained from different parts of the lake bed is, at least qualitatively. Depending on the lake, microfossils may be deposited unevenly and may or may not be redistributed by water movements. Davis [151] found that relatively heavy oak pollen grains were evenly distributed over the sediment of a Michigan lake, yet lighter ragweed pollen settled slowly and was carried to the windward edges before reaching the sediment. On the other hand she found that the surface few millimetres of sediment were resuspended during mixing periods and that the material could be recycled upwards into the water perhaps four times before becoming part of the permanent sediment. This process tended to even out irregularities in previous sedimentation, both in space and in season.

Benthic animals also may progressively mix the surface sediments as they form, thus evening out annual variations into a sort of moving average and smoothing long-term trends. Davis [152] found that tubificid (Oligochaeta) worms, whose feeding was centred at 3–4 cm depth, displaced sediments upwards and displaced larger particles, e.g. of pollen grains, less than smaller ones. The worms fed with their mouths downwards and their anuses upwards.

Much evidence comes from diatoms and in general these are preserved well. There is some selectivity, however. In deep lakes only a fraction of the diatom silica polymerized during wall formation reaches the permanent sediment [623] and dissolution may occur in the surface sediments of shallow lakes [25]. Long, thin diatoms, for example *Synedra, Asterionella*, tend to break easily by abrasion with inorganic particles or in invertebrate guts, and may then more easily be redissolved. Some very thin-walled genera, for example *Rhizosolenia* are rarely preserved at all. Centric and littoral diatoms seem to persist best, compared with planktonic Araphidineae, though intact specimens, even of these, are to be found in very old lake muds.

In interpretation, much depends on knowledge of the current autecology of particular species, yet unfortunately this is usually poorly known. Much more confidence can be placed on the detection of whole arrays of species commonly associated with particular habitats than on the occurrence of particular species. A further problem of reliance on single 'indicator' species of diatoms is that the diagnostic patterning of the wall may be changed as dissolved silicate levels fall [43]. One 'species' may thus change into another. Sufficient laboratory work has not been done yet to determine how widespread is this phenomenon.

Perhaps the greatest problem, and possibly the most misleading feature of sediment core interpretation, however, is the way in which the results are expressed. Early workers often calculated the percentage representation of a particular species or group of fossils in their total count from a sediment sample of undetermined weight. This gave relative changes in frequency. Thus a decrease in frequency of a species A, relative to that of a species B would be interpreted as a decline in the incidence of A and an increase in B. The true situation might have been an increase or decrease in both species, with B increasing at a greater rate or decreasing at a lesser rate than A.

An improvement has been the expression of results on the basis of per unit weight of sediment, because this gives some indication of absolute changes if the sedimentation rate has been constant. However, where sedimentation rates have changed greatly, an absolute increase in a species may have been accompanied by a decrease in its numbers per unit weight of sediment if the rate of sedimentation of inorganic or organic matter has increased at an even greater rate. The only reliable way of expressing results of counts of fossils or of chemical analyses is thus in terms of amount laid down per unit area of lake bed per year and means that comprehensive dating of the core must be carried out. This is expensive and has not been done for most cores. All of the above limitation should be borne in mind in the following accounts of results from particular lakes, which will illustrate some of the features of lake development.

10.5 EXAMPLES OF LAKE DEVELOPMENT

10.5.1 Blea Tarn, English Lake District

Hills of hard, Borrowdale volcanic rocks up to 600 m high surround the small, upland lake called Blea Tarn in the English Lake District. It is 3.4 ha in area, 8 m in maximum depth and is surrounded by tundra-like moorland of grasses and sedges and by *Sphagnum* bogs. There are some plantations of conifers but the only human disturbance of the catchment is a small sheep farm, and its effect on the lake is likely to be slight. Habitation of the area in the past is also known to have been only sporadic so Blea Tarn provides an example of a lake and its catchment perhaps largely under control of 'natural' events.

A core taken from the lake [332] was a little over 350 cm long (Table 10.1) and was dated from changes in the pollen content. Distinct phases of vegetation change, registered by pollen, have been recorded in north-western

Table 10.1. Palaeolimnology of Blea Tarn, Langdale (after Haworth [332])

Depth (cm)	Zone	Date BP	Sediments	Pollen	Sedimentation rate (mm year^{-1})	Diatoms and interpretation
0–147	VIIB	Sub-boreal 4900	Brown muds, detritus gyttja	Ash, elm declines	0.32	Very infertile conditions
147–227	VIIA	Atlantic 7400		Oak, elm, birch, alder (forest period)		*Eunotia* (infertile water)
227–327	VI V IV	Boreal 10 200	Clay gyttja	Hazel peak		Acidophils increase *Asterionella, Fragilaria, Melosira*; fertile, alkaline indicators increase
		POST GLACIAL			0.36	
327–343	III	LATE GLACIAL	Pink clay	*Artemisia Empetrum Lycopodium*		Diatoms scarce, cold? tarn completely frozen over
343–349.5	II/III	10 700		*Empetrum* and *Artemisia*, juniper, birch		
349.5–353.5	II	Allerød 11 900	Dark, mostly organic silt			*Campylodiscus noricus* (alkaliphilic)
353.5–357 Below 357	I	Pre-14 900	Unvarved pink clay Varved clay	*Salix*, grasses, sedges no pollen	0.07	*Pinnularia, Cymbella, Fragilaria* No plankters

Europe and dated originally by radiocarbon measurement. These phases are numbered I-VIIb (or VIII in more southerly areas) and fall into two groups, the late-glacial (late Weichselian) phases I-III of tundra vegetation, and the post-glacial (Flandrian) IV-VIIIb in which various sorts of woodland developed in north-western Europe as the climate warmed.

Blea Tarn was formed in an ice-carved depression in the debris eroded from the hills by the glaciers and had water in it by about 14 000 BP (before present). For 4000–5000 years its biological production was low—relatively few diatom frustules and little organic matter are present in the clay and silt late-glacial sediments. Around 11 900 BP there was a brief period of warming, the Allerød interphase (named after a site in Scandinavia where it was first detected in deposits) in which the sedge, grass and dwarf willow tundra was briefly diversified with birch trees and juniper bushes. The lake sediments were then slightly richer in organic matter and diatoms; the diatoms present in the late-glacial period included few plytoplankters and were of a mixed collection, some characteristic of slightly alkaline water.

Final melting of the ice, a little more than 10 000 years ago in this area, led to marked changes in the general area and lake. Hazel (*Corylus avellana*) colonized, followed by birch (*Betula*), oak (*Quercus*), elm (*Ulmus*) and, in the wetter phase of VIIa (called the Atlantic period), alder (*Alnus glutinosa*). The lake sediments became less inorganic and ultimately were of largely organic, brown lake muds. The diatom fossils indicate a post-glacial phase of high fertility, with a well developed plankton. Species such as *Asterionella formosa* and *Rhopalodia gibba*, and the genera *Fragilaria* and *Melosira* were common and indicate an increase in nutrient loading. This may be attributed to the presence of much-pulverized (and hence easily leached) rock left by glacial action and an increase in run off, as temperatures increased, to transport the nutrients to the lake. The organic matter of the sediment was probably catchment-derived from the deepening soils of the woodlands near the lake. As phases IV and V progressed from around 10 500 BP to 7400 BP the fertile lake diatom flora did not persist. Progressively it was replaced by one of less fertile, more acid waters characterized by *Eunotia* and *Pinnularia*, which has persisted to the present day with little change in 7000 years.

The decrease in fertility of the lake is attributed to exhaustion of the readily-available nutrients in the glacial debris and subsequent dependence on the slow weathering of fresh basal rock at the base of the soil profiles. This is extremely slow and studies on the composition of present day waters of upland lake district tarns show that it differs little from that of rainfall (Mackereth [497]).

10.5.2 **Esthwaite**

A glacier flowing down a small river valley, also in the English Lake District, carved out the basin of Esthwaite and dammed the lake with a moraine at the foot of the valley. It is a larger lake (100 ha) than Blea Tarn and lies among more subdued scenery of softer Silurian rocks which are more readily weathered than the Borrowdale volcanic rocks around Blea Tarn. For all but the last few thousand years, however, Esthwaite had a similar history to that of the smaller lake, and rather more information is available on the chemistry and animal remains of the sediment as well as on diatoms and pollen.

The late-glacial phases (I–III) (Table 10.2) supported tundra around the lake and clays were deposited in the presumably rather cloudy water as the fine debris of the glacial till was eroded. The warming in phase II was marked by the appearance of Cladocera in the water, particularly species of Chydoridae, a group usually associated with the shallow littoral zone (Chapter 8), but diatoms were not well preserved, perhaps partly because of abrasion by the silt and clay. The concentrations of sodium, potassium, calcium and magnesium in the sediments were around 10, 30, 5–10 and less than 10 mg g^{-1} dry wt respectively.

With the start of the post-glacial period the surge in production noted for Blea Tarn was recorded in the sediments. The leaching of broken rock fragments led to increases particularly of calcium (22 mg g^{-1} dry wt) and phosphorus in the sediments. The latter increased from about 3.5 to 5 mg g^{-1} dry wt as phase VI was succeeded by the early part of phase VIIa. Diatoms (*Melosira arenaria, Epithemia, Fragilaria*) of a fertile phase and planktonic Cladocera numbers increased markedly. Analysis of the organic component of sediments laid down in phase VIIa has shown a ratio of n-alkanoic to branched and cyclic monocarboxylic acids similar to that found currently in the surface sediments of fertile lakes.

As phase VIIb was entered about 5000 years ago, however, the regression to less fertile, more acidic conditions found in Blea Tarn took place. Calcium concentrations fell to about 5 mg g^{-1} dry wt and those of phosphorus to about 2 mg g^{-1} dry wt. *Cyclotella, Tabellaria, Eunotia, Anomoeoneis, Gomphonema*, and *Cymbella* in the diatom flora collectively indicate a decline in fertility. Increases in sodium and potassium concentrations were recorded in sediments of zones VIIa and VIIb perhaps reflecting increased rainfall. Much of the sodium, in particular, of English Lake District waters is provided from droplets picked up by the prevailing winds from the sea only a few tens of km to the west.

In Esthwaite, in contrast to Blea Tarn, the unproductive phase has not continued to the present day. During the latter part of phase VIIb, or the last 3000 years there has again been an increase in fertility. At first this was moderate but very recently, in the nineteenth century, it has been marked. *Asterionella* and *Melosira* have reappeared in the diatom flora, and chironomids became abundant about 1600 years ago. At about the same time the cladoceran *Bosmina longirostris* appeared and partly replaced the previously recorded *B. coregoni*. Sedimentation rates have increased in the

Table 10.2. Palaeolimnology of Esthwaite. (Compiled from Cranwell [122], Goulden [293], Mackereth [499] and Round [691])

Depth (cm)	Sedimentation rate (mm year^{-1})	Date BP	Sediment/pollen	Diatoms	Animals	Organic compounds	Ca (mg g^{-1})	Phosphorus (mg g^{-1})
0–150	Current 5.6–11	VIIb (upper)	Plantain	*Asterionella, Melosira,* fertile phase	Chironomids increase from 100 cm up, also *Bosmina longirostris*	0–5 cm 82% n-alkanoic 18% branched cyclic acids	4–5	2 Falls with dilution by increasing sediment
150–300	0.55	VIIb	Oak, birch, alder	*Cyclotella, Gomphonema, Cymbella, Eunotia, Anomoeoneis.* Acidic indicators: infertile	*Bosmina coregoni* predominant			
		5100 BP						
300–410	0.46	VIIa	Alder increase Brown mud from here up	*Melosira, Stephanodiscus,* fertile phase		336–382 cm 82% *n*-alkanoic 13% branched cyclic acids		3.5
410–437	0.19	8900 BP	Banded clay/mud	*Melosira arenaria, Epithemia, Fragilaria,* fertile phase				
437–460		V	Pine, hazel					
460–478	0.16	IV	Grey clay	Poor flora			22	
		10 300 BP III						
478–498					Littoral arctic forms		5–10	
498–515		II	Grey silt/clay *Salix,* sedges					
515		I	Laminated clay					

nineteenth century to 5.6–11 mm year^{-1} from a mean post-glacial rate of only 0.48 mm year^{-1}

The reason for these changes is to be found in human activity in the catchment area (Macan [491]). Pollen of a common agricultural weed, the plantain, *Plantago lanceolata* is found in sediments younger than about 3000 years, and around 3700 BP, neolithic peoples are known, from archaeological evidence, to have moved into the Esthwaite catchment area, though not to any extent into that of Blea Tarn. Forest clearance for fuel and agriculture leads to release of nutrients previously stored in the woodland biomass (Chapter 3), to greater soil erosion and perhaps to greater leachability of nutrients from the excreta of domestic stock. The clearances were intensified by a new wave of colonization around 900 BP by Norsemen and Norse Irish, perhaps retreating from political disturbances in the Isle of Man. Many place names in the English Lake District are of Norse derivation.

During the nineteenth century the English Lake District became popularized particularly by the poetry of Southey and Wordsworth, who both lived there, and since the building of a railway, completed in 1847, the populations of both residents and tourists have increased. In turn this has led to improved mains sanitation and it is the sewage effluent from the popular tourist village of Hawkshead, on the inflow river to Esthwaite, which has led to the latest phase of eutrophication of the lake in the last 30 years or so.

10.5.3 Pickerel Lake

The Prairies of N.E. Dakota, USA, form a great contrast to the uplands of the English Lake District, though their lakes also were formed by ice and are no more than 10 000 or 11 000 years old. A large ice block, calved from the continental glacier as it melted back, was buried amid the tons of rubble of limestone and shale washed out from under the glacier. As this block itself finally melted it left in its place a basin up to 25 m deep which now holds Pickerel Lake. The lake sediments occupy about 8 m of the basin and provide a record of lake development [333] in an area which is now generally drier than north-western Europe and which lies in the grassland rather than woodland biome.

Four phases (Table 10.3) in the surrounding vegetation have been deduced by pollen analysis. After the retreat of ice, spruce (*Picea*) and tamarack (*Larix*) forest had developed by 10 500 BP, and by 9400 BP this had diversified to a forest of oak, elm, birch, alder, fir (*Abies*), sycamore (*Acer*) and ash (*Fraxinus*), probably with some grassy openings in it. The climate then became progressively drier for the forest was replaced by prairie grassland, with grasses including *Andropogon* (blue-stem), Compositae (*Ambrosia* (ragweed) and *Artemisia*) and annual herbs of the family Chenopodiaceae. This vegetation has persisted since 9400 BP, though since 4200 BP there has been some patchy deciduous woodland development.

Table 10.3. Palaeolimnology of Pickerel Lake, Dakota. (Based on Haworth [333])

Depth in core (cm)	¹⁴C dates	Sedimentation rate (mm year⁻¹)	Pollen	Diatoms	Interpretation
0				Fewer, varied, alkalibiontic, increase in halophiles. Some *Melosira granulata*	Some enrichment (increase in *Fragilaria crotonensis* and in *Ambrosia* pollen). Agricultural fertilization and disturbance
100			Prarie grasses. Some increase in woodland		
200	2700 BP	1.05		Many, no halophobes, % planktonic decreases	Transition
300		0.91			
400	4200 BP		Blue stem grass prairie. *Ambrosia, Artemisia,* composites, chenopods. Some wood. Seeds of wet-mud plants	Many, alkaliphilous, no halophobes, more plankton *Melosira granulata, Stephanodiscus.* Some sandy layers	Eutrophic (*M. granulata*) Transition to open prairie (summer droughts), fluctuation in lake level. Increased erosion (sand layers). Evaporative concentration—some brackish taxa
600	9400 BP		Mixed deciduous woodland. *Quercus Ulmus, Betula Alnus, Abies Acer, Fraxinus*	Fewer. *Fragilaria.* Benthic (molluscs appear). Acidophilous species decrease. Many benthic, alkaliphilous *Fragilaria, Epithemia.* Some halophilic	Change to more alkaline
700		3.82			
800	10 500 BP		Forest. *Picea, Larix*	Acid/neutral, acidophilous, halophobe *Eunotia*	Spruce forest, acid soils, low nutrient loading
825	Glacial till				

The drying of the climate, and its effect on the catchment vegetation is reflected also in events in Pickerel Lake. In the early forest phases the effect of the raw conifer forest humus was to neutralize the alkaline ground water of the calcareous glacial deposits, for the diatom flora was then a mixed one. There were some indicators of neutral to acid waters, such as *Eunotia* and *Tabellaria*, which do not appear in the younger sediments, while genera now found in fertile, alkaline waters (*Fragilaria, Epithemia, Mastigloia*) were also present. Diatoms of strongly alkaline habitats were absent and the mainly benthic flora, coupled with the remains of aquatic plants (*Potamogeton pectinatus, Najas flexilis, Najas marina*) testifies to poor plankton development.

As the deciduous woodland replaced the conifer forest and then itself was succeeded by grassland, the diatom flora underwent a transition and molluscs appeared in quantity in the lake. Diatom indicators of neutral to acid water disappeared and the flora was generally sparser. Then, as the prairie developed, a major increase in diatoms took place. There was a high production of phytoplankters of fertile lakes—*Melosira ambigua, M. granulata, Fragilaria crotonensis* and *Stephanodiscus niagarae*—and some species characteristic of brackish waters were found. The sort of climate that favours prairie has summer droughts, giving increased mineralization of nutrients, great fluctuations in water level and in ionic content of the water.

The fluctuations in levels are indicated by layers of sand, eroded from the marginal high beaches, in the sediment profile and by the presence of seeds of annual higher plants characteristic of exposed wet mud. Evaporative concentration of ions in summer was demonstrated by brackish diatom taxa and the increased nutrient loading by the characteristic plankton species.

Again this burst of production was not sustained, for after 4200 BP some woodland development may have stabilized the soil and decreased the erosion rate. Presumably this was associated with a less extreme, somewhat damper climate, and was linked with a mixed diatom flora less dominated by indicators of high fertility. Towards the top of the sediment profile a further enrichment, though not as great as that between 9400 and 4200 BP has been noted. *Melosira granulata* and *Fragilaria crotonensis* have again increased and coupled with an increase in ragweed pollen suggest that this is associated with agricultural activity and fertilization of the land during the past two centuries.

10.5.4 Lago di Monterosi

Not all lakes are as young as those formed by ice action, but few of the older ones have been studied by palaeolimnologists. The Lago di Monterosi in the Campagna, 40 km from Rome, in Italy, is one. It lies in an old volcanic crater which is almost circular and about 600 m in diameter. At present the lake is a moderately fertile one with extensive beds of aquatic plants (*Nymphaea,*

Myriophyllum, Ceratophyllum) and a plankton not untypical of soft lowland waters, but it has changed greatly during its existence [383].

The basin was formed by a volcanic explosion a little more than 26 000 years ago, when northern European was still heavily glaciated. At that time the Roman Campagna was probably a cold, dry steppe with large mammals like the mammoth grazing over it. Sediments from the lake in its first 3000 years (period A, Table 10.4) indicate an alkaline, rather fertile lake with remains of *Gloeotrichia* (a blue-green alga), *Chaoborus* and *Glyptotendipes* (Chironomidae) and of aquatic plants of fertile habitats. Sponge spicules,

Table 10.4. Palaeolimnology of the Lago di Monterosi. (After Hutchinson *et al.* [383])

Period	Years BP	Features
	0	As in C_2.
C_2		Productivity less than during C_1, but more than in B. Diatoms, some blue-green algae and chironomids, but *Sphagnum* increased. More intensive farming (pigs, grain) and soil erosion
	1200	Decline in productivity and eutrophy, decrease in chironomids. Fewer blue-green algae but eutrophic diatoms present
C_1	AD 10– BC 340	In this period, dramatic changes. Increase in Ca^{2+}, rate of sedimentation, chironomids and chydorids. Marked increased in eutrophy. Blue-green algae very common
	2800	Surroundings more wooded (*Abies* and *Quercus*). Less *Sphagnum*. Probably a little less acid. Human activity in the basin probably began increasing. Chenopodiaceae pollen
B	5000	Lake shallow, but still with low sedimentation rate. Increase in acidophil diatoms and *Sphagnum*. Water plant flora decreased. Chrysophycean cysts indicate oligotrophic conditions, chironomids and cladocerans very rare
	13 000	Sedimentation rate low and little organic matter deposited. Probably local plant cover well developed, with grasses. More water-plant pollen—lake shallower
	23 000	Early lake dilute, slightly alkaline (due to Na and K bicarbonates). Moderate sedimentation rate. Few chironomids, diatoms of wide tolerance. Sponges and pollen of *Myriophyllum*, Nymphaeaceae and *Potamogeton*. Surroundings cold, dry, steppe-like with *Artemisia*
A		
	26 000 BP	Origin of lake by volcanic explosion

also of species from such habitats, remain from what was probably a richly developed invertebrate fauna in the weed-beds. Such conditions might be expected. The climate was cold—the Wurm II glacial maximum in the north had still to be approached—but the crater rim and surrounding permeable catchment of fresh ash, lapilli and scoria provided fresh rock surfaces for leaching of nutrients.

This early productive phase, seen also in other lakes discussed above, did not last beyond 23 000 BP, again presumably as the fresh deposits became leached out and the run-off waters less rich in dissolved ions. There followed a very long period (B) of reduced production and low sedimentation rates (0.0044–0.0075 cm year^{-1}) which lasted for 20 000 years with little change. *Sphagnum*, a plant of acid to neutral infertile waters, colonized. The sponges of alkaline waters declined and chironomids became scarce. Woodland of oak, fir and hazel developed in the surrounding area, and stabilization of the soils by the forest probably also reduced nutrient loading on the lake. Some minor changes could be ascribed to changing water levels related to climatic fluctuations.

Then, dramatically, the lake became again very fertile. There were maxima of blue-green algae. Other algae (*Pediastrum, Melosira granulata*), characteristic of very fertile lake plankton, appeared and there was a rich cladoceran and chironomid fauna. Calcium was deposited heavily in the sediments and the sedimentation rate increased nearly sixfold to at least 0.038 cm year^{-1}. *Sphagnum* disappeared. The date for this transition given by^{14}C analysis is 2220 ± 120 years BP and after corrections have been applied for fluctuations in the rate of formation of ^{14}C and for variability in the method, the date most likely falls between 285 and 150 BC.

No evidence of natural events could be found to explain the changes which occurred but there was a major interference by man in the catchment at the time. The Roman consul Lucius Cassius Longinius had a road built around 171 BC. This was to give rapid passage between Rome, eastern Tuscany and the upper Arno valley, where unrest, which was to culminate in the second Punic war, was about to break out. The ancient Via Cassia runs around the southern edge of the catchment of the Lago di Monterosi, and the disturbance of road building in what is quite a small catchment (little larger than the lake itself) seems to be the only sensible explanation of changes in the lake. The roadworks seem to have disturbed the vegetation cover and altered the drainage pattern slightly so that water previously drained away from the lake burst through as springs (the Fontana di Papa Leone) near its edge. This water percolates through calcareous deposits and explains the increased calcium deposition in the lake during the period in question. Soil erosion caused by removal of vegetation explains the increase in sedimentation rate.

The immediate effects of the road building were short and the lake settled back into a less productive phase soon afterwards. It has never quite returned to its state in period B, however, because of the permanent alteration in the hydrology caused by the springs, and because, in the post-imperial period, agriculture and deforestation caused further changes and disturbance. *Sphagnum* did reappear in the fifteenth century and the numbers of chironomids and crops of blue-green algae have greatly declined. There was a nadir in the fortunes of the Monterosi area in the fifteenth century and these seem related to a period of decreased production sandwiched between the post road-building period, when pollen of nettles (*Urtica*) and Chenopodiaceae indicate cultivation and stock rearing, and the last 200 years of farming propserity.

10.6 FILLING IN OF SHALLOW LAKES

Sediment is deposited in all lakes, so that theoretically all lakes should eventually be made so shallow that they are converted to areas of reed swamp, with a river flowing through them, if they were fed originally by surface water. In a very deep lake, however, at sedimentation rates, where man has not disturbed the catchment, of the order of a millimetre or so per year, this process may take a very long time. L. Tanganyika in Africa, for example, has a maximum depth of 1500 m and a current sedimentation rate of around 0.5 mm year^{-1}. This should give it a future span of about 3 million years—enough for geological events even to disrupt the basin before such a state is reached.

Lakes of only a few metres depth, however, may fill in not only by sedimentation, but also by peat accumulation at the margins and encroachment of swamp from the edges towards the middle. An example will illustrate this process.

10.6.1 **Tarn Moss, Malham**

Malham Tarn, in Yorkshire, England, is a small lake situated in glacial drift and fed by nutrient-rich spring water. In the past it has been larger, but encroachment of aquatic plants has now completely filled about half of its original area [637, 638]. About 8 m of peat have been laid down in parts where encroachment has been complete (Fig. 10.6). The earliest deposit, from the late glacial phase (see above), is boulder clay deposited as outwash from a glacier, and immediately overlying this are several thin layers of inorganic materials consisting of clay with stones up to 1 cm diameter. These were laid down in the period about 12–13 000 BP, just after the lake formed, and represent the washing in of loose material from the surrounding, probably not well-vegetated catchment. Pollen analyses show that a tundra vegetation then predominated (pollen phases I–III). The lake water was probably turbid with suspended clay, and light penetration limited. Aquatic

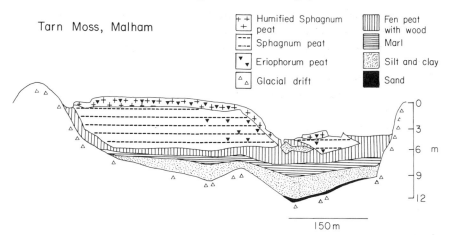

Fig. 10.6. Profile reconstruction from peat borings of part of Tarn Moss, which occupies part of the basin of Malham Tarn, a small lake in Yorkshire, England. (Based on Pigott and Pigott [637, 638].)

plant remains are therefore not plentiful. The water cleared as the clay was deposited and the sediments become silty and of marl, which is calcareous. The oospores of *Chara* (stoneworts) were abundant. Shell fragments of at least six snail species also testify that animal and plant communities appropriate to highly calcareous water (from the inflow springs) were present. The *Chara* probably formed most of the sediment with the marl deposited on its surface.

In the remaining lake similar conditions still prevail. However, in the shallower water emergent plants colonized the calcareous sediment and laid down peat. Grass and sedge remains have been found but the peat is relatively well decomposed and is amorphous and black in parts. Such peats are characteristic of nutrient rich conditions and are formed by a diverse flora known collectively as fen in the UK, alkaline bog in the USA, and as minerotrophic mire elsewhere. Fen conditions persisted for several thousand years while concomitantly the vegetation of the catchment had changed from tundra to birch forest to hazel scrub. The water table in fen is approximately at the surface of the peat. Temporarily drier phases allowed some drying out of the peat surface and invasion by birch trees in parts. The remains of these are to be found in woody (brush wood) peat within the fen peat. Fen vegetation, laying down fen peat, is still present around the tarn edges where nutrient-rich water still has access. However, some few thousands of years ago in the middle of the fen mat the peat built up to such an extent that even after compression by new peat laid on top, it was above the level of the groundwater table. At this point rain water became the main source of water to the peat surface and leached it. Acid conditions were thus created for bog plants, which, by definition, occupy acid peat soils, to colonize and *Sphagnum* species took over from the fen plants. Many *Sphagnum* species are able to maintain the water around them acid by ion exchange. They absorb cations such as Mg^{2+} and Ca^{2+} and release H^+ ions. The *Sphagnum* peat, which is easily recognizable, sometimes even to species level, from the distinctive structure of *Sphagnum* leaves, is about 3–4 m thick in the centre of the encroachment area of Malham Tarn. As it grew thicker it dried out and heather (*Calluna vulgaris*) colonized. Its woody remains and leathery flowers are well preserved. In recent centuries a general drying out has allowed cotton grass (*Eriophorum vaginatum*) to invade and cotton grass peat forms the most recent peat layer.

All small temperate lakes where vegetation is encroaching have not behaved exactly as Malham Tarn has (Fig. 10.7). If local climate is dry, capillary action may provide sufficient salts to the fen peat surface to prevent acidification and bog may never invade. Alder and willow woodland (carr) may invade instead, as happens in the drier areas of eastern England and western Wisconsin. If the groundwater is acid, bog may invade immediately without the intervention of a fen stage, as happens around many American and Scandinavian lakes set in sandy glacial drift. Either bog or fen stages may dry out sufficiently to allow colonization of hardwood forest, or in very

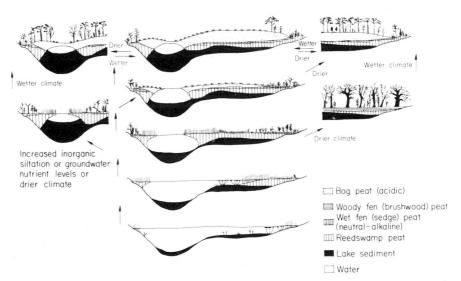

Increased inorganic
siltation or groundwater
nutrient levels or
drier climate

☐ Bog peat (acidic)

▤ Woody fen (brushwood) peat

▥ Wet fen (sedge) peat
(neutral - alkaline)

▥ Reedswamp peat

■ Lake sediment

☐ Water

Fig. 10.7. Some possible developmental sequences of vegetation in small lake basins. The central column shows a sequence from open water through reed swamp and fen to domed bog, which might occur in a moderately wet climate and result in floating bog in the deeper parts of the basin. To each side are variations dependent on long-term climatic changes, reflected in woodland development or in erosional changes in the catchment. These might increase the ionic content of the groundwater sufficiently to retard or prevent the vegatation surface becoming sufficiently leached for acid enough conditions to develop and favour bog formation. (Based on Sinker [744].)

wet climates the bog may form the climax vegetation as happens in the maritime seaboards in Maine, British Columbia and Sweden. Indeed the bog may grow so well that it forms a floating mat or quaking bog (Fig. 10.7).

10.7 CONSENSUS—NATURAL EUTROPHICATION

Shallow lakes generally have higher total phosphorus levels in the water than deep ones. They also give an appearance of high fertility because of their plant beds and the rich invertebrate fauna these often support. The higher fertility of shallow lakes is, of course, a feature of the nature of their catchments which have fertile, easily eroded soils or human activities, which *inter alia* have favoured also rapid filling in. The catchments are also generally of subdued relief and the lake basins formed in them may consequently never have been very deep.

The apparent correlation between deepness and infertility, shallowness and fertility, has led to an association with the process of filling in of one of steady increasing fertility, or eutrophication. This has been embodied in a long-held idea that lakes therefore become naturally more fertile with time. One most important result of palaeolimnological studies has been to counter this idea.

Sediments, themselves the agents of filling in, contain the evidence (Blea Tarn, Esthwaite, Lago di Monterosi) that over long periods lakes may become less fertile as they become more shallow. This 'oligotrophication' is a much more reasonable general expectation than one of increasing fertility. The upheavals which make lake basins might be expected initially to expose readily weatherable minerals, but these are in finite supply and must eventually become leached out. Some parallel can be found in the conversion from fen to bog and indeed the words eutrophic and oligotrophic were originally coined to describe this process.

On the other hand, just as 'natural' eutrophication is certainly not inevitable, neither is 'natural' oligotrophication. Pickerel Lake became naturally more fertile as climatic conditions changed to favour prairie rather than woodland in the surrounding catchment. The important point, however, is that this change in lake fertility was not an internal change in an undisturbed catchment. It was a change consequent on major disturbance of the catchment by climate. All of which underlies the general principle that what happens in a lake is determined by what happens in its catchment. Often related to climate, the direction of change may reverse, perhaps several times in the development of an ancient lake. Table 10.5 illustrates this for a Venezuelan lake, Lake Valencia. It is an ancient lake formed in a tectonic depression some time in the mid-Tertiary period perhaps 50 million years ago.

Table 10.5. Changes over the last 12 000 years in the ancient Lake Valencia, Venezuela. (After Bradbury *et al.* [65])

Date years BP	Selected evidence	Interpretation	Comment
0	High chlorophyll *a* and diatoms (*Nitzschia amphibia*) not previously recorded	No overflow	Drier climate and evidence of artificial eutrophication
10000		Overflow but increasing salinity	
2000	Freshwater littoral and planktonic diatoms. Chenopodiaceae & grass & tree pollen. Little emergent aquatic plant pollen. Increase in		
3000	organic matter		
4000		Overflow, decreasing salinity	Maximum freshness
5000			
6000	Around 5500 BP, stromatolites	No continuous overflow; higher salinity	
7000	Around 7100 BP stromatolites (layered deposits formed by shallow water algae)	Overflow present; low salinity	Start of pluvial period
8000			
9000			Climate becomes wetter

Freshwater lake with variable salinity $<5‰$ ↑

	Between 8700 and 10 500 BP, diatoms and ostracods of saline water; evaporite minerals. Laminae, little benthos, some		End of interpluvial dry period
10 000	tree pollen		

Saline lake ↑

11 000	Absence of diatoms, ostracods, cladocerans but pollen of emergent aquatic plants. Little tree pollen. Low organic content and chlorophyll *a* in sediment		
12 000			

Marsh ↑

During the last 12 000 years or so, Lake Valencia has been, successively, a shallow marsh, a closed basin saline lake, and a freshwater lake of fluctuating salinity. Its sediments 12 000 years ago were of clay, containing few diatoms, but pollen of *Alternanthera, Ludwigia* and *Typha*, which are all emergent aquatic plants.

With the general retreat of the polar glaciers this tropical lake became deeper, around 10 500 BP as the climate became warmer and wetter but it was very saline. Diatoms of salt water (e.g. *Cyclotella striata*) were present and the minerals aragonite and dolomite, which are laid down at salinities greater than about $10 \, g \, l^{-1}$ of total salts if Ca and Mg are abundant, are present. After about 8700 BP the lake became deeper and fresher, reaching its maximum freshness about 3000 BP. It had periods of low water, indicated by layered deposits (stromatolites) formed in the shallows about 5500 and 7100 BP, but was still a freshwater lake with an outflow when it was observed discharging to the R. Orinoco in the early eighteenth century. At present it seems to be drying out again and has no outflow but it is not known whether this is a reflection of a further natural climatic change, or changes in the local hydrology due to agricultural development of the catchment. The flooding of the L. Valencia marsh around 10 500 BP coincides with the beginning of the latest interglacial (or pluvial) period, the Holocene.

There does not seem, therefore, to be any common pattern in the development of lakes. What happens depends on the nature of the original basin, local circumstances in the catchment and climatic change. The one feature common to most is the recent major change brought about by human activity.

10.8 FINALE

The last paragraph of the previous section will come as no surprise. Lakes have been no less affected by man than upland rivers and floodplain ecosystems. Sometimes, the word 'dying' is applied to lakes undergoing severe eutrophication. It is a graphic, but silly term because such lakes are highly productive. Nonetheless, fresh waters remain perhaps the most vulnerable of habitats. This is for several reasons. First, water is subconsciously disliked by many people, perhaps through fear of drowning, or past associations with once vast and inhospitable stretches of marsh. Secondly, lakes act as sinks for many of the products of human activity in their catchments, while rivers are naturally-provided drains for the removal of waste to the sea. Thirdly, water is nevertheless a valuable and essential resource, often scarcest where it is required most and hence needing to be stored. Reservoirs have frequently replaced riverine habitats and their previously more diverse aquatic ecosystem. Lastly, as with other ecosystems, man has had the notion that by some form of management he can 'improve' them.

Natural ecosystems represent the current equilibria in a continuous process of change and adjustment. All chemical systems (including living ones) tend always to adjust towards a state of maximum homoeostasis when confronted with changing external conditions. This means that they develop mechanisms which tend to annul or minimize the effects of random external changes (e.g. short-term weather), or to anticipate and perhaps make use of regularly predictable ones (e.g. tidal and photoperiodic changes). This is achieved in ecosystems through the continuous operation of natural selection on the organisms comprising them. It has resulted in production of arrays of species best fitted, in the prevailing environment, to maintain a system of maximal homoeostasis while simultaneously retaining a capacity, through genetic mechanisms, to keep on adjusting to inevitable continuing change. It is impossible to 'improve' an ecosystem formed and regulated in this way. Though we may not fully understand them there are always good reasons, in undisturbed ecosystems, why, for example, fish growth is moderate, or aquatic plant beds are extensive. Attempts to increase fish growth or clear the beds will always result in a chain of repercussions which will inevitably have to be counteracted, perhaps at considerable expense. Geomorphological systems may also be viewed in the same terms. A river floodplain may only be under water once every few years, but it is still part of the river bed, essential for the efficient and natural disposal of occasional high discharges. Attempts to change its use from this purpose to farming, for example, must mean expensive engineering works to increase the height of the river banks to accommodate the flood water. Objective economic cost–benefit analyses have rarely been produced for such schemes. In the long run one suspects that, so far as the general public good is concerned, the balance might be very unfavourable.

Inevitably, however, aquatic ecosystems have been changed either from a humanitarianly defensible need for water-borne disease control or for the establishment of farmland, from simple ignorance of the consequences of pollution and eutrophication, or from the need to prevent flood damage to property built on natural floodplains through stupidity or greed. There is now a belated trend in public opinion that they should be 'conserved'.

Conservation of natural ecosystems can be defined formally in several ways. One is that it is a process of maintaining the maximum diversity of organisms and ecosystems; another is that it is 'the total management of the rural areas . . . for the fair and equal benefit of all groups which have a direct interest in their use' [874]. What conservation of fresh water usually amounts to is making the best of what is left after waterways and their catchments have been used for the more obvious needs of society. It is true that most developed countries have legislation permitting the reservation of particularly interesting areas, and that attitudes towards the retention of as much natural habitat as possible are emerging. It is also true that enlightened countries are developing ways of restoring much changed lakes and swamps and of limiting water pollution, and that powerful allies may be found in the large numbers of people who enjoy angling, wildfowling or birdwatching.

Nonetheless, on balance, more waterways are being degraded and lost than created and restored. The agents of change are varied, but three—drainage and river management, acidification and eutrophication—seem of widespread significance. Any careful observer of the freshwater scene, freed of shackles of political or occupational discretion, must, I believe, be no less than appalled at what has been done and is still being done.

> What would the world be, once bereft
> Of wet and wildness? Let them be left,
> O let them be left, wildness and wet;
> Long live the weeds and the wilderness yet
> Gerald Manley Hopkins (1844–1889)
> *'Inversnaid'*.

Further Reading

The scientific literature, particularly for such a wide subject area as dealt with in this book, is very large. References quoted in the text give leads into the literature and in this section I have quoted books, reviews and papers which themselves give good lists of references or introduce topics of relevance for which there was no space in the main text. Journals with papers relevant to each chapter are also quoted. The following journals are widely relevant: *Limnology and Oceanography, Canadian Journal of Fisheries and Aquatic Sciences, Freshwater Biology, Internationale Revue der gesamten Hydrobiologie, Archiv für Hydrobiologie, Hydrobiologia, Internationale Verhandlungen der Vereinigung theoretische und angewandte Limnologie, Journal of Ecology, Journal of Animal Ecology, Journal of Applied Ecology, Ecology, Ecological Monographs, Oecologia, Oikos, Holarctic Ecology.*

Chapter 1

General aspects of water management: 434, 615, 622.

Chapter 2

Properties of water: 230, 379. Osmoregulation: 650, 659, 725. Respiration: 725. Classification of phyla: 34, 519. African fishes: 246, 422.

Chapter 3

Water chemistry: 270, 292, 379, 497, 658, 782, 859. Acidification: 121, 177, 203, 236, 471, 788. Catchment processes: 469, 541, 605, 695, 696. Water pollution: 523. Analytical methods: 9, 286, 501, 701. Journals: *Tellus, Journal of Atmospheric Chemistry.*

Chapter 4

Basic hydrology: 302, 466, 553. General stream ecology: 148, 386, 813, 862, 863. Hot springs: 98. Micro-ecology: 783. Stream invertebrate biology and identification: 133, 192, 262, 489, 533, 534, 664, plus keys published by the Freshwater Biological Association, Ambleside, Cumbria LA22 0LP, UK.

365

Sampling methods for stream animals: 178, 197, 200, 502, 625, 683. Salmon: 400, 401, 536, 537, 579, 847. Acidification: 205, 606, 608. Effects of hydroelectric installations and river regulation: 269, 453, 839.

Chapter 5

Aquatic plants: 30, 118, 127, 301, 382, 521, 707, 710, 731. Aquatic plant control: 355, 590, 852. Blanket weed: 57, 861. Swamps: 95, 211, 232, 248, 287, 345, 363, 419, 437, 475, 476, 517, 594, 703, 844, 845, 846. Nutrient exchanges in swamps: 39, 68, 69, 260, 261, 366. Diseases: 487, 574, 632, 636. Drainage: 15, 142, 146, 273, 279, 789. Aquatic plant identification: 212, 329, 525. Pollution monitoring: 343. Journals: *Aquatic Botany, Journal of the Water Pollution Control Federation, Water Resources Research, Water Research.*

Chapter 6

Basic limnology: 38, 41, 183, 280, 284, 379, 380, 460, 699, 855. Light: 435, 436. Eutrophication and lake restoration: 285, 585, 821, 822, 824, 843. Temporary waters: 865. Meromictic lakes: 20. Saline lakes: 70, 312, 835.

Chapter 7

General: 380. Phytoplankton: 58, 145, 227, 320, 405, 575, 668, 669, 693. Zooplankton and fish: 3, 76, 181, 310, 376, 431, 432, 448, 681, 799, 827, 880. Biomanipulation: 12, 13, 736, 748. Identification: 172, 192, 264, 315, 373, 388, 641, 653, 729. Journals: *Journal of Plankton Research, Microbial Ecology, Archiv für Mikrobiologie, Journal of General Microbiology.*

Chapter 8

General: 370. Epiphyte and associated communities: 91, 131, 317, 692, 810, 811, 854, 857. Journals: *Aquatic Botany, Journal of Phycology, British Phycological Journal.*

Chapter 9

Fish biology: 265, 444, 589, 827. Management: 182, 402, 403, 528. Journals: *Environmental Biology of Fishes, Journal of Fish Biology, Transactions of the American Fisheries Society, Journal of Fish Diseases, Journal of Fisheries Management.*

Chapter 10

Weed problems: 355, 543. Palaeolimnological methods: 48, 233, 547. Lake development: 278, 307, 334, 654, 743. Conservation: 180, 459, 465, 515, 549. Journals: *Boreas, Biological Conservation.*

References

1 Alexander, W. J. R. (1982) *Water requirements of the Pongolo floodplain ecosystem and recommended operating rules for the Pongolapoort Dam.* Report of the Department of Environmental Affairs, Pretoria.

2 Algeus, S. (1950) The utilization of aspartic acid, succinamide, and asparagine by *Scenedesmus obliquous. Physiol. Plant.,* 3, 225-35.

3 Allan, J. D. (1976) Life history patterns in zooplankton. *Am. Nat.,* 110, 165-80.

4 Allanson, B. R. (1973) The fine structure of the periphyton of *Chara* sp. and *Potamogeton natans* from Wytham Pond, Oxford and its significance to the macrophyte-periphyton metabolic model of R. G. Wetzel and H. L. Allen. *Freshwat Biol.,* 3, 535-42.

5 Allen, E. D. and Spence, D. H. N. (1981) The differential ability of aquatic plants to utilize the inorganic carbon supply in fresh waters. *New Phytol.,* 87, 269-83.

6 Allen, H. L. (1967) Acetate utilization by hetcrotrophic bacteria in a pond. *Hydrologiai Kozlony.* 1967, 295-7.

7 Allen, H. L. (1971) Primary productivity, chemo-organotrophy, and nutritional interactions of epiphytic algae and bacteria on macrophytes in the littoral of a lake. *Ecol. Monogr.,* 41, 97-127.

8 Allen, K. R. (1951) The Horokiwi stream. A study of a trout population. *Fish. Bull. New Zealand Mar. Depart.,* 10, 1-231.

9 American Public Health Association (1976) *Standard Methods for the Examination of Water and Wastewater, including Bottom Sediments and Sludge,* 14th edn. APHA, New York.

10 Anderson, A. M. (1961) Further observations concerning the proposed introduction of Nile perch into Lake Victoria. *East Afr. Agric. Forest. J.,* 26, 195-201.

11 Andersson, F. and Olsson, B. (1985) *Lake Gårdsjön. An Acid Forest lake and its catchment. Ecol. Bull.,* 32, Stockholm.

12 Andersson, G. (1985) The influence of fish on eutrophic lake ecosystems. *Lake Pollution and Recovery,* pp. 50-3. European Water Pollution Control Association, Rome.

13 Andersson, G., Berggren, H., Cronberg, G. and Gelin, C. (1978) Effects of planktivorous and benthivorous fish on organisms and water chemistry in eutrophic lakes. *Hydrobiologia,* 59, 9-15.

14 Andreae, M. O. and Barnard, W. R. (1984) The marine chemistry of dimethylsulfide. *Mar. Chem.,* 14, 267-79.

15 Armentano, T. V. and Menges, E. S. (1986) Patterns of change in the carbon balance of organic soil-wetlands of the temperate zone. *J. Ecol.,* 74, 753-77.

16 Armillas, P. (1971) Gardens on swamps. *Science,* 174, 653-61.

17 Armitage, P. D. (1978) The impact of Cow Green reservoir on invertebrate populations in the River Tees. *Ann. Rep. Freshwat. Biol. Assoc.,* 46, 47-56.

18 Arnold, D. E. (1971) Ingestion, assimilation, survival, and reproduction by *Daphnia pulex* fed seven species of blue-green algae. *Limnol. Oceanogr.,* 16, 906-20.

19 Azam, F. and others (1983) The ecological role of water-column microbes in the sea. *Mar. Ecol. Progr. Series,* 10, 257-63.

20 Bagenal, T. B. (1978) *Methods for Assessment of Fish Production in Freshwaters.* Blackwell Scientific Publications, Oxford.

21 Bailey-Watts, A. E. (1976) Planktonic diatoms and some diatom-silica relations in a shallow eutrophic Scottish loch. *Freshwat. Biol.,* 6, 69-80.

22 Baker, A. L., Baker, K. K. and Tyler, P. A. (1985) Fine layer depth relationships of lakewater chemistry, planktonic algae, and photosynthetic bacteria in meromictic Lake Fidler, Tasmania. *Freshwat. Biol.,* 15, 735-47.

367

23 Baker, J. H. and Farr, I. S. (1977) Origin, characterization and dynamics of suspended bacteria in two chalk streams. *Arch. Hydrobiol.,* **80**, 308–26.

24 Baker, J. H. and Farr, I. S. (1987) Importance of dissolved organic matter produced by duckweed (*Lemna minor*) in a southern English river. *Freshwat. Biol.,* **17**, 325–30.

25 Baker, J. H. and Orr, D. R. (1986) Distribution of epiphytic bacteria on freshwater plants. *J. Ecol.,* **74**, 155–65.

25a Baldock, B. M. (1986) Peritrich ciliates epizoic on larvae of *Brachycentrus subnubilis* (Trichoptera): importance in relation to the total protozoan population in streams. *Hydrobiologia,* **132**, 125–31.

26 Baldock, B. M., Baker, J. H. and Sleigh, M. A. (1983) Abundance and productivity of protozoa in chalk streams. *Hol. Ecol.,* **6**, 238–46.

27 Balon, E. K. and Bruton, M. N. (1986) Introduction of alien species or why scientific advice is not heeded. *Env. Biol. Fishes,* **16**, 225–30.

28 Balon, E. K. and Stewart, D. J. (1983) Fish assemblages in a river with an unusual gradient (Luongo, Africa-Zaire system), reflections on river zonation, and description of another new species. *Env. Biol. Fishes,* **9**, 225–52.

29 Barel, C. D. N., Dorit, R., Greenwood, P. H., Fryer, G., Hughes, N., Jackson, P. B. N., Kawanabe, H., Lowe-McConnell, E., Witte, F. and Yamaoka, K. (1985) Destruction of fisheries in Africa's lakes. *Nature,* **315**, 19–20.

30 Barko, J. W., Murphy, P. G. and Wetzel, R. G. (1977) An investigation of primary production and ecosystem metabolism in a Lake Michigan dune pond. *Arch. Hydrobiol.,* **81**, 155–87.

31 Barko, J. W. and Smart, R. M. (1980) Mobilization of sediment phosphorus by submersed freshwater macrophytes. *Freshwat. Biol.,* **10**, 229–38.

32 Bärlocher, F. and Kendrick, B. (1973a) Fungi and food preferences of *Gammarus pseudolimnaeus. Arch. Hydrobiol.,* **72**, 501–16.

33 Bärlocher, F. and Kendrick, B. (1973b) Fungi in the diet of *Gammarus pseudolimnaeus* (Amphipoda). *Oikos,* **24**, 295–300.

34 Barnes, R. S. K. (Ed.) (1984) *A Synoptic Classification of Living Organisms.* Blackwell Scientific Publications, Oxford.

35 Baross, J. A. and Deming, J. W. (1983) Growth of 'black smoker' bacteria at temperatures of at least 250°C. *Nature,* **303**, 423–6.

36 Barrett, C. F., Fowler, D., Irving, J. G., Kallend, A. S., Martin, A., Scriven, R. A. and Tuck, A. F. (1982) *Acidity of Rainfall in the United Kingdom.* A preliminary report. Warren Spring Laboratory, Stevenage.

37 Bayless, J. and Smith, W. B. (1967) The effects of channelization upon the fish population of lotic waters in eastern North Carolina. *Proc. Ann. Conf. Southeast Assoc. Game Fish Commissions,* **18**, 230–8.

38 Bayley, I. A. E. and Williams, W. D. (1973) *Inland Waters and their Ecology.* Longman, London.

39 Bayley, S. E., Zollek, Jr, J., Hermann, A. J., Dolan, T. J. and Tortora, L. (1985) Experimental manipulation of nutrients and water in a freshwater marsh: effects on biomass, decomposition, and nutrient accumulation. *Limnol. Oceanogr.,* **30**, 500–12.

40 Beadle, L. C. (1943) Osmotic regulation and the faunas of inland waters. *Biological Reviews,* **18**, 172–83.

41 Beadle, L.C. (1981) *The Inland Waters of Tropical Africa* (2nd edn.). Longman, London.

42 Beauchamp, R. S. A. (1964) The rift valley lakes of Africa. *Verh. int. Verein. theor. angew. Limnol.,* **15**, 91–9.

43 Belcher, J. H., Swale, E. M. F. and Heron, J. (1966) Ecological and morphological observations on a population of *Cyclotella pseudostelligera* Hustedt. *J. Ecol.,* **54**, 335–40.

44 Bellamy, L. S. and Reynoldson, T. B. (1974) Behaviour in competition for food amongst lake-dwelling triclads. *Oikos,* **25**, 356–64.

45 Bengtsson, L., Fleischer, S., Lindmark, G. and Ripl, W. (1975) Lake Trummen restoration project I. Water and sediment chemistry. *Ver. int. Verein. theor. angew. Limnol.,* **19**, 1080–7.

46 Bengtsson, L. and Gelin, C. (1975) Artificial aeration and suction dredging methods for controlling water quality. *The Effects of Storage on Water Quality,* 313–42. Water Research Centre, Medmenham.

47 Berg, K. (1938) Studies on the bottom animals of Esrom Lake. *K. Danske vidensk Selsk. skr.,* **7**, 1–255.

48 Berglund, B. E. (Ed.) (1986) *Handbook of Holocene Palaeoecology and Palaeohydrology.* Wiley, Chichester.

49 De Bernardi, R. and Giussani, G. (1975) Population dynamics of three Cladocerans of Lago Maggiore related to predation pressure by a planktophagous fish. *Verh. int. Verein. theor. angew. Limnol.*, 19, 2906–12.

50 Bird, D. F. and Kalff, J. (1986) Bacterial grazing by planktonic lake algae. *Science*, 231, 493–5.

51 Bird, D. F. and Kalff, J. (1987) Algal phagotrophy: regulating factors and importance relative to photosynthesis in *Dinobryon* (Chysophyeae). *Limnol. Oceanogr.*, 32, 277–84.

52 Birdsey, E. C. and Lynch, V. H. (1962) Utilization of nitrogen compounds by unicellular algae. *Science*, 137, 763–4.

53 Björk, S. (1972) Swedish lake restoration program gets results. *Ambio*, 1(5), 153–65.

54 Björk, S. (1985) Scandinavian lake restoration activities. In *Lakes Pollution and Recovery*, pp. 293–301. European Water Pollution Control Association, Rome.

54a Blackie, J. R., Ford, E. D., Horne, J. E. M., Kinsman, D. J. J., Last, F. T. and Moorhouse, P. (1980) Environmental effects of deforestation. An annotated bibliography. *Occ. Publ. Freshwat. Biol. Assoc.*, 10.

55 Boar, R. R. and Crook, C. E. (1985) Investigations into the causes of reedswamp regression in the Norfolk Broads. *Verh. int. Verein. theor. angw. Limnol.*, 22, 2916–19.

56 Bogdan, K. G. and Gilbert, J. J. (1982) Seasonal patterns of feeding by natural populations of *Keratella, Polyarthra* and *Bosmina*: clearance rates, selectivities and contributions to community grazing. *Limnol. Oceanogr.*, 27, 918 34.

57 Bolas, P. M. and Lund, J. W. G. (1974) Some factors affecting the growth of *Cladophora glomerata* in the Kentish Stour. *Water Treat. Examin.*, 23, 25–51.

58 Bold, H. C. and Wynne, M. J. (1978) *Introduction to the Algae*. Prentice-Hall, New Jersey.

59 Booker, M. J. and Walsby, A. E. (1981) Bloom formation and stratification by a planktonic blue-green alga in an experimental water column. *Br. Phycol. J.*, 16, 411–21.

60 Borgstrom, G. (1978) The contribution of freshwater fish to human food. *Ecology of Freshwater Fish Production* (Ed. by S. D. Gerking), pp. 469–91. Blackwell Scientific Publications, Oxford.

61 Boston, H. L. and Adams, M. S. (1983) Evidence of crassulacean acid metabolism in two North American isoetids. *Aquat. Bot.*, 15, 381–6.

62 Boston, H. L. and Adams, M. S. (1986) The contribution of crassulacean acid metabolism to the annual productivity of two aquatic vascular plants. *Oecologia*, 68, 615–22.

63 Bothwell, M. L. (1985) Phosphorus limitation of lotic periphyton growth rates: An intersite comparison using continuous-flow troughs (Thompson River System, British Columbia). *Limnol. Oceanogr.*, 30, 527–42.

64 Bradbury, I. K. and Grace, J. (1983) Primary production in wetlands. *Ecosystems of the World 4A: Mires: Swamp, Bog, Fen and Moor* (Ed. by A. J. P. Gore), pp. 285–310. Elsevier, Amsterdam.

65 Bradbury, J. P., Leyden, B., Salgado-Labouriau, M., Lewis, W. M. Jr., Schubert, C., Binford, M. W., Frey, D. G., Whitehead, D. R. and Weibezahn, F. H. (1981). Late Quaternary environmental history of Lake Valencia, Venezuela. *Science*, 214, 1299–305.

66 Brimblecome, P. and Stedman, D. H. (1982) Historical evidence for a dramatic increase in the nitrate component of acid rain. *Nature*, 298, 460–2.

67 Brinkhurst, R. O. (1974) *The Benthos of Lakes*. Macmillan, London.

68 Brinson, M. M. (1977) Decomposition and nutrient exchange of litter in an alluvial swamp forest. *Ecology*, 58, 601–9.

69 Brinson, M. M., Bradshaw, H. D., Holmes, R. N. and Elkins, J. B. Jr. (1980) Litterfall, stemflow, and throughfall nutrient fluxes in an alluvial swamp forest. *Ecology*, 6, 827–35.

70 Brock, M. A. (1982) Biology of the salinity tolerant genus *Ruppia* L. in saline lakes in South Australia. 1. Morphological variation within and between species and ecophysiology. *Aquat. Bot.*, 13, 219–48.

71 Brock, T. D. (1967) Life at high temperatures. *Science*, 158, 1012–19.

72 Brock, T. D. (1978) *Thermophilic Microorganisms and Life at High Temperatures*. Springer-Verlag, New York.

73 Brooker, M. P. (1981) The impact of impounds on the downstream fisheries and general ecology of rivers. *Adv. Appl. Biol.*, 4, 91–152.

74 Brooker, M. P. and Edwards, R. W. (1975) Aquatic herbicides and the control of water weeds. *Water Res.*, 9, 1–15.

75 Brookes, A., Gregory, K. J. and Dawson,

F. H. (1983) An assessment of river channelization in England and Wales. *Sci. Total Env.*, **27**, 97–111.

76 Brookes, J. L. (1965) Predation and relative helmet size in cyclomorphic *Daphnia*. *Proc. Nat. Acad. Sci., U.S.*, **53**, 119–26.

77 Brooks, J. L. and Dodson, S. I. (1965) Predation, body size and composition of plankton. *Science*, **150**, 28–35.

78 Brown, A. W. A. (1962) A survey of *Simulium* control in Africa. *Bull. World Health Org.*, **27**, 511.

79 Brown, S. and Colman, B. (1963) Oscillaxanthin in lake sediments. *Limnol. Oceanogr.*, **8**, 352–3.

80 Bryant, M. D. (1983) The role and management of woody debris in west coast salmonid nursery streams. *N. Am. J. Fish. Man.*, **3**, 322–30.

81 Brylinsky, M. (1980) Estimating the productivity of lakes and reservoirs. *The Functioning of Freshwater Ecosystems* (Ed. by E. D. Le Cren and R. H. Lowe-McConnell), pp. 411–54. Cambridge University Press, Cambridge.

82 Brylinsky, M. and Mann, K. H. (1973) An analysis of factors governing productivity in lakes and reservoirs. *Limnol. Oceanogr.*, **18**, 1–14.

83 Burgis, M. J. and Morris, P. (1987) *The Natural History of Lakes*. Cambridge University Press, Cambridge.

84 Burns, C. W. and Rigler, F. H. (1967) Comparison of filtering rates of *Daphnia rosea* in lake water and suspensions of yeast. *Limnol. Oceanogr.*, **12**, 492–502.

85 Burton, T. M., King, D. L. and Ervin, J. L. (1979) Aquatic plant harvesting as a lake restoration technique. *Lake Restoration*. U.S.E.P.A., Washington.

86 Campbell, J. I. and Meadows, P. S. (1972) An analysis of aggregations formed by the caddis fly larva *Potamophylax latipennis* in its natural habitat. *J. Zool.*, **167**, 133–47.

87 Canter, H. M. (1979) Fungal and protozoan parasites and their importance in the ecology of phytoplankton. *Ann. Rep. Freshwat. Biol. Assoc.*, **47**, 43–50.

88 Canter, H. M. (1980) Observations on the ameoboid protozoan *Asterocaelum* (Protomyxida) which ingests algae. *Protistologica*, **16**, 475–83.

89 Cantrell, M. A. and McLachlan, A. J. (1982) Habitat duration and dipteran larvae in tropical rain pools. *Oikos*, **38**, 343–8.

90 Carignan, R. and Kalff, J. (1982) Phosphorus release by submerged macrophytes: significance to epiphyton and phytoplankton. *Limnol. Oceanogr.*, **27**, 419–27.

91 Carpenter, S. R. (1983) Submersed macrophyte community structure and internal loading: relationship to lake ecosystem productivity and succession. *Lake Restoration, Protection and Management* (Ed. by J. Taggart), pp. 105–11. U.S.E.P.A. Washington.

92 Carpenter, S. R. and Lodge, D. M. (1986) Effects of submersed macrophytes on ecosystem processes. *Aquat. Bot.*, **26**, 341–70.

93 Carpenter, S. R. and McCreary, N. J. (1985) Effects of fish nests on pattern and zonation of submersed macrophytes in a softwater lake. *Aquat. Bot.*, **22**, 21–32.

94 Carter, G. S. (1955) *The Papyrus Swamps of Uganda*. Heffer, Cambridge University Press, Cambridge.

95 Carter, G. S. and Beadle, L. C. (1931) The fauna of the swamps of the Paraguayan Chaco in relation to its environment. II Respiratory adaptations in the fishes. *J. Linn. Soc. (Zool.)*, **37**, 327–68.

96 Carter, J. C. H. and Goudie, K. A. (1986) Diel vertical migrations and horizontal distributions of *Limnocalanus macrurus* and *Senecella calanoides* (Copepoda, Calanoida) in lakes of Southern Ontario in relation to planktivorous fish. *Can. J. Fish. Aquat. Sci.*, **43**, 2508–14.

97 Casey, H. C. and Downing, A. (1976) Levels of inorganic nutrients in *Ranunculus penicillatus* var. *calcareus* in relation to water chemistry. *Aquat. Bot.*, **2**, 75–80.

98 Castenholz, R. W. (1973) Ecology of blue-green algae in hot springs. *Biology of Blue-green Algae* (Ed. by N. G. Carr and B. A. Whitton), pp. 379–414. Blackwell Scientific Publications, Oxford.

99 Caswell, H. (1972) On instantaneous and finite birth rates. *Limnol. Oceanogr.*, **17**, 787–91.

100 Cattaneo, A. and Kalff, J. (1979) Primary production of algae growing on natural and artificial aquatic plants: a study of interactions between epiphytes and their substrate. *Limnol. Oceanogr.*, **24**, 1031–7.

101 Cattaneo, A. and Kalff, J. (1980) The relative contribution of aquatic macrophytes and their epiphytes to the production of macrophyte beds. *Limnol. Oceanogr.*, **25**, 280–9.

102 Caufield, C. (1982a) *Tropical Moist-For-*

ests. International Institute for Environment and Development, London.

103 Caufield, C. (1982b) Brazil, energy and the Amazon. *New Sci.*, 28 Oct. 1982, 240–3.

104 Caufield, C. (1985a) The Yangtze beckons the Yankee dollar. *New Sci.*, 5 Dec. 1985, 26–27.

105 Caufield, C. (1985b) *In the Rainforest.* Heinemann, London.

106 Caulfield, P. (1971) *Everglades.* Sierra Club, Ballantyne, New York.

107 Cederholm, C. J. and Peterson, N. P. (1985) The retention of coho salmon (*Oncorhynchus kisutch*) carcasses by organic debris in small streams. *Can. J. Fish. Aquat. Sci.*, 42, 1222–5.

108 Chambers, P. A. (1987) Light and nutrients in the control of aquatic plant community structure. II. In situ observations. *J. Ecol.*, 75, 621–8.

109 Charlson, R. J. and Rodhe, H. (1982) Factors controlling the acidity of natural rainwaters. *Nature*, 295, 683–5.

110 Charnock, A. (1983) A new course for the Nile. *New Sci.*, 27 October 1983, 285–8.

111 Christiansen, R., Friis, N. J. S. and Søndergaard, M. (1985) Leaf production and nitrogen and phosphorus tissue content of *Littorella uniflora* (L.) Aschers in relation to nitrogen and phosphorus enrichment of the sediment in oligotrophic Lake Hampen, Denmark. *Aquat. Bot.*, 23, 1–11.

112 Christie, W. J. (1974) Changes in the fish species composition of the Great Lakes. *J. Fish. Res. Board Canada*, 31, 827 54.

113 Chudbya, H. (1965) *Cladophora glomerata* and accompanying algae in the Skawa River. *Acta Hydrobiol.*, 7, 93–126.

114 Codd, G. A. and Bell, S. G. (1985) Eutrophication and toxic Cyanobacteria in freshwaters. *Water Poll. Contr.*, 84, 225–32.

115 Cogbill, C. V. and Likens, G. E. (1974) Acid precipitation in the northeastern United States. *Water Res. Research*, 10, 1133–7.

116 Coleman, M. J. and Hynes, H. B. N. (1970) The vertical distribution of the invertebrate fauna in the bed of a stream. *Limnol. Oceanogr.*, 15, 31–40.

117 Collingwood, R. W. (1977) *A survey of eutrophication in Britain and its effects on water supplies.* Technical Report TR40. Water Research Centre, Medmenham.

118 Cook, C. D. K., Gut, B, J., Rix, T. M., Schneller, J. and Seitz, M. (1974) *The*

Waterplants of the World. Junk, The Hague.

119 Cooke, G. D., McComas, M. R., Waller, D. W. and Kennedy, R. H. (1977) The occurrence of internal phosphorus loading in two small eutrophic glacial lakes in northeastern Ohio. *Hydrobiologia*, 56, 129–35.

120 Coulter, G. W., Allanson, B. R., Bruton, M. W., Greenwood, P. H., Hart, R. C., Jackson, P. B. N. and Ribbink, A. J. (1986) Unique qualities and special problems of the African Great lakes. *Env. Biol. Fishes*, 17, 161–84.

121 Cowling, E. B. (1982) Acid precipitation in historical perspective. *Env. Sci. Technol.*, 16, 110A–23A.

122 Cranwell, P. A. (1973) Branched chain and cyclopropanoid acids in a recent sediment. *Chem. Geol.*, 11, 307–13.

123 Cranwell, P. A. (1974) Monocarboxylic acids in late sediments: indicators derived from terrestrial and aquatic biota of palaeoenvironmental trophic levels. *Chem. Geol.*, 14, 1–14.

124 Cranwell, P. A. (1976) Organic geochemistry of lake sediments. In *Environmental Biogeochemistry* Vol. 1 *Carbon, Nitrogen, Phosphorus, Sulphur and Selenium Cycles,* (Ed. by J. O. Nriagu), pp. 75–88. Ann Arbor Science, Ann Arbor, Michigan.

125 Crawford, R. M. M. (1966) The control of anaerobic respiration as a determining factor in the distribution of the genus *Senecio. J. Ecol.*, 54, 403–13.

126 Crawford, R. M. M. (1982) Root survival in flooded soils Part A. Analytical Studies. *Mires: Swamp, Bog, Fen and Moor* (Ed. by A. J. P. Gore), pp. 257–83. Elsevier, Amsterdam.

127 Crawford, R. M. M. (Ed.) (1987) *Plant Life in Aquatic and Amphibious Habitats.* Blackwell Scientific Publications, Oxford.

128 Crawford, R. M. M. and McManmon, M. (1968) Inductive responses of alcohol and malic dehydrogenase in relation to flooding tolerance in roots. *J. Exp. Bot.*, 19, 435–41.

128a Crawford, R. M. M. and Tyler, P. (1969) Organic acid metabolism in relation to flooding tolerance in roots. *J. Ecol.*, 57, 235–44.

129 Crisp, D. T. (1984) Effects of Cow Green reservoir upon downstream fish populations. *Ann. Rep. Freshwat. Biol. Assoc.*, 52, 47–62.

130 Croome, R. L. (1986) Biological studies on meromictic lakes. *Limnology in Aus-*

tralia (Ed. by P. De Deckker and W. D. Williams), pp. 113–30. Junk, Dordrecht.

131 Crowder, L. B. and Cooper, W. E. (1982) Habitat structural complexity and the interaction between bluegills and their prey. *Ecology*, **63**, 1802–13.

132 Cummins, K. W. (1974) Structure and function of stream ecosystems. *Bioscience*, **24**, 631–41.

133 Cummins, K. W. and Klug, M. J. (1979) Feeding ecology of stream invertebrates. *Ann. Rev. Ecol. Syst.*, **10**, 147–72.

134 Cummins, K. W., Petersen, R. C., Howard, F. O., Wuycheck, J. C. and Holt, V. I. (1973) The utilization of leaf litter by stream detritivores. *Ecology*, **54**, 336–45.

135 Currie, D. J. (1986) Does orthophosphate uptake supply sufficient phosphorus to phytoplankton to sustain their growth? *Can. J. Fish. Aquat. Sci.*, **43**, 1482–7.

136 Cushing, C. E., McIntyre, C. D., Cummins, K. W., Minshall, G. W., Petersen, R. C., Sedell, J. R. and Vannote, R. L. (1983) Relationships among chemical, physical and biological indices along river continua based on multivariate analyses. *Arch. Hydrobiol.*, **98**, 317–26.

137 Cushing, D. H. (1968) *Fisheries Biology*. University of Wisconsin Press, Madison.

138 Cushing, D. H. (1975) *Marine Ecology and Fisheries*. Cambridge University Press, Cambridge.

139 Cushing, D. H. (1977) *Science and the Fisheries*. Arnold, London.

140 Dacey, J. W. H. (1980) Internal wind in water lilies: an adaptation for life in anaerobic sediments. *Science*, **210**, 1017–19.

141 Dacey, J. W. H. (1981) Pressurized ventilation in the yellow water lily. *Ecology*, **62**, 1137–47.

142 Darby, H. C. (1983) *The Changing Fenland*. Cambridge University Press, Cambridge.

143 Darch, J. P. (1983) *Drained Field Agriculture in Central and South America*. British Archaeological Reports International Series, 189.

144 Darch, J. P. (1988) Drained field agriculture: parallels from past to present. *J. Biogeogr.*, (in press).

145 Darley, W. M. (1982) *Algal Biology: A Physiological Approach*. Blackwell Scientific Publications, Oxford.

146 Darnell, R. M. (1976) *Impacts of Construction Activities in Wetlands of the United States*. U.S. Environmental Protection Agency. EPA-600/3-76-041. Corvallis, Oregon.

147 Davies, B. R., Hall, A. and Jackson, P. B. N. (1975) Some ecological aspects of the Cabora Bassa Dam. *Biol. Cons.*, **8**, 189–201.

148 Davies, B. R. and Walker, K. F. (1986) *The Ecology of River Systems*. Junk, Dordrecht.

149 Davies, B. R. and Walmsley, R. D. (1985) *Perspectives in Southern Hemisphere Limnology*. Junk, Dordrecht. (Also published in *Hydrobiologia*, **125**.)

150 Davies, R. W. and Reynoldson, T. B. (1971) The incidence and intensity of predation on lake-dwelling triclads in the field. *J. Anim. Ecol.*, **40**, 191–214.

151 Davis, M. B. (1968) Pollen grains in lake sediments: redeposition caused by seasonal water circulation. *Science*, **162**, 796–9.

152 Davis, R. B. (1974) Stratigraphic effects of tubificids in profoundal lake sediments. *Limnol. Oceanogr.*, **19**, 466–88.

153 Dawson, F. H. (1976) The annual production of the aquatic macrophyte. *Ranunculus penicillatus* var *calcarcus* (R. W. Butcher) C. D. K. Cook. *Aquat. Bot.*, **2**, 51–74.

154 Dawson, F. H. (1978) Aquatic plant management in semi-natural streams: the role of marginal vegetation. *J. Env. Man.*, **6**, 213–21.

155 Dawson, F. H. (1979) *Ranunculus calcareus* and its role in lowland streams. *Ann. Rep. Freshwat. Biol. Assoc.*, **47**, 60–9.

156 Dawson, F. H., Castellano, E. and Ladle, M. (1978) Concept of species succession in relation to river vegetation and management. *Verh. int. Verein. theor. angew, Limnol.*, **20**, 1429–34.

157 Dawson, F. H. and Haslam, S. M. (1983) The management of river vegetation with particular reference to shading effects of marginal vegetation. *Landscape Planning*, **10**, 147–69.

158 Dawson, F. H. and Kern-Hansen, V. (1979) The effect of natural and artificial shade on the macrophytes of lowland streams and the use of shade as a management technique. *Int. Rev. ges. Hydrobiol.*, **64**, 437–55.

159 Deevey, E. S. (1942) The biostratonomy of Linsley Pond. *Am. J. Sci.*, **240**, 233–64, 313–24.

160 Deevey, E. S. (1969) Cladoceran populations of Rogers Lake, Connecticut during late- and post-glacial time. *Mitt. int. Ver. theor. angew. Limnol.*, **17**, 56–63.

161 Degens, E. T., Von Herzen, R. P. and

Wong, H-K. (1971) Lake Tanganyika: Water chemistry, sediments, geological structure. *Naturwissenschaften,* **58**, 229–41.

162 De Mott, W. R. (1982) Feeding selectivities and related ingestion rates of *Daphnia* and *Bosmina*. *Limnol. Oceanogr.,* **27**, 518–27.

163 Denny, P. (1980) Solute movement in submerged angiosperms. *Biol. Rev.,* **50**, 65–92.

164 Dierberg, F. E. and Brezonik, D. L. (1983) Nitrogen and phosphorus mass balances in natural and sewage-enriched cypress domes. *J. Appl. Ecol.,* **20**, 323–37.

165 Dillon, P. J. (1975) The application of the phosphorus-loading concept to eutrophication research. *Env. Canada Sci. Ser.,* **46**, Burlington, Ontario.

166 Dillon, P. J. and Rigler, F. H. (1974a) A test of a simple nutrient budget model predicting the phosphorus concentration in lake water. *J. Fish. Res. Bd. Canada,* **31**, 1771–8.

167 Dillon, P. J. and Rigler, F. H. (1974b) A test of a simple method for predicting the capacity of a lake for development based on lake trophic status. *J. Fish. Res. Bd. Canada,* **32**, 1519–31.

168 Dodson, S. I. (1974a) Adaptive changes in plankton morphology in response to size-selective predation: a new hypothesis of cyclomorphosis. *Limnol. Oceanogr.,* **19**, 721–9.

169 Dodson, S. I. (1974b) Zooplankton competition and predation: an experimental test of the size-efficiency hypothesis. *Ecology,* **55**, 605–13.

170 Dodson, S. I. and Egger, D. L. (1980) Selective feeding of red phalaropes on zooplankton of Arctic ponds. *Ecology,* **61**, 755–63.

171 Dokulil, M. (1974) Der Neusiedler See (Österreich). *Ber. Naturhist. Ges.,* **118**, 205–11.

172 Donner, J. (1966) *Rotifers.* Warne, London.

173 Douglas, M. S. (1947) *The Everglades: River of Grass.* Ballantine, New York.

174 Downing, J. A. (1986) A regression technique for the estimation of epiphytic invertebrate populations. *Freshwat. Biol.,* **16**, 161–73.

175 Downing, J. A. and Cyr, H. (1985) Quantitative estimation of epiphytic invertebrate populations. *Can. J. Fish. Aquat. Sci.,* **42**, 1570–9.

176 Downing, J. A. and Rigler, F. H. (1984) *A Manual on Methods for the Assessment of Secondary Productivity in Freshwaters.* Blackwell Scientific Publications, Oxford.

177 Drablos, D. and Tollan, A. (Eds) (1980) *Proceedings of an International Conference on the Ecological Impact of Acid Precipitation.* SNSF Project, Sandefjord, Norway.

178 Drake, C. M. and Elliott, J. M. (1983) A new quantitative airlift sampler for collecting macroinvertebrates on stony bottoms in deep rivers. *Freshwat. Biol.,* **13**, 545–60.

179 Drake, J. C. and Heaney, S. I. (1987) Occurrence of phosphorus and its potential remobilization in the littoral sediments of a productive English lake. *Freshwat. Biol.,* **17**, 513–24.

180 Duffey, E. and Watt, A. S. (Eds) (1971) *The Scientific Management of Animal and Plant Communities for Conservation.* Blackwell Scientific Publications, Oxford.

181 Duncan, A. (1984) Assessment of factors influencing the composition, body size and turnover rate of zooplankton in Parakrama Samudra, an irrigation reservoir in Sri Lanka. *Hydrobiologia,* **113**, 201–15.

182 Dunn, I. G. (1972) The commercial fishery of L. George, Uganda (E. Africa). *Afr. J. Trop. Hydrobiol. Fish.,* **2**, 109–20.

183 Duthie, J. R. (1972) Detergents: nutrient considerations and total assessment. *Nutrients and Eutrophication* (Ed. by G. E. Likens), pp. 205–16. American Society for Limnology and Oceanography, Lawrence, Ka.

184 Eaton, J. W., Murphy, K. J. and Hyde, T. M. (1981) Comparative trials of herbicidal and mechanical control of aquatic weeds in canals. *Aquatic Weeds and their Control.* 105–16. Association of Applied Biologists, Oxford.

185 Eccles, D. H. (1985) Lake flies and sardines—a cautionary note. *Biol. Cons.,* **33**, 309–33.

186 Edmondson, W. T. (1960) Reproductive rates of rotifers in natural populations. *Mem. Ist. Ital. Idrobiol.,* **12**, 21–77.

187 Edmondson, W. T. (1970) Phosphorus, nitrogen and algae in Lake Washington after diversion of sewage. *Science,* **169**, 690–1.

188 Edmondson, W. T. (1972) The present condition of Lake Washington. *Verh. int. Verein. theoret. angew. Limnol.,* **18**, 284–91.

189 Edmondson, W. T. (1974) Secondary production. *Mitt. Verein. int. theoret. angew. Limnol.,* **20**, 229–72.

190 Edmondson, W. T. (1979) Lake Washington and the predictability of limnological events. *Erg. Limnol. Archiv. Hydrobiol. Beih.,* **13**, 234–41.

191 Edmondson, W. T., Anderson, G. C. and Peterson, D. R. (1956) Artificial eutrophication of Lake Washington. *Limnol. Oceanogr.,* **1**, 47–53.

192 Edmondson, W. T., Ward, H. B. and Whipple, G. C. (Eds) (1959) *Freshwater Biology.* Wiley, New York.

193 Edwards, A. M. C. (1971) *Aspects of the chemistry of four East Anglian rivers.* Ph.D. Thesis, University of East Anglia, Norwich.

194 Edwards, R. W., Densem, J. W. and Russell, P. A. (1979) An assessment of the importance of temperature as a factor controlling the growth rate of brown trout in streams. *J. Anim. Ecol.,* **48**, 501–8.

195 Elliott, J. M. (1971) The distances travelled by drifting invertebrates in a Lake District stream. *Oecologia,* **6**, 350–79.

196 Elliott, J. M. (1976) The energetics of feeding, metabolism and growth of brown trout (*Salmo trutta* L.) in relation to body weight, water temperature and ration size. *J. Anim. Ecol.,* **45**, 923–48.

197 Elliott, J. M. (1977a) Some methods for the statistical analysis of samples of benthic invertebrates. *Sci. Pub. Freshwat. Biol., Assoc.,* **25**, 160 pp.

198 Elliott, J. M. (1977b) Feeding, metabolism and growth of brown trout. *Ann. Rep. Freshwat. Biol. Assoc.,* **45**, 70–7.

199 Elliott, J. M. (1984) Hatching time and growth of *Nemurella pictetii* (Plecoptera: Nemouridae) in the laboratory and a Lake District stream. *Freshwat. Biol.,* **14**, 491–500.

200 Elliott, J. M. and Tullett, P. A. (1978) *A Bibliography of Samplers for Benthic Invertebrates. Occ. Publ. Freshwat. Biol. Assoc.,* **4**.

200a Emerson, J. W. (1971) Channelization, a case study. *Science,* **173**, 325–6.

201 Eminson, D. F. and Moss, B. (1980) The composition and ecology of periphyton communities in freshwaters. I. The influence of host type and external environment on community composition. *Br. Phycol. J.,* **15**, 429–46.

202 Entz, B. A. G., DeWitt, J. W., Massoud, A. and Khallaf, E. S. (1971) Lake Nasser, United Arab Republic. *Afr. J. Trop. Hydrobiol. Fish.,* **1**, 69–83.

203 Environmental Resources Ltd (1983) *Acid Rain: A Review of the Phenomenon in the EEC and Europe.* Graham and Trotman, Guildford.

204 Eriksson, F., Hörnström, E., Mossberg, P. and Nyberg, P. (1983) Ecological effects of lime treatment of acidified lakes and rivers in Sweden. *Hydrobiologia,* **101**, 145–64.

205 Eriksson, M. O. G. (1984) Acidification of lakes: Effects on waterbirds in Sweden. *Ambio,* **13**, 260–2.

206 Estep, K. W., Davis, P. G., Keller, M. D. and Sieburth, J. McN. (1986) How important are oceanic algal nannoflagellates in bacterivory? *Limnol. Oceanogr.,* **31**, 646–50.

207 Etherington, J. R. (1982) *Environment and Plant Ecology.* Wiley, Chichester.

208 Etherington, J. R. (1983) *Wetland Ecology.* Arnold, London.

209 Ettlinger, M. and Ferch, H. (1978) Synthetic zeolites as new builders for detergents. *Manuf. Chem. Aerosol News,* October 1978, 51–66.

210 Evans-Pritchard, E. E. (1940) *The Nuer.* Oxford University Press, Oxford.

211 Ewel, K. C. and Odum, H. T. (Eds) (1984) *Cypress Swamps.* University of Florida Press, Gainesville.

212 Fassett, N. C. (1957) *A Manual of Aquatic Plants,* 2nd edn. University of Wisconsin Press, Madison.

213 Fay, P., Stewart, W. D. P., Walsby, A. E. and Fogg, G. E. (1968) Is the heterocyst the site of nitrogen fixation in blue-green algae? *Nature,* **220**, 810–12.

214 Finlay, B. J. (1980) Temporal and vertical distribution of ciliophoran communities in the benthos of a small eutrophic loch with particular reference to the redox profile. *Freshwat. Biol.,* **10**, 15–34.

215 Finlay, B. J. (1982) Effects of seasonal anoxia on the community of benthic ciliated protozoa in a productive lake. *Arch. Protist.,* **15**, 215–22.

216 Finlay, B. J. and Ochsenbein-Gattlen, A. (1982) Ecology of free-living protozoa (a bibliography). *Occ. Publ. Freshwat. Biol. Assoc.,* **17**.

217 Finlay, B. J., Span, A. S. W. and Harman, J. P. (1983) Nitrate respiration in primitive eukaryotes. *Nature,* **303**, 333–6.

218 Fisher, S. G. and Likens, G. E. (1973) Energy flow in Bear Brook, New Hampshire: an integrative approach to stream ecosystem metabolism. *Ecol. Monogr.,* **43**, 421–39.

219 Fitkau, E. J. (1970) Role of caimans in the nutrient regime of mouth-lakes in

Amazon effluents (a hypothesis). *Biotropica*, **2**, 138–42.

220 Fitkau, E. J. (1973) Crocodiles and the nutrient metabolism of Amazonian water. *Amazoniana*, **4**, 101–33.

221 Fitzgerald, G. P. (1969) Some factors in the competition or antagonism among bacteria, algae and aquatic weeds. *J. Phycol.*, **5**, 351–9.

222 Fletcher, M., Gray, T. R. G. and Jones, J. G. (Eds) (1987) *Ecology of Microbial Communities*. Cambridge University Press, Cambridge.

223 Flower, R. J. (1987) The relationship between surface sediment diatom assemblages and pH in 33 Galloway Lakes. *Hydrobiologia*, **143**, 93–104.

224 Flower, R. J. and Battarbee, R. N. (1983) Diatom evidence for recent acidification of two Scottish Lochs. *Nature*, **305**, 130–2.

225 Flower, R. J., Battarbee, R. W. and Appleby, P. G. (1987) The recent palaeolimnology of acid lakes in Gallaway, southwest Scotland: Diatom analysis, pH trends, and the role of afforestation. *J. Ecol.*, **75**, 797–824.

226 Fogg, G. E. (1971) Extracellular products of algae in freshwater. *Arch. Hydrobiol. Beih Erg. Limnol.*, **5**, 1–25.

227 Fogg, G. E. (1975) *Algal Cultures and Phytoplankton Ecology*. 2nd edn. University of Wisconsin Press, Madison.

228 Fogg, G. E. and Westlake, D. F. (1955) The importance of extracellular products of algae in freshwater. *Verh. int. Verein. theoret. angew. Limnol.*, **12**, 219–32.

229 Foy, R. H. (1985) Phosphorus inactivation in a eutrophic lake by the direct addition of ferric aluminium sulphate: impact on iron and phosphorus. *Freshwat. Biol.*, **15**, 613–30.

230 Franks, F. (1983) *Water*. Royal Society of Chemistry, London.

231 Frantz, T. C. and Cordone, A. J. (1967) Observations on deep water plants in lake Tahoe, California and Nevada. *Ecology*, **48**, 709–14.

232 Freeman, B. J. and Freeman, M. C. (1985) Production of fishes in a subtropical blackwater ecosystem: The Okefenokee Swamp. *Limnol. Oceanogr.*, **30**, 686–92.

233 Frey, D. G. (Ed.) (1969) *Symposium on Palaeolimnology. Mitt. int. Ver. Limnol.*, **17**, 448 pp.

234 From, P. O. (1980) A review of some physiological and toxicological responses of freshwater fish to acid stress. *Env. Biol. Fishes*, **5**, 79–93.

235 Frost, W. E. and Brown, M. E. (1967) *The Trout*. Collins, London.

236 Fry, G. L.A. and Cooke, A. S. (1984) *Acid Deposition and its Implications for Nature Conservation in Britain*. Nature Conservancy Council, Shrewsbury.

237 Fryer, G. (1959) The trophic interrelationships and ecology of some littoral communities of Lake Nyasa with especial reference to the fishes and a discussion of the evolution of a group of rock-frequenting Cichlidae. *Proc. Zool. Soc. London*, **132**, 153–281.

238 Fryer, G. (1960) Concerning the proposed introduction of Nile perch into Lake Victoria. *E. Afr. Agric. J.*, **25**, 267–70.

239 Fryer, G. (1968) Evolution and adaptive radiation in the Chydoridae (Crustacea: Cladocera): A study in comparative functional morphology and ecology. *Phil. Trans. R. Soc.*, (B) **254**, 221–385.

240 Fryer, G. (1971) Functional morphology and niche specificity in chydorid and macrothricid cladocerans. *Trans. Am. Micros. Soc.*, **90**, 103–4.

241 Fryer, G. (1973) The Lake Victoria fisheries: some facts and fallacies. *Biol. Cons.*, **5**, 304–8.

242 Fryer, G. (1977a) The atyid prawns of Dominica. *Ann. Rep. Freshwat. Biol. Assoc.*, **45**, 48–54.

243 Fryer, G. (1977b) Studies on the functional morphology and ecology of the atyid prawns of Dominica. *Phil. trans. R. Soc.*, (B) 57–129.

244 Fryer, G. (1987) Quantitative and qualitative: numbers and reality in the study of living organisms. *Freshwat. Biol.*, **17**, 177–90.

245 Fryer, G. and Iles, T. D. (1969) Alternative routes to evolutionary success as exhibited by African cichlid fishes of the genus *Tilapia* and the species flocks of the Great Lakes. *Evolution*, **23**, 359–69.

246 Fryer, G. and Iles, T. D. (1972) *The Cichlid Fishes of the Great Lakes of Africa*. Oliver and Boyd, Edinburgh.

247 Furch, K. (1984) Water chemistry of the Amazon Basin: the distribution of chemical elements among freshwaters. *The Amazon* (Ed. by H. Sioli), pp. 167–200. Junk, Dordrecht.

248 Furtado, J. I. and Mori, S. (Eds) (1982) *Tasek Bera. The Ecology of a Freshwater Swamp*. Junk, The Hague.

249 Gaarder, T. and Gran, H. H. (1927)

Investigations of the production of plankton in the Oslo Fjord. *J. du Conseil*, **42**, 1–48.

250 Ganf, G. G. (1983) An ecological relationship between *Aphanizomenon* and *Daphnia pulex*. *Austr. J. Mar. Freshwat. Res.*, **34**, 755–73.

251 Ganf, G. G. and Blazka, P. (1974) Oxygen uptake, ammonia and phosphate excretion by zooplankton of a shallow equatorial lake (Lake George, Uganda). *Limnol. Oceanogr.*, **19**, 313–25.

252 Ganf, G. G. and Oliver, R. L. (1982) Vertical separation of light and available nutrients as a factor causing replacement of green algae by blue-green algae in the plankton of a stratified lake. *J. Ecol.*, **70**, 529–844.

253 Ganf, G. G. and Shiel, R. J. (1985) Particle capture by *Daphnia carinata*. *Austr. J. Mar. Freshwat. Res.*, **36**, 69–86.

254 Ganf, G. G. and Viner, A. B. (1973) Ecological stability in a shallow equatorial lake (Lake George, Uganda). *Proc. R. Soc.*, (B) **184**, 321–46.

255 Gardner, W. S. and Lee, G. F. (1975) The role of amino acids in the nitrogen cycle of Lake Mendota. *Limnol. Oceanogr.*, **20**, 379–88.

256 Garrick, L. D. and Lang, J. W. (1977) The alligator revealed. *Nat. Hist.*, **86**, 54–61.

257 Garrod, D. J. (1961a) The rational exploitation of the *Tilapia esculenta* stock of the North Buvuma island area, Lake Victoria. *E. Afr. Agric. Forest. J.*, **27**, 69–76.

258 Garrod, D. J. (1961b) The history of the fishing industry of Lake Victoria, East Africa, in relation to the expansion of marketing facilities. *E. Afr. Agric. Forest. J.*, **27**, 95–9.

259 Gash, J. H. C., Oliver, H. R., Stuttleworth, W. J. and Stewart, J. B. (1978) Evaporation from forests. *J. Inst. Water Eng. Sci.*, **32**, 104–10.

260 Gaudet, J. J. (1977) Uptake, accumulation and loss of nutrients by papyrus in tropical swamps. *Ecology*, **58**, 415–22.

261 Gaudet, J. J. (1979) Seasonal changes in nutrients in a tropical swamp: North Swamp, Lake Naivasha, Kenya. *J. Ecol.*, **67**, 953–81.

262 Gee, J. H. R. (1982) Resource utilization by *Gammarus pulex* (Amphipoda) in a Cotswold stream: a micro distribution study. *J. Anim. Ecol.*, **51**, 817–32.

263 Geller, W. and Müller, H. (1981) The filtration apparatus of Cladocera: Filter mesh sizes and their implications on food selectivity. *Oecologia*, **49**, 316–21.

264 George, E. A. (1976) A guide to algal keys (excluding seaweeds). *Br. Phycol. J.*, **11**, 49–55.

265 Gerking, S. D. (1978) *Ecology of Freshwater Fish Production*. Blackwell Scientific Publications, Oxford.

266 Gerloff, G. C. and Krombholz, P. H. (1966) Tissue analysis as a measure of nutrient availability for the growth of aquatic plants. *Limnol. Oceanogr.*, **11**, 529–37.

267 Gersberg, R. M., Elkins, B. W. and Goldman, C. R. (1983) Nitrogen removal in artificial wetlands. *Water Res.*, **17**, 1009–14.

268 Gessner, F. (1952) Der Druck in seiner Bedeutung für das Wachstum submerser Wasserpflanzen. *Planta*, **40**, 391–7.

269 Gibbons, J. W. and Sharitz, R. R. (1974) *Thermal Ecology*. U.S. Atomic Energy Commission, Augusta, Georgia.

270 Gibbs, R. J. (1970) Mechanisms controlling world water chemistry. *Science*, **170**, 1088–90.

271 Gilbert, J. J. (1973) Induction and ecological significance of gigantism in the rotifer *Asplanchna sieboldi*. *Science*, **181**, 63–6.

272 Gilbert, P. A. and De Jong, A. L. (1978) The use of phosphate in detergents and possible replacements for phosphate. *Phosphorus in the environment: its chemistry and biochemistry*. CIBA Foundation Symposium 57. Elsevier, Amsterdam.

273 Gilette, R. (1972) Stream channelization: conflict between ditchers, conservationists. *Science*, **176**, 890–3.

274 Gliwicz, M. S. and Sieniawska, A. (1986) Filtering activity of *Daphnia* in low concentrations of a pesticide. *Limnol. Oceanogr.*, **31**, 1132–7.

275 Gliwicz, Z. M. (1977) Food size selection and seasonal succession of filter feeding zooplankton in an eutrophic lake. *Ekol. Polsk.*, **25**, 179–225.

276 Gliwicz, Z. M. (1980) Filtering rates, food size selection and feeding rates in Cladocerans—another aspect of interspecific competition in filter feeding zooplankton. *Evolution and Ecology of Zooplankton Communities* (Ed. by W. C. Kerfoot) pp. 282–91. University of New England Press, Hanover.

277 Godmaire, H. and Planas, D. (1986) Influence of *Myriophyllum spicatum* L. on the species composition, biomass and

primary productivity of phytoplankton. *Aquat. Bot.*, **23**, 299–308.

278 Godwin, H. (1975) *The History of the British Flora*. 541 pp. Cambridge University Press, Cambridge.

279 Godwin, H. (1978) *Fenland: its Ancient Past and Uncertain Future*. Cambridge University Press, Cambridge.

280 Goldman, C. R. and Horne, A. J. (1983) *Limnology*. McGraw Hill, New York.

281 Goldman, J. C., McCarthy, J. J. and Peavey, D. G. (1979) Growth rate influence on the chemical composition of phytoplankton in oceanic waters. *Nature*, **179**, 210–15.

282 Goldsmith, E. and Hildyard, N. (1985) *The Social and Environmental Effects of Large Dams*. 1. Overview. Wadebridge Ecological Centre, Wadebridge.

283 Goldsmith, F. and Hildyard, N. (Eds) (1986) *The Social and Environmental Effects of Large Dams*. 2. Case Studies. Wadebridge Ecological Centre, Wadebridge.

284 Goltermann, H. L. (1975) *Physiological Limnology*. Elsevier, Amsterdam.

285 Golterman, H. L. (Ed.) (1977) *Interactions Between Sediments and Freshwater*. Junk, The Hague.

286 Golterman, H. L., Clymo, R. S. and Ohnstad, M. A. M. (1978) *Methods for Physical and Chemical Analysis of Freshwaters*. IBP Handbook No. 8. Blackwell Scientific Publications, Oxford.

287 Good, R. E., Whigham, D. F., Simpson, R. L. and Jackson, C. G. Jr. (1978) *Freshwater Wetlands*. Academic Press, New York.

288 Goode, D. A. (1981) *Lead Poisoning in Swans*. Report of the Nature Conservancy Council's Working Group. NCC, London.

289 Goodwin, P. (1976) Volta ten years on. *New Sci.*, 596–7.

290 Gophen, M. and Geller, W. (1984) Filter mesh size and food particle uptake by *Daphnia*. *Oecologia*, **64**, 408–12.

291 Gorham, E. (1958) The influence and importance of daily weather conditions in the supply of chloride, sulphate and other ions to freshwater from atmospheric precipitation. *Phil. Trans. R. Soc.*, (B) **247**, 147–78.

292 Gorham, E. (1961) Factors influencing supply of major ions to inland waters with special reference to the atmosphere. *Geol. Soc. Am. Bull.*, **72**, 795–840.

293 Goulden, C. E. (1964) The history of the Cladoceran fauna of Esthwaite Water (England) and its limnological significance. *Arch. Hydrobiol.*, **60**, 1–52.

294 Goulding, M. (1980) *The Fishes and the Forest: Explorations in Amazonian Natural History*. University of California Press, Los Angeles.

295 Goulding, M. (1981) *Man and Fisheries on an Amazon frontier*. Junk, The Hague.

296 Graham, R. (1986) Ghana's Volta resettlement scheme. *The Social and Environmental Effects of Large Dams* (Ed. by F. Goldsmith and N. Hildyard), pp. 131–9. Wadebridge Ecological Centre, Wadebridge.

297 Green, J. (1960) Zooplankton of the River Sokoto. The Rotifers. *Proc. Zool. Soc. London*, **135**, 491–523.

298 Green, J. (1967) The distribution and variation of *Daphnia lumholtzi* (Crustacea: Cladocera) in relation to fish predation in Lake Albert, East Africa. *J. Zool.*, **151**, 181–97.

299 Green, J., Corbet, S. A. and Betney, E. (1973) Ecological studies on crater lakes in West Cameroon. The blood of endemic cichlids in Barombi Mbo in relation to stratification and their feeding habits. *J. Zool.*, **170**, 299–308.

300 Green, J., Corbet, S. A., Watts, E. and Lan, O. B. (1976) Ecological studies on Indonesian lakes. Overturn and restratification of Ranu Lamongan. *J. Zool.*, **180**, 315–54.

301 Gregg, W. W. and Rose, F. L. (1982) The effects of aquatic macrophytes on the stream microenvironment. *Aquat. Bot.*, **14**, 309–24.

302 Gregory, K. J. and Walling, D. E. (1973) *Drainage Basin Form and Process*. Arnold, London.

303 Griffiths, M. and Edmondson, W. T. (1975) Burial of oscillaxanthin in the sediment of Lake Washington. *Limnol. Oceanogr.*, **20**, 945–52.

304 Griffiths, M., Perrott, P. S. and Edmondson, W. T. (1969) Oscillaxanthin in the sediment of Lake Washington. *Limnol. Oceanogr.*, **14**, 317–26.

305 Gulland, J. A. (1969) *Manual of Methods for Fish Stock Assessment*. Part 1. *Fish Population Analysis*. FAO, Rome.

306 Gulland, J. A. (1970) The fish resources of the oceans. FAO Fish. Tech. Paper 97, 1–425.

307 Haberyan, K. A. (1987) Fossil diatoms and the palaeolimnology of Lake Rukwa, Tanzania. *Freshwat. Biol.*, **17**, 429–36.

308 Hall, C. A. S. and Moll, R. (1975) Methods of assessing aquatic primary productivity. *Primary Productivity of the Biosphere* (Ed. by H. Leith and R. H. Whittaker), pp. 19–54. Springer-Verlag, New York.

309 Hall, D. J. (1964) An experimental approach to the dynamics of a natural population of *Daphnia galeata mendotae*. *Ecology*, **45**, 94–111.

310 Hall, D. J., Threlkeld, S. T., Burns, C. W. and Crowley, P. H. (1976) The size and efficiency hypothesis and the size structure of zooplankton communities. *A. Rev. Ecol. Syst.*, **7**, 177–208.

311 Hall, R. J., Likens, G. E., Fiance, S. B. and Hendry, G. R. (1980) Experimental acidification of a stream in the Hubbard Brook Experimental Forest, New Hampshire. *Ecology*, **61**, 976–89.

312 Hammer, U. T. (1986) *Saline Lake Ecosystems of the World*. Junk, Dordrecht.

313 Haney, J. F. (1971) An *in situ* method for the measurement of zooplankton grazing rates. *Limnol. Oceanogr.*, **16**, 970–6.

314 Harding, D. (1966) Lake Kariba, the hydrology and development of fisheries. *Man-made Lakes* (Ed. by R. Lowe-McConnell), pp. 7–20. Academic Press, London.

315 Harding, J. P. and Smith, W. A. (1974) *A Key to British Freshwater Cyclopoid and Calanoid Copepods. Sci. Publ. Freshwat. Boil. Assoc.*, **18**.

316 Harding, J. P. C., Burrows, I. G. and Whitton, B. A. (1981) Heavy metals in the Derwent reservoir catchment, Northern England. *Heavy Metals in Northern England: Environmental and Biological Aspects* (Ed. by P. J. Say and B. A. Whitton), pp. 73–86. University of Durham Press, Durham.

317 Harlin, M. M. (1975) Epiphyte–host relations in sea grass communities. *Aquat. Bot.*, **1**, 125–31.

318 Harriman, R. and Morrison, B. R. S. (1981) Forestry, fisheries and acid rain in Scotland. *Scott. Forest.*, **36**, 89–95.

319 Harriman, R. and Morrison, B. R. S. (1982) Ecology of streams draining forested and non-forested catchments in an area of central Scotland subject to acid precipitation. *Hydrobiologia*, **88**, 251–63.

320 Harris, G. P. (1985) The answer lies in the nesting behaviour. *Freshwat. Biol.*, **15**, 375–80.

321 Harris, G. P. (1986) *Phytoplankton Ecology*. Chapman and Hall, London.

322 Hart, D. (1968) *The Volta River Project*. Edinburgh University Press, Edinburgh.

323 Hart, R. C. (1986a) Zooplankton abundance, community structure and dynamics in relation to inorganic turbidity, and their implications for a potential fishery in subtropical Lake Le Roux, South Africa. *Freshwat. Biol.*, **16**, 351–72.

324 Hart, R. C. (1986b) *Plankton, Fish and Man—a Triplet in Limnology*. Inaugural Speech, Rhodes University, Grahamstown.

325 Hartman, R. T. and Brown, D. L. (1967) Changes in internal atmosphere of submersed vascular hydrophytes in relation to photosynthesis. *Ecology*, **48**, 252–8.

326 Hartmann, J. (1977) Fischereiliche Veränderungen in kulturbedingt eutrophierenden Seen. *Schweiz. Zh. Hydrol.*, **39**, 243–54.

327 Harvey, H. H., Dillon, P. J., Kramer, J. R., Pierce, R. C. and Whelpdale, D. M. (1981) *Acidification in the Canadian aquatic environment. Scientific criteria for an assessment of the effects of acidic deposition on aquatic ecosystems*. National Research Council of Canada, Ottawa.

328 Haslam, S. M. (1978) *River Plants*. Cambridge University Press, Cambridge.

329 Haslam, S. M., Sinker, C. A. and Wolseley, P. A. (1975) British water plants. *Field Studies*, **4**, 243–351.

330 Havel, J. E. and Dodson, S. I. (1984) *Chaoborus* predation on typical and spined morphs of *Daphnia pulex*: Behavioural observations. *Limnol. Oceanogr.*, **29**, 487–94.

331 Havel, J. E. and Dodson, S. I. (1985) Environmental cues for cyclomorphosis in *Daphnia retrocurva* Forbes. *Freshwat. Biol.*, **15**, 469–78.

332 Haworth, E. Y. (1969) The diatoms of a sediment core from Blea Tarn, Langdale. *J. Ecol.*, **57**, 429–39.

333 Haworth, E. Y. (1972) Diatom succession in a core from Pickerel Lake, Northeastern South Dakota. *Geol. Soc. Am. Bull.*, **83**, 157–72.

334 Haworth, E. Y. and Lund, J. W. G. (1984) *Lake Sediments and Environmental History*. Leicester University Press, Leicester.

335 Hayes, C. R. and Greene, C. A. (1984) The evolution of eutrophication impact in public water supply reservoirs in East Anglia. *Water Poll. Cont.*, **1984**, 42–51.

336 Healey, F. P. and Hendzel, L. L. (1980) Physiological indicators of nturient defi-

ciency in lake phytoplankton. *Can. J. Fish. Aquat. Sci.*, **37**, 442–53.

337 Heaney, S. I. (1971) The toxicity of *Microcystis aeruginosa* Kratz from some English reservoirs. *Water Treat. Exam.*, **20**, 235–44.

338 Hebert, P. D. N. (1978) The adaptive significance of cyclomorphosis in *Daphnia*: more possibilities. *Freshwat. Biol.*, **8**, 313–20.

339 Hecky, R. E. and Kling, H. J. (1981) The phytoplankton and protozooplankton of the euphotic zone of Lake Tanganyika: species composition, biomass, chlorophyll content, and spatio-temporal distribution. *Limnol. Oceanogr.*, **26**, 548–64.

340 Heeg, J. and Breen, C. M. (1982) *Man and the Pongolo*. South African National Scientific Programmes Report 56. Council for Scientific and Industrial Research, Pretoria.

341 Hellawell, J. M. (1977) Change in natural and managed ecosystems: detection, measurement and assessment. *Proc. R. Soc. London*, (B) **197**, 31–57.

342 Hellawell, J. M. (1978) *Biological Surveillance of Rivers*. Water Research Centre, Stevenage.

343 Hellawell, J. M. (1986) *Biological Indicators of Freshwater Pollution and Environmental Management*. Elsevier, London.

344 Hessen, D. O. (1985) Filtering structures and particle size selection in co-existing Cladocera. *Oecologia*, **60**, 368–72.

345 Hickley, P. and Bailey, R. G. (1986) Fish communities in the perennial wetland of the Sudd, southern Sudan. *Freshwat. Biol.*, **16**, 695–709.

346 Hickling, C. F. (1961) *Tropical Inland Fisheries*. Longman, London.

347 Hickling, C. F. (1966) On the feeding processes of the white amur, *Ctenopharyngodon idella*, Val. *J. Zool.*, **148**, 408–19.

348 Hildebrand, S. G. (1974) The relation of drift to benthos density and food level in an artificial stream. *Limnol. Oceanogr.*, **19**, 951–7.

349 Hildrew, A. G., Townsend, C. R. and Francis, J. (1984) Community structure in some southern England streams: the influence of species interactions. *Freshwat. Biol.*, **14**, 297–310.

350 Hinton, H. E. (1968) Reversible suspension of metabolism and the origin of life. *Proc. R. Soc.*, (B) **171**, 43–7.

351 Hobbie, J. E., Crawford, C. C. and Webb, K. L. (1968) Amino acid flux in an estuary. *Science*, **159**, 1463–4.

352 Holdren, G. C. Jr. and Armstrong, D. E. (1980) Factors affecting phosphorus release from intact lake sediment cores. *Envir. Sci. Technol.*, **14**, 79–87.

353 Holdway, P. A., Watson, R. A. and Moss, B. (1978) Aspects of the ecology of *Prymnesium parvum* (Haptophyta) and water chemistry in the Norfolk Broads, England. *Freshwat. Biol.*, **8**, 295–311.

354 Holland, D. G. and Harding, J. P. C. (1984) Mersey. *Ecology of European Rivers* (Ed. by B. A. Whitton). Blackwell Scientific Publications, Oxford.

355 Holm, L. G., Weldon, L. W. and Blackburn, R. D. (1969) Aquatic weeds. *Science*, **166**, 699–708.

356 Holm, N. P., Ganf, G. G. and Shapiro, J. (1983) Feeding and assimilation rates of *Daphnia pulex* fed *Aphanizomenon flosaquae*. *Limnol. Oceanogr.*, **28**, 677–87.

357 Hopson, A. J. (1972) A study of the Nile perch in L. Chad. *Overseas Research Publications*, 19. H.M.S.O., London.

358 Horne, A. J. and Fogg, G. E. (1970) Nitrogen fixation in some English lakes. *Proc. R. Soc.*, **B175**, 351–66.

359 Horne, A. J. and Goldman, C. R. (1972) Nitrogen fixation in Clear Lake, California, I. Seasonal variation and the role of heterocysts. *Limnol. Oceanogr.*, **17**, 678–92.

360 Hornung, M. (1984) The impact of upland pasture improvement on solute outputs in surface waters. *Agriculture and the Environment* (Ed. by D. Jenkins), pp. 150–5. Institute of Terrestrial Ecology, Cambridge.

361 Hossell, J. C. and Baker, J. H. (1979a) Estimation of the growth rates of epiphytic bacteria and *Lemna minor* in a river. *Freshwat. Biol.*, **9**, 319–27.

362 Hossell, J. C. and Baker, J. H. (1979b) Epiphytic bacteria of the freshwater plant *Ranunculus penicillatus*: enumeration, distribution and identification. *Arch. Hydrobiol.*, **86**, 332–7.

363 Howard-Williams, C. (1977) Swamp ecosystems. *Malay. Nat. J.*, **31**, 113–25.

364 Howard-Williams, C. (1981) Studies on the ability of a *Potamogeton pectinatus* community to remove dissolved nitrogen and phosphorus compounds from lake water. *J. Appl. Ecol.*, **18**, 619–37.

365 Howard-Williams, C. (1983) Wetlands and watershed management: The role of aquatic vegetation. *J. Limnol. Soc. South. Afr.*, **9(2)**, 54–62.

366 Howard-Williams, C. (1985) Cycling and

retention of nitrogen and phosphorus in wetlands: a theoretical and applied perspective. *Freshwat. Biol.*, **15**, 391–431.

367 Howard-Williams, C., Davies, J. and Pickmere, S. (1982) The dynamics of growth, the effects of changing area and nitrate uptake by watercress *Nasturtium officinale* R. Br. in a New Zealand Stream. *J. Appl. Ecol.*, **19**, 589–601.

368 Howard-Williams, C., Davis, B. R. and Cross, R. H. M. (1978) The influence of periphyton on the surface structure of a *Potamogeton pectinatus* L. leaf (an hypothesis). *Aquat. Bot.*, **5**, 87–91.

369 Howard-Williams, C. and Downes, M. T. (1984) Nutrient removal by stream bank vegetation. *Land Treatment of Wastes* (Ed. by R. J. Wilcock), pp. 409–22. National Water and Soil Conservation Authority, Wellington, New Zealand.

370 Howard-Williams, C. and Lenton, G. M. (1975) The role of the littoral zone in the functioning of a shallow tropical lake ecosystem. *Freshwat. Biol.*, **5**, 445–59.

371 Howell, A. D. (1932) *Florida Bird Life.* Florida Department of Game and Freshwater Fish, Talahassee, Florida.

372 Hrbacek, J., Bvorakova, K., Korinek, V. and Prochazkova, L. (1961) Demonstration of the effect of the fish stock on the species composition of the zooplankton and the intensity of metabolism of the whole plankton association. *Verh. int. Verein. theoret. angew. Limnol.*, **14**, 192–5.

373 Huber-Pestalozzi, G. (1938) *Das Phytoplankton des Süsswassers, Systematik und Biologie.* E. Schweitzerbartische Verlagsbuchhandlung, Stuttgart.

374 Hubschman, J. H. (1971) Lake Erie: Pollution abatement, then what? *Science*, **171**, 536–640.

375 Hughes, J. C. and Lund, J. W. G. (1962) The rate of growth of *Asterionella formosa* Hass in relation to its ecology. *Arch. Mikrobiol.*, **42**, 117–29.

376 Hurlbert, S. H., Loayza, W. and Moreno, T. (1986) Fish–flamingo–plankton interactions in the Peruvian Andes. *Limnol. Oceanogr.*, **31**, 457–68.

377 Hurlbert, S. H., Zedler, J. and Fairbanks, D. (1971) Ecosystem alteration by mosquito fish (*Gambusia affinis*) predation. *Science*, **175**, 639–41.

378 Hutchinson, G. E. (1937) A contribution to the limnology of arid regions. *Trans. Conn. Acad. Arts Sci.*, **33**, 47–132.

379 Hutchinson, G. E. (1957) *A Treatise on Limnology. Vol. I. Geography, Physics, Chemistry.* Wiley, New York.

380 Hutchinson, G. E. (1967) *A Treatise on Limnology.* Vol. II. Wiley, New York.

381 Hutchinson, G. E. (1973) Eutrophication. *Amer. Sci.*, **61**, 269–79.

382 Hutchinson, G. E. (1975) *A Treatise on Limnology. Vol. 3. Limnological Botany.* Wiley, New York.

383 Hutchinson, G. E., Bonatti, E., Cowgill, U. M., Goulden, C. E., Leventhal, E. A., Mallett, M. E., Margaritora, F., Patrick, R., Racek, A., Roback, S. A., Stella, E., Ward-Perkins, J. B. and Wellman, T. R. (1970) Ianula: an account of the history and development of the Lago di Monterosi, Latium, Italy. *Trans. Am. Phil. Soc.*, **60**, 1–178.

384 Hutchinson, G. E. and Löffler, H. (1956) The thermal classification of lakes. *Proc. Nat. Acad. Sci.*, **42**, 84–6.

385 Hynes, H. B. N. (1970) *The Biology of Polluted Waters.* Liverpool University Press, Liverpool.

386 Hynes, H. B. N. (1979) *The Ecology of Running Waters.* Liverpool University Press, Liverpool.

387 Hynes, H. B. N. and Kaushik, N. K. (1969) The relationship between dissolved nutrient salts and protein production in submerged autumnal leaves. *Verh. int. Verein. theor. angew. Limnol.*, **17**, 95–103.

387a Hynes, H. B. N., Kaushik, N. K., Lock, M. A., Lush, D. L., Stocker, Z. S. J., Wallace, R. R. and Williams, D. D. (1974) Benthos and allochthonous organic matter in streams. *J. Fish. Res. Bd. Canada*, **31**, 545–63.

388 Illies, J. (Ed.) (1978) *Limnofauna Europaea. A Checklist of the Animals Inhabiting European Inland Waters with Accounts of their Distribution and Ecology.* Gustav-Fischer Verlag, Stuttgart.

389 Imevbore, A. M. A. and Adegoke, O. S. (Eds) (1975) *The Ecology of Lake Kainji.* University of Ife Press, Nigeria.

390 Imhof, G. (1973) Aspects of energy flow by different food chains in a reed bed—a review. *Polsk. Arch. Hydrobiol.*, **20**, 165–8.

391 Imhof, G. and Burian, K. (1972) *Energy flow studies in a wetland ecosystem.* Special Publication. Austrian Academy of Sciences. Springer-Verlag, Vienna.

392 Infante, A. and Abella, S. E. B. (1985) Inhibition of *Daphnia* by *Oscillatoria* in Lake Washington. *Limnol. Oceanogr.*, **30**, 1046–52.

393 Infante, A. and Litt, A. H. (1985) Differences between two species of *Daphnia* in the use of 10 species of algae in Lake Washington. *Limnol. Oceanogr.*, **30**, 1053–9.

394 Ingold, C. T. (1966) The tetraradiate fungal spore. *Mycologia*, **58**, 43–56.

395 Ivlev, V. S. (1961) *Experimental Ecology of the Feeding of Fishes.* Yale University Press, Newhaven.

396 Jacobs, J. (1967) Untersuchungen zur Funktion und Evolution der Zyklomorphose bei *Daphnia* mit besonderer Berücksichtigung der Selektion durch Fische. *Arch. Hydrobiol.*, **62**, 467–541.

397 Jackson, P. B. N. (1971) The African Great Lakes fisheries: past, present and future. *Afr. J. Trop. Hydrobiol. Fish.*, **1**, 35–49.

398 Jawed, M. (1969) Body nitrogen and nitrogenous excretion in *Neomysis rayii* Murdoch and *Euphausia pacifica* Hansen. *Limnol. Oceanogr.*, **14**, 748–54.

399 Jenkin, P. M. (1942) Seasonal changes in the temperature of Windermere (English Lake District). *J. Anim. Ecol.*, **11**, 248–69.

400 Jenkins, D. (Ed.) (1985) *The Biology and Management of the River Dee.* Institute of Terrestrial Ecology, Cambridge.

401 Jenkins, D. and Shearer, W. M. (1986) *The Status of the Atlantic Salmon in Scotland.* Institute of Terrestrial Ecology, Huntingdon.

402 Jenkins, R. M. (1976) Prediction of fish production in Oklahoma reservoirs on the basis of environmental variables. *Ann. Okl. Acad. Sci.*, **5**, 11–20.

403 Jenkins, R. M. (1977) Prediction of fish biomass, harvest, and prey-predator relations in reservoirs. *Proc. Conf. Assess. Effects Power-Plant-Induced Mortality on Fish Populations* (Ed. by W. Van Winkle), pp. 282–93. Pergamon, New York.

404 Jewson, D. H. (1976) The interaction of components controlling net phytoplankton photosynthesis in a well-mixed lake (Lough Neagh) Northern Ireland. *Freshwat. Biol.*, **6**, 551–76.

405 Jewson, D. H. (1977) A comparison between *in situ* photosynthetic rates determined using ^{14}C uptake and oxygen evolution methods in Lough Neagh, Northern Ireland. *Proc. R. Irish Acad.*, (B) **77**, 87–99.

406 Jhingran, V. G. (1975) *Fish and Fisheries of India.* Hindustan Publishing Corporation, Delhi.

407 Johannes, R. E. (1965) Influence of marine protozoa on nutrient regeneration. *Limnol. Oceanogr.*, **10**, 434–42.

407a Johanson, K. and Nyberg, P. (1981) *Acidification of surface waters in Sweden— effects and extent 1980.* Publications of the Institute of Freshwater Research, Drottningholm 6. 118 pp. Swedish with English summary.

408 Johnson, M. G. and Brinkhurst, R. O. (1971a) Associations and species diversity in benthic macroinvertebrates of Bay of Quinte and Lake Ontario. *J. Fish. Res. Bd. Canada*, **28**, 1683–1697.

409 Johnson, M. G. and Brinkhurst, R. O. (1971b) Production of benthic macroinvertebrates of Bay of Quinte and Lake Ontario. *J. Fish. Res. Bd. Canada*, **28**, 1699–714.

410 Johnson, M. G. and Brinkhurst, R. O. (1971c) Benthic community metabolism in Bay of Quinte and Lake Ontario. *J. Fish. Res. Bd. Canada*, **28**, 1715–25.

411 Jonasson, P. M. (1972) Ecology and production of the profundal benthos in relation to phytoplankton in Lake Esrom. *Oikos*, suppl. 14, 1–148.

412 Jonasson, P. M. (1977) Lake Esrom research, 1867–1977. *Folia Limnol. Scand.*, **17**, 67–90.

413 Jonasson, P. M. (1978) Zoobenthos of lakes. *Verh. int. Verein. theor. angew. Limnol.*, **20**, 13–37.

414 Jones, J. G. (1979a) Microbial activity in lake sediments with particular reference to electrode potential gradients. *J. Gen. Microbiol.*, **115**, 19–26.

415 Jones, J. G. (1979b) *A Guide to Methods for Estimating Microbial Numbers and Biomass in Freshwater. Sci. Publ. Freshwat. Biol. Assoc.*, **39**.

416 Jones, J. G. (1985) Microbes and microbial processes in sediments. *Phil. Trans. R. Soc. London*, (A)**315**, 3–17.

417 Jones, J. R. E. (1940) A study of the zinc-polluted river Ystwyth in north Cardiganshire, Wales. *Ann. Appl. Biol.*, **27**, 368–78.

418 Jones, J. R. E. (1964) *Fish and River Pollution.* Butterworth, London.

419 Junk, W. J. (1983) Ecology of swamps on the middle Amazon. *Ecosystems of the World 4B Mires: Swamp, Bog, Fen and Moor Regional Studies* (Ed. by A. J. P. Gore) pp. 269–94. Elsevier, Amsterdam.

420 Kadlec, R. H. and Tilton, D. L. (1979) The use of fresh-water wetlands as a tertiary wastewater treatment alternative. *Crit. Rev. Environ. Control*, **9**, 185–212.

421 Kairesalo, T. and Koskimies, I. (1987) Grazing by oligochaetes and snails on epiphytes. *Freshwat. Biol.*, **17**, 317–24.

422 Kalk, M., McLachlan, A. and Howard-Williams, C. (1979) *Lake Chilwa: Studies of Change in a Tropical Ecosystem.* Junk, The Hague.

423 Kamp-Nielsen, L. (1974) Mud-water exchange of phosphate and other ions in undisturbed sediment cores and factors affecting the exchange rates. *Arch. Hydrobiol.*, **73**, 218–37.

424 Kaushik, N. K. and Hynes, H. B. N. (1968) Experimental study on the role of autumn shed leaves in aquatic environments. *J. Ecol.*, **56**, 229–43.

425 Kaushik, N. K. and Hynes, H. B. N. (1971) The fate of dead leaves that fall into streams. *Arch. Hydrobiol.*, **68**, 465–515.

426 Keating, K. I. (1978) Blue-green algal inhibition of diatom growth: transition from mesotrophic to eutrophic community structure. *Science*, **199**, 971–3.

427 Keeley, J. E. (1979) Population differentiation along a flood frequency gradient: physical adaptations to floods in *Nyssa sylvatica*. *Ecol. Monogr.*, **49**, 89–108.

428 Keeley, J. E. (1982) Distribution of diurnal acid metabolism in the genus *Isoetes*. *Am. J. Bot.*, **69**, 254–7.

429 Keeley, J. E. and Morton, B. A. (1982) Distribution of diurnal acid metabolism in submerged plants outside the genus *Isoetes*. *Photosynthetica*, **16**, 546–53.

430 Kerfoot, W. C. (1977) Competition in Cladoceran communities: the cost of evolving defenses against copepod predation. *Ecology*, **58**, 303–13.

431 Kerfoot, W. C. (Ed.) (1980) *Evolution and Ecology of Zooplankton Communities.* University Press of New England, Hanover.

432 Kerfoot, W. C. and Sih, A. (1987) *Predation—Direct and Indirect Impacts on Aquatic Communities.* University Press of New England, Hanover.

433 Ketchum, B. H. (1972) *The Water's Edge: Critical Problems of the Coastal Zone.* M. I. T. Press, Cambridge, Massachusetts.

434 Kirby, C. (1984) *Water in Great Britain.* Penguin, Harmondsworth.

435 Kirk, J. T. O. (1983) *Light and Photosynthesis in Aquatic Ecosystems.* Cambridge University Press, Cambridge.

436 Kirk, J. T. O. (1986) Optical limnology—a manifesto. *Limnology in Australia* (Ed. by P. de Deckker and W. D. Williams), pp. 33–62. CSIRO, Canberra.

437 Kramer, D. L. and McClure, M. (1982) Aquatic surface respiration, a widespread adaptation to hypoxia in tropical freshwater fishes. *Env. Biol. Fishes*, **7**, 47–55.

438 Kramer, J. and Tessier, A. (1982) Acidification of aquatic systems: a critique of chemical approaches. *Env. Sci. Technol.*, **16**, 606A–615A.

439 Krokhin, E. M. (1975) Transport of nutrients by salmon migrating from the sea into lakes. *Coupling of Land and Water Systems* (Ed. by A. D. Hasler), 153–6. Springer-Verlag, New York.

440 Kuznetsov, S. I. (1977) Trends in the development of ecological microbiology. *Adv. Aquat. Microbiol.*, **1**, 1–48.

441 Lack, T. J. (1981) Advances in the management of eutrophic reservoirs. *Notes on Water Res.*, **27**, 1–4. Water Research Centre.

442 Ladle, M. and Welton, J. S. (1984) The ecology of chalk-stream invertebrates studied in a recirculating stream. *Ann. Rep. Freshwat. Biol. Assoc.*, **52**, 63–74.

443 Lagler, K. F. (1969) *Man-made Lakes—Planning and Development.* U.N. Development Programme, F.A.O., Rome.

444 Lake, P. S., Barmuta, L. A., Boulton, A. J., Campbell, I. C. and St. Clair, R. M. (1985) Australian streams and Northern Hemisphere stream ecology comparisons and problems. *Proc. Ecol. Soc. Australia*, **14**, 61–82.

445 Lamarra, V. (1975) Digestive activities of carp as a major contributor to the nutrient loading of lakes. *Verh. Int. Verein. theoret. angew. Limnol.*, **19**, 2461–8.

446 Lampert, W. (1978) A field study on the dependence of the fecundity of *Daphnia* spec. on food concentration. *Oecologia*, **36**, 363–9.

447 Lampert, W. (1981) Inhibiting and toxic effects of blue-green algae on *Daphnia*. *Int. Rev. ges. Hydrobiol.*, **66**, 285–98.

448 Lampert, W. (Ed.) (1985) *Food Limitation and the Structure of Zooplankton Communities.* E. Schweizerbart'sche, Stuttgart.

449 Lampert, W., Fleckner, W., Rai, H. and Taylor, B. E. (1986) Phytoplankton control by grazing zooplankton: a study on the spring clear-water phase. *Limnol. Oceanogr.*, **31**, 478–90.

450 Lampert, W. and Taylor, B. E. (1984) In situ grazing rates and particle selection by zooplankton: effects of vertical migration. *Verh. int. Verein. theor. angew. Limnol.*, **22**, 943–6.

451 Lampert, W. and Taylor, B. E. (1985)

Zooplankton grazing in a eutrophic lake: implications of diel vertical migration. *Ecology*, **66**, 68–82.

452 Lane, P. A. (1979) Vertebrate and invertebrate predation intensity on freshwater zooplankton communities. *Nature*, **280**, 391–2.

453 Langford, T. E. (1983) *Electricity Generation and the Ecology of Natural Waters*. Liverpool University Press, Liverpool.

454 Lasenby, D. C., Northcote, T. G. and Fürst, M. (1986) Theory, practice and effects of *Mysis relicta* introductions to North American and Scandinavian lakes. *Can. J. Fish. Aquat. Sci.*, **43**, 1277–84.

455 Lavergne, M. (1986) The seven deadly sins of Egypt's Aswan High Dam. *The Social and Environmental Effects of Large Dams* (Ed. by F. Goldsmith and N. Hildyard), pp. 181–3.

456 Lawson, G. W. (1970) Lessons of the Volta—a new man-made lake in tropical Africa. *Biol. Cons.*, **2**, 90–6.

457 Leah, R. T., Moss, B. and Forrest, D. E. (1980) The role of predation in causing major changes in the limnology of a hypereutrophic lake. *Int. Rev. ges. Hydrobiol.*, **65**, 223–47.

458 Lean, D. R. S. (1973) Phosphorus dynamics in lake waters. *Science*, **179**, 678–80.

458a Le Blanc, R. (1980) Closing address. *Atlantic Salmon: its Future* (Ed. by A. J. Went), pp. 234–46. Fishing News Books, Farnham.

459 Le Cren, E. D. (1964) The interactions between freshwater fisheries and nature conservation. *Proc. MAR Conf.* IUCN Publications, 431–7.

460 Le Cren, E. D. and Lowe-McConnell, R. H. (1980) *The Functioning of Freshwater Ecosystems*. Cambridge University Press, Cambridge.

461 Lee, G. F. (1977) Significance of oxic vs. anoxic conditions for Lake Mendota sediment phosphorus release. *Interactions between Sediments and Freshwater* (Ed. by H. L. Golterman), 294–306. Junk, The Hague.

462 Leedale, G. F. (1967) *Euglenoid Flagellates*. Prentice Hall, Englewood Cliffs, New Jersey.

462a Lehman, J. T. (1976) Ecological and nutritional studies on *Dinobryon* Ehrenb. Seasonal periodicity and the phosphate toxicity problem. *Limnol. Oceanogr.*, **21**, 646–58.

463 Lehman, J. T. and Scavia, D. (1982) Microscale patchiness of nutrients in plankton communities. *Science*, **216**, 729–30.

464 Lennox, L. J. (1984) Lough Ennell: laboratory studies on sediment phosphorus release under varying mixing, aerobic and anaerobic conditions. *Freshwat. Biol.*, **14**, 183–7.

465 Leopold, A. (1949) *A Sandy County Almanac*. Oxford University Press, Oxford.

466 Leopold, L. B. (1974) *Water, A Primer*. Freeman, San Francisco.

467 Likens, G. E. and Bormann, F. H. (1974) Acid rain: a serious regional environmental problem. *Science*, **184**, 1176–9.

468 Likens, G. E. and Bormann, F. H. (1975) An experimental approach in New England landscapes. *Coupling of Land and Water Systems* (Ed. by A. D. Hasler), pp. 7–29. Springer-Verlag, New York.

469 Likens, G. E., Bormann, F. H., Pierce, R. S. Eaton, J. S. and Johnson, N. M. (1977) *Biogeochemistry of a Forested Ecosystem*. Springer-Verlag, New York.

470 Linacre, E. T., Hicks, B. B., Sainty, G. R. and Grauze, G. (1970) The evaporation from a swamp. *Agric. Meteorol.*, **7**, 375–86.

471 Linthurst, R. H. (1983) *The Acidic Deposition Phenomenon and its Effects*. Critical Assessment Review Papers II. U.S. Environmental Protection Agency, Washington D.C.

472 Lodge, D. M. (1985) Macrophyte-gastropod associations: observations and experiments on macrophyte choice by gastropods. *Freshwat. Biol.*, **15**, 695–708.

473 Lodge, D. M. (1986) Selective grazing on periphyton: a determinant of freshwater gastropod distributions. *Freshwat. Biol.*, **16**, 831–41.

474 Lowe-McConnell, R. H. (Ed.) (1966) Man-made lakes. *Symp. Inst. Biol. London*, **15**, Academic Press, New York.

475 Lowe-McConnell, R. H. (1975) *Fish Communities in Tropical Freshwaters*. Longman, London.

476 Lowe-McConnell, R. H. (1987) *Ecological Studies in Tropical Fish Communities*. Cambridge University Press, Cambridge.

477 Lund, J. W. G. (1949) Studies on *Asterionella*. I. The origin and nature of the cells producing seasonal maxima. *J. Ecol.*, **37**, 389–419.

478 Lund, J. W. G. (1950) Studies on *Asterionella formosa* Hass. II. Nutrient depletion and the spring maximum. *J. Ecol.*, **38**, 1–14, 15–35.

479 Lund, J. W. G. (1954) The seasonal cycle

of the plankton diatom *Melosira italica*
(Ehr.) Kütz susp. *subarctica* O. Müll. *J.
Ecol.,* **42**, 151–79.

480 Lund, J. W. G. (1964) Primary productivity and periodicity of phytoplankton.
Verh. int. Verein. theoret. angew. Limnol.,
15, 37–56.

481 Lund, J. W. G. (1971) The seasonal
periodicity of three planktonic desmids
in Lake Windermere. *Mitt. int. Verein.
theor. angew. Limnol.,* **19**, 3–25.

482 Lund, J. W. G. (1975) The uses of large
experimental tubes in lakes. *The Effects of
Storage on Water Quality.* 291–311. Water
Research Centre, Medmenham.

483 Lund, J. W. G. and Reynolds, C. S. (1982)
The development and operation of large
limnetic enclosures in Blelham Tarn,
English Lake District, and their contri-
bution to phytoplankton ecology. *Prog.
Phycol. Res.,* **1**, 2–65.

484 Lynch, M. (1979) Predation, competition,
and zooplankton community structure:
an experimental study. *Limnol. Oceanogr.,*
24, 253–72.

485 Lynch, M. (1980) *Aphanizomenon* blooms:
alternate control and cultivation by
*Daphnia pulex. Evolution and Ecology of
Zooplankton Communities* (Ed. by W. C.
Kerfoot), 299–304. University Press of
New England, Hanover.

486 Lynch, M. and Shapiro, J. (1980) Predation, enrichment and phytoplankton community structure. *Limnol. Oceanogr.,* **26**,
86–102.

487 Lyons, K. M. (1978) *The Biology of
Helminth Parasites.* Arnold, London.

488 Maberley, S. C. and Spence, D. H. N.
(1983) Photosynthetic inorganic carbon
use by freshwater plants. *J. Ecol.,* **71**, 705–
24.

489 Macan, T. T. (1959) *A Guide to Freshwater
Invertebrate Animals.* Longman, London.

490 Macan, T. T. (1963) *Freshwater Ecology.*
Longman, London.

491 Macan, T. T. (1970) *Biological Studies of
the English Lakes.* American Elsevier.

492 Macan, T. T. (1973) *Ponds and Lakes.*
Allen and Unwin, London.

493 Macan, T. T. (1976) A twenty-one-year
study of the water-bugs in a moorland
fishpond. *J. Anim. Ecol.,* **45**, 913–22.

494 Macan, T. T. (1977) The influence of
predation on the composition of freshwater animal communities. *Biol. Rev.,* **52**,
45–70.

495 Macan, T. T. and Kitching, A. (1972)
Some experiments with artificial sub-

strata. *Verh. int. Verein. theor. angew.
Limnol.,* **18**, 213–20.

496 Mackenzie, D. (1986) Geology of Cameroon's gas catastrophe. *New Sci.,* 4 September 1986, 26–7.

497 Mackereth, F. J. H. (1957) Chemical
analysis in ecology illustrated from lake
district tarns and lakes. 1. Chemical
analysis. *Proc. Linn. Soc. London,* **67**,
(1954–1955), 159–64.

498 Mackereth, F. J. H. (1958) A portable
core sampler for lake deposits. *Limnol.
Oceanogr.,* **3**, 181–91.

499 Mackereth, F. J. H. (1965) Chemical
investigations of lake sediments and their
interpretation. *Proc. R. Soc.,* (B) **161**, 293–
375.

500 Mackereth, F. J. H. (1966) Some chemical
observations on post-glacial lake sediments. *Phil. Trans. R. Soc.,* (B) **250**, 165–
213.

501 Mackereth, F. J. H., Heron, J. and
Talling, J. F. (1978) *Water Analysis. Sci.
Publ. Freshwat. Biol. Assoc.,* **36**.

502 Mackey, A. P., Cooling, D. A. and Berrie,
A. D. (1984) An evaluation of sampling
strategies for qualitative surveys of macro-
invertebrates in rivers, using pond nets.
J. Appl. Ecol., **21**, 515–34.

503 McCauley, E. and Kalff, J. (1981) Empirical relationships between phytoplankton
and zooplankton biomass in lakes. *Can.
J. Fish. Aquat. Sci.,* **38**, 458–63.

504 McCauley, E. and Kalff, J. (1987) Effect
of changes in zooplankton on orthophosphate dynamics of natural phytoplankton
communities. *Can. J. Fish. Aquat. Sci.,*
44, 176–82.

505 McLachlan, A. J. (1974a) Recovery of the
mud substrate and its associated fauna
following a dry phase in a tropical lake.
Limnol. Oceanogr., **19**, 74–83.

506 McLachlan, A. J. (1974b) Development
of some lake ecosystems in tropical Africa,
with special reference to the invertebrates. *Biol. Rev.,* **49**, 365–97.

507 McLachlan, A. J. (1981a) Food sources
and foraging tactics in tropical rain pools.
Zool. J. Linn. Soc., **71**, 265–77.

508 McLachlan, A. J. (1981b) Interaction
between insect larvae and tadpoles in
tropical rain pools. *Ecol. Ent.,* **6**, 175–82.

509 McLachlan, A. J. (1983) Life history
tactics of rain-pool dwellers. *J. Anim.
Ecol.,* **52**, 545–61.

510 McLachlan, A. J. and Cantrell, M. A.
(1980) Survival strategies in tropical rain
pools. *Oecologia,* **47**, 344–51.

511 McLachlan, A. J., Pearce, L. J. and Smith, J. A. (1979) Feeding interactions and cycling of peat in a bog lake. *J. Anim. Ecol.*, **48**, 851–61.

512 McQueen, D. G., Post, J. R. and Mills, E. L. (1986) Trophic relationships in freshwater pelagic ecosystems. *Can. J. Fish. Aquat. Sci.*, **43**, 1571–81.

513 Mahan, D. C. and Cummings, K. W. (undated) *A profile of Augusta Creek in Kalamazoo and Barry Counties, Michigan*. Technical Report No. 3. W. K. Kellogg Biological Station, Michigan State University, Michigan.

514 Maitland, P. S. (1972) *Key to British Freshwater Fishes. Sci. Publ. Freshwat. Biol. Assoc.*, **27**, 139 pp.

515 Maitland, P. S. (1974) The conservation of freshwater fishes in the British Isles. *Biol. Cons.*, **6**, 7–14.

516 Malley, D. F. (1980) Decreased survival and Ca uptake by the crayfish, *Orconectes virilis* in low pH. *Can. J. Fish. Aquat. Sci*, **37**, 364–72.

517 Maltby, E. (1986) *Waterlogged Wealth*. Earthscan, London.

518 Manny, B. A., Miller, M. C. and Wetzel, R. G. (1971) Ultraviolet combustion of dissolved organic compounds in lake waters. *Limnol. Oceanogr.*, **16**, 71–85.

519 Margulis, L. and Schwartz, K. V. (1982) *Five Kingdoms: an Illustrated Guide to the Phyla of Life on Earth*. Freeman, San Francisco.

520 Marshall, B. E. and Langerman, J. D. (1979) The Tanganyika sardine in Lake Kariba. *Rhodesia Sci. News.*, **13**, 104–5.

521 Marshall, E. J. P. and Westlake, D. F. (1975) Recent studies on the role of aquatic macrophytes in their ecosystem. *Proc. European Weed Ress Socs Symps No. 5 on Aquatic Weeds*, 43–51.

521a Marshall, T. K. (1977) Morphological, physiological and ethological differences between walleye (*Stizostedion vitreum*) and pikeperch (*S. lucioperca*). *J. Fish Res. Bd Canada*, **34**, 1515–23.

522 Martens, E., von (1858) On the occurrence of marine animal forms in freshwater. *Ann. Nat. Hist. (Series 3)* **1**, 50–63.

523 Mason, C. F. (1981) *Biology of Freshwater Pollution*. Longman, London.

524 Mason, C. F. and Bryant, R. J. (1975) Periphyton production and grazing by chironomids in Alderfen Broad, Norfolk. *Freshwat. Biol.*, **5**, 271–7.

525 Mason, H. L. (1957) *A Flora of Marshes of California*. University of California Press, Berkeley and Los Angeles.

526 Mathews, C. P. and Westlake, D. F. (1969) Estimation of production by populations of higher plants subject to high mortality. *Oikos*, **20**, 156–60.

527 Maxwell, G. (1957) *A Reed Shaken by the Wind*. Longmans, London. (1983) Penguin, Harmondsworth.

528 May, R. M., Beddington, J. R., Clarke, C. S., Holt, S. J. and Laws, R. M. (1979) Management of multispecies fisheries. *Science*, **205**, 267–77.

529 Melack, J. M. (1976) Primary productivity and fish yields in tropical lakes. *Trans. Am Fish. Soc.*, **105**, 575–80.

530 Melack, J. M. (1979) Temporal variability of phytoplankton in tropical lakes. *Oecologia*, **44**, 1–7.

531 Melack, J. M. and Kilham, P. (1974) Photosynthetic rates of phytoplankton in East African alkaline, saline lakes. *Limnol. Oceanogr.*, **19**, 743–55.

532 Melack, J. M., Kilham, P. and Fisher, T. R. (1982) Responses of phytoplankton to experimental fertilization with ammonium and phosphate in an African soda lake. *Oecologia*, **52**, 321–6.

533 Mellanby, H. (1963) *Animal Life in Freshwater*. Chapman and Hall, London.

534 Merritt, R. W. and Cummins, K. W. (1978) *An Introduction to the Aquatic Insects of North America*. Kendall/Hunt, Dubuque, Iowa.

535 Milbrink, G. (1977) On the limnology of two alkaline lakes (Nakuru and Naivasha) in the east rift valley system in Kenya. *Int. Rev. ges. Hydrobiol.*, **62**, 1–17.

536 Mills, D. H. (1967) A study of trout and young salmon populations in forest streams with a view to management. *Forestry*, **40**, 85–90.

537 Mills, D. H. (1971) *Salmon and Trout: a Resource, its Ecology, Conservation and Management*. Oliver and Boyd, Edinburgh.

538 Mills, D. H. (1980) Scottish salmon rivers and their future management. *Atlantic Salmon: its Future* (Ed. by A. E. J. Went), 70–81. Fishing News Books, Farnham.

539 Mills, S. (1982a) Salmon: demise of the landlord's fish. *New Sci.*, 364–7.

540 Mills, S. (1982b) Britain's native trout is floundering. *New Sci.*, 498–501.

541 Ministry of Agriculture, Fisheries and Food (1976) *Agriculture and Water Quality*. Technical Bulletin, 32.

542 Mitchell, D. S. (1972) The Kariba weed:

Salvinia molesta. Brit. Fern. Gaz., 10, 251–2.

543 Mitchell, D. S. (Ed.) (1974) *Aquatic Vegetation and its Use and Control.* UNESCO, Paris.

544 Moncur, A. (1986) Volcanic gas kills 1500 villagers. *The Guardian* August 126, 1986, p. 1.

545 Monod, J. (1942) *Recherches sur la croissance des cultures bacteriennes.* Hermann, Paris.

546 Moore, J. W. (1980) Zooplankton and related phytoplankton cycles in a eutrophic lake. *Hydrobiologia,* 74, 99–104.

547 Moore, P. D. and Webb, J. A. (1978) *An Illustrated Guide to Pollen Analysis.* Hodder & Stoughton, London.

548 Morgan, A. and Kalk, M. (1970) Seasonal changes in the waters of Lake Chilwa (Malawi) in a drying phase 1966–1968. *Hydrobiologia,* 36, 81–103.

549 Morgan, N. C. (1972) Problems of the conservation of freshwater ecosystems. *Symp. zool. Soc. London,* 29, 135–54.

550 Moriarty, D. J. W. (1973) The physiology of digestion of blue-green algae in the cichlid fish *Tilapia nilotica. J. Zool.,* 171, 25–39.

551 Moriarty, D. J. W., Darlington, J. P. E. C., Dunn, I. G., Moriarty, C. M. and Tevlin, M. P. (1973) Feeding and grazing in Lake George, Uganda. *Proc. R. Soc. London* (B) 184, 227–346.

552 Moriarty, D. J. W. and Moriarty, C. M. (1973) The assimilation of carbon from phytoplankton by two herbivorous fishes: *Tilapia nilotica* and *Haplochromis nigripinnis. J. Zool.,* 171, 41–55.

553 Morisawa, M. (1985) *Rivers.* Longman, London.

554 Mortimer, C. H. (1941–1942) The exchange of dissolved substances between mud and water in lakes. *J. Ecol.,* 29, 280–329, 30, 147–201.

555 Mortimer, C. H. (1973) The Loch Ness Monster—limnology or paralimnology. *Limnol. Oceanogr.,* 18, 343–4.

556 Morton, W. (1982) Comparative catches and food habits of dolly varden and arctic charrs, *Salvelinus malma* and *S. alpinus* at Karluk, Alaska, in 1939–1941. *Env. Biol. Fishes,* 7, 7–29.

557 Moss, B. (1969) Limitation of algal growth in some Central African waters. *Limnol. Oceanogr.,* 14, 591–601.

558 Moss, B. (1972) Studies on Gull Lake, Michigan. I. Seasonal and depth distribution of phytoplankton. *Freshwat. Biol.,* 2, 289–307.

559 Moss, B. (1973a) The influence of environmental factors on the distribution of freshwater algae: an experimental study. II. The role of pH and the carbon dioxide-bicarbonate system. *J. Ecol.,* 61, 157–77.

560 Moss, B. (1973b) The influence of environmental factors on the distribution of freshwater algae: an experimental study. IV. Growth of test species in natural lake waters, and conclusion. *J. Ecol.,* 61, 193–211.

561 Moss, B. (1977) Adaptations of epipelic and epipsammic freshwater algae. *Oecologia,* 27, 103–8.

562 Moss, B. (1981) The composition and ecology of periphyton communities in freshwaters. II Inter-relationships between water chemistry, phytoplankton populations, and periphyton populations in a shallow lake and associated experimental reservoirs ('Lund Tubes'). *Br. Phyc. J.,* 16, 59–76.

563 Moss, B. (1983) The Norfolk Broadland: experiments in the restoration of a complex wetland. *Biol. Rev.,* 58, 521–61.

564 Moss, B. (1986) Restoration of lakes and lowland rivers. *Ecology and Design in Landscape* (Ed. by A. D. Bradshaw, D. A. Goode and E. Thorp) pp. 399–415. Blackwell Scientific Publications, Oxford.

565 Moss, B. (1987a) The art of lake restoration. *New Sci.,* 1550, 41–5.

566 Moss, B. (1987b) The Broads. *Biologist,* 34, 7–13.

567 Moss, B., Balls, H., Booker, I., Manson, K. and Timms, M. (1984) The River Bure, United Kingdom: patterns of change in chemistry and phytoplankton in a slow-flowing fertile river. *Verh. int. Verein. theor. angew. Limnol.,* 22, 1959–64.

568 Moss, B., Balls, H. R. and Irvine, K. (1985) Management of the consequences of eutrophication in lowland lakes in England—engineering and biological solutions. *Management Strategies for Phosphorus in the Environment* (Ed. by J. N. Lester and P. W. W. Kirk), 180–5. Selper, London.

569 Moss, B., Balls, H., Irvine, K. and Stansfield, J. (1986) Restoration of two lowland lakes by isolation from nutrient-rich water sources with and without removal of sediment. *J. Appl. Ecol.,* 23, 391–414.

570 Moss, B. and Leah, R. T. (1982) Changes in the ecosystem of a guanotrophic and

brackish shallow lake in Eastern England: potential problems in its restoration. *Int. Rev. ges. Hydrobiol., 67*, 625–59.

571 Moss, B., Wetzel, R. G. and Lauff, G. H. (1980) Annual productivity and phytoplankton changes between 1969 and 1974 in Gull Lake, Michigan. *Freshwat. Biol.,* **10**, 113–121.

572 Moss, C. E. (1913) *Vegetation of the Peak District.* Cambridge University Press, Cambridge.

573 Müller, K. (1974) Stream drift as a chronobiological phenomenon in running water ecosystems. *Ann. Rev. Ecol. Syst.,* **5**, 309–23.

574 Muller, R. (1975) *Worms and Diseases.* Heinemann, London.

575 Munawar, M. and Talling, J. F. (Eds) (1986) *Seasonality of Freshwater Phytoplankton: A Global Perspective.* Junk, Dordrecht (reprinted from *Hydrobiologia*, **138**, 1986).

576 Munk, W. H. and Riley, G. A. (1952) Absorption of nutrients by aquatic plants. *J. Mar. Res.,* **11**, 215–40.

577 Munro, A. L. S. and Brock, R. S. (1968) Distinction between bacterial and algal utilization of soluble substances in the sea. *J. Gen. Microbiol.,* **51**, 35–42.

578 Murphy, K. J., Hanbury, R. G. and Eaton, J. W. (1981) The ecological effects of 2-methylthiotriazine herbicides used for aquatic weed control in navigable canals. 1. Effects on aquatic flora and water chemistry. *Arch. Hydrobiol.,* **91**, 294–331.

579 Murphy, M. L., Heiftez, J., Johnson, S. W., Koski, K. V. and Thedinga, J. F. (1986) Effects of clear-cut logging with and without buffer strips on juvenile salmonids in Alaskan streams. *Can. J. Fish. Aquat. Sci.,* **43**, 1521–33.

580 Murphy, P. M. and Learner, M. A. (1982) The life history and production of the leech *Helobdella stagnalis* (Hirudinea: Glossiphonidae) in the River Ely, South Wales. *Freshwat. Biol.,* **12**, 321–30.

581 Murtaugh, P. A. (1981a) Size-selective predation on *Daphnia* by *Neomysis mercedis. Ecology,* **62**, 894–900.

582 Murtaugh, P. A. (1981b) Selective predation by *Neomysis mercedis* in Lake Washington. *Limnol. Oceanogr.,* **26**, 445–53.

583 Myers, N. (Ed.) (1985) *The Gaia Atlas of Planet Management.* Pan, London.

584 Nalewajko, C. and Paul, P. (1985) Effects of manipulations of aluminium concentrations and pH on phosphate uptake and photosynthesis of planktonic communities in two Precambrian shield lakes. *Can. J. Fish. Aquat. Sci.,* **42**, 1946–53.

585 National Academy of Sciences (1969) *Eutrophication: causes, consequences, correctives.* N. A. S., Washington.

586 National Water Council (1981) *River Quality—the 1980 Survey and Future Outlook.* National Water Council, London.

587 Nature Conservancy Council (1984) *Nature Conservation in Great Britain.* Nature Conservancy Council, Shrewsbury.

588 Nelson, J. S. (1976) *Fishes of the World.* Wiley, New York.

589 Nelson, J. S. (1984) *Fishes of the World,* 2nd edn. Wiley, New York.

590 Newbold, C. (1975) Herbicides in aquatic systems. *Biol. Cons.,* **7**, 97–118.

591 Newbold, C., Purseglove, J. and Holmes, N. (1983) *Nature Conservation and River Engineering.* Nature Conservancy Council, Shrewsbury.

592 Noordwijk, M. van (1984) *Ecology Textbook for the Sudan.* Khartoum University Press.

593 Novitski, R. P. (1978) Hydrologic characteristics of Wisconsin's wetlands and their influence on floods, stream flow and sediment. *Wetland Functions and Values: the State of our Understanding.* American Water Resources Association, Minneapolis.

594 Nteta, N. D. (Ed.) (1979) *Proceedings of the Symposium on the Okavango Delta and its Future Utilisation.* Botswana Society, Gaborone.

595 Nyholm, N. E. I. (1981) Evidence of involvement of aluminium in causation of defective formation of eggshells and of impaired breeding in wild passerine birds. *Env. Res.,* **26**, 363–71.

596 Nyholm, N. E. I. and Myhrberg, H. E. (1977) Severe eggshell defects and impaired reproductive capacity in small passerines in Swedish Lappland. *Oikos,* **29**, 336–41.

597 Obeng, L. E. (Ed.) (1969) *Man-made Lakes, the Accra Symposium.* Accra, Ghana University Press.

598 Obeng, L. E. (1977) Should dams be built? The Volta Lake example. *Ambio,* **6**, 46–50.

599 O'Brien, W. J. and Vinyard, G. L. (1978) Polymorphine predation: the effect of invertebrate predation on the distribution of two varieties of *Daphnia carinata* in South India ponds. *Limnol. Oceanogr.,* **23**, 452–60.

600 Odum, H. T. (1956) Primary production in flowing waters. *Limnol. Oceanogr.*, **1**, 102–17.

601 Office of Technology Assessment (1984) *Wetlands—their Use and Regulation.* Publication OTA-o-207. U. S. Congress.

602 O'Hop, J., Wallace, J. B. and Haefner, J. D. (1984) Production of a stream shredder, *Peltoperla maria* (Plecoptera: Peltoperlidae) in disturbed and undisturbed hardwood catchments. *Freshwat. Biol.*, **14**, 13–22.

603 Ohwada, K. and Taga, N. (1973) Seasonal cycles of vitamin B_{12}, thiamine and biotin in Lake Sagami. Patterns in their distribution and ecological significance. *Int. Rev. ges. Hydrobiol.*, **58**, 851–71.

604 Olsen, S. (1964) Phosphate equilibrium between reduced sediments and water. Laboratory experiments with radioactive phosphorus. *Verh. int. Verein. theor. angew. Limnol.*, **15**, 333–41.

605 Omernik, J. M. (1976) The influence of land use on stream nutrient levels. U. S. Environmental Protection Agency Report EPA-600/3-76-014. Corvallis, Oregon.

606 Ormerod, S. V., Allinson, N., Hudson, D. and Tyler, S. J. (1986) The distribution of breeding dippers (*Cinclus cinclus* (L.)) Aves in relation to stream acidity in upland Wales. *Freshwat. Biol.*, **16**, 501–8.

607 Ormerod, S. V., Boole, P., McCahon, C. P., Weatherly, N. S., Pascoe, D. and Edwards, R. W. (1987) Short term experimental acidification of a Welsh stream: comparing the biological effects of hydrogen ions and aluminium. *Freshwat. Biol.*, **17**, 341–56.

608 Ormerod, S. V., Tyler, S. J. and Lewis, J. M. S. (1985) Is the breeding distribution of dippers influenced by stream acidity. *Bird Study*, **32**, 32–9.

609 Orth, R. J. and van Montfrans, J. (1984) Epiphyte-sea grass relationships with an emphasis on the role of micrograzing: a review. *Aquat. Bot.*, **18**, 43–69.

610 Osborne, P. L. (1980) Prediction of phosphorus and nitrogen concentrations in lakes from both internal and external loading rates. *Hydrobiologia*, **69**, 229–33.

611 Osborne, P. L. and McLachlan, A. J. (1985) The effect of tadpoles on algal growth in temporary rain-filled rock pools. *Freshwat. Biol.*, **15**, 77–88.

612 Osborne, P. L. and Moss, B. (1977) Palaeolimnology and trends in the phosphorus and iron budgets of an old man-made lake, Barton Broad, Norfolk. *Freshwat. Biol.*, **7**, 213–34.

613 Osborne, P. L. and Phillips, G. L. (1978) Evidence for nutrient release from the sediments of two shallow and productive lakes. *Verh. int. Verein. theor. angew, Limnol.*, **20**, 654–8.

614 Ostrofsky, M. L. and Zettler, E. R. (1986) Chemical defences in aquatic plants. *J. Ecol.*, 279–88.

615 Overman, M. (1968) *Water—Solutions to a Problem of Supply and Demand.* Open University Press, Milton Keynes.

616 Owens, M., Garland, J. H. N., Hart, I. C. and Wood, G. (1972) Nutrient budgets in rivers. *Symp. Zool. Soc. London*, **29**, 21–40.

617 Paerl, H. W. and Ustach, J. F. (1982) Blue-green algae scums: an explanation for their occurrence during freshwater blooms. *Limnol. Oceanogr.*, **27**, 212–17.

618 Paffenhöfer, G-A., Strickler, J. R. and Alcaroz, M. (1982) Suspension-feeding by herbivorous calanoid copepods: a cinematographic study. *Mar. Biol.*, **67**, 193–9.

619 Page, P., Ouellet, M., Hillaire-Marcel, C. and Dickman, M. (1984) Isotopic analysis (^{18}O, ^{13}C, ^{14}C) of two meromictic lakes in the Canadian Arctic archipelago. *Limnol. Oceanogr.*, **29**, 564–73.

620 Pain, S. (1987) Australian invader threatens Britain's waterways. *New Sci.*, 23 July 1987, 26.

621 Paloheimo, J. E. (1974) Calculation of instantaneous birth rate. *Limnol. Oceanogr.*, **19**, 692–4.

622 Parker, D. J. and Penning-Rowsell, E. C. (1980) *Water Planning in Britain.* Allen and Unwin, London.

623 Parker, J. I. and Edgington, D. N. (1976) Concentration of diatom frustules in Lake Michigan sediment cores. *Limnol. Oceanogr.*, **21**, 887–93.

624 Paton, A. (1976) Dams and their interfaces. *Proc. R. Soc. London*, (A) **351**, 1–17.

625 Peckarsky, B. L. (1984) Sampling the stream benthos. *A Manual on Methods for the Assessment of Secondary Productivity in Fresh Waters* (Ed. by J. A. Downing and F. H. Rigler) pp. 131–160. Blackwell Scientific Publications, Oxford.

626 Penhale, P. A. and Thayer, G. W. (1980) Uptake and transfer of carbon and phosphorus by eelgrass (*Zostera marina* L.) and its epiphytes. *J. Exper. Mar. Biol. Ecol.*, **42**, 113–23.

627 Pennington, W. (1969) *The History of*

British Vegetation. English Universities Press, London.

628 Pennington, W. (1984) Long-term natural acidification of upland sites in Cumbria: evidence from post-glacial lake sediments. *Ann. Rep. Freshwat. Biol. Assoc.,* **52**, 28–46.

629 Pennington, W. and Lishman, J. P. (1971) Iodine in lake sediments in Northern England and Scotland. *Biol. Rev.,* **46**, 279–313.

630 Peters, R. H. (1984) Methods for the study of feeding, grazing and assimilation by zooplankton. *A Manual on Methods for the Assessment of Secondary Productivity in Freshwaters* (Ed. by J. A. Downing and F. H. Rigler), pp. 336–412. Blackwell Scientific Publications, Oxford.

631 Peters, R. H. and Rigler, F. H. (1973) Phosphorus release by *Daphnia. Limnol. Oceanogr.,* **18**, 821–39.

632 Peters, W. and Gilles, H. M. (1977) *A Colour Atlas of Tropical Medicine and Parasitology.* Wolfe Medical Publications, London.

633 Peterson, R. H., Daye, P. G. and Metcalfe, J. L. (1980) Inhibition of Atlantic Salmon (*Salmo salar*) hatching at low pH. *Can. J. Fish. Aquat. Sci.,* **37**, 770–4.

634 Petr, T. (1969) Fish population changes in Volta Lake over the period January 1965–September 1966. In *Man-made Lakes, the Accra Symposium* (Ed. by L. E. Obeng). Ghana University Press, Accra.

635 Phillips, G. L., Eminson, D. F. and Moss, B. (1978) A mechanism to account for macrophyte decline in progressively eutrophicated freshwaters. *Aquat. Bot.,* **4**, 103–26.

636 Phillips, R. S. (1983) *Malaria.* Arnold, London.

637 Pigott, C. D. and Pigott, M. E. (1963) Late-glacial and post-glacial deposits at Malham, Yorkshire. *New Phytol.,* **62**, 317–34.

638 Pigott, M. E. and Pigott, C. D. (1959) Stratigraphy and pollen analysis of Malham tarn and Tarn Moss. *Fld. Stud.,* **1**, 17 pp.

639 Pister, E. P. (1985) Desert pupfishes: reflections on reality, desirability and conscience. *Env. Biol. Fishes,* **12**, 3–12.

640 Pitcher, T. J. and Hart, P. J. B. (1982) *Fisheries Ecology.* Croom Helm, London.

641 Pontin, R. M. (1978) *A Key to the Freshwater Planktonic and Semi-Planktonic Rotifera of the British Isles. Sci. Publ. Freshwat. Biol. Assoc.,* 38.

642 Porter, K. G. (1973) Selective grazing and differential digestion of algae by zooplankton. *Nature,* **244**, 179–80.

643 Porter, K. G. (1976) Enhancement of algal growth and productivity by grazing zooplankton. *Science,* **192**, 1332–4.

644 Porter, K. G., Feig, T. S. and Vetter, E. F. (1983) Morphology, flow regimes, and filtering rates of *Daphnia, Ceriodaphnia* and *Bosmina* fed natural bacteria. *Oecologia,* **58**, 156–63.

645 Porter, K. G. and McDonough, R. (1984) The energetic cost of response to blue-green algal filaments by cladocerans. *Limnol. Oceanogr.,* **29**, 365–9.

646 Porter, K. G. and Orcutt, J. D. Jr. (1980) Nutritional adequacy, manageability and toxicity as factors that determine the food quality of green and blue-green algae for *Daphnia. Evolution and Ecology of Zooplankton Communities* (Ed. by W. C. Kerfoot), pp. 258–81. University of New England Press, Hanover.

647 Porter, K. G., Pace, M. L. and Battey, J. F. (1979) Ciliate protozoans as links in freshwater planktonic food chains. *Nature,* **277**, 563–5.

648 Post, J. R. and McQueen, D. J. (1987) The impact of planktivorous fish on the structure of a plankton community. *Freshwat. Biol.,* **17**, 79–90.

649 Potts, W. T. W. (1954) The energetics of osmotic regulation in brackish- and freshwater animals. *J. Exp. Biol.,* **31**, 618–30.

650 Potts, W. T. W. and Parry, G. (1964) *Osmotic and Ionic Regulations in Animals.* Pergamon, Oxford.

651 Poynton, S. L. and Bennett, C. E. (1985) Parasitic infections and their interactions in wild and cultured brown trout and cultured rainbow trout from the River Itchen, Hampshire. *Fish and Shellfish Pathology* (Ed. by A. E. Ellis) pp. 353–7.

652 Prepas, E. and Rigler, F. H. (1978) The enigma of *Daphnia* death rates. *Limnol. Oceanogr.,* **23**, 970–88.

653 Prescott, G. W. (1962) *Algae of the Western Great Lakes Area.* W. C. Brown, Dubuque, Iowa.

654 Priddle, J. and Heywood, R. B. (1980) Evolution of Antarctic lake ecosystems. *Biol. J. Linn. Soc.,* **14**, 51–66.

655 Prowse, G. A. (1959) Relationships between epiphytic algal species and their macrophyte hosts. *Nature,* **183**, 1204–5.

656 Pullen, R. (1985) Tilapias: 'Everyman's Fish'. *Biologist,* **32**, 84–8.

657 Qureshi, A. A. and Patel, J. (1976)

References

Adenosine triphosphate (ATP) levels in microbial cultures and a review of the ATP biomass estimation technique. *Env. Canada Sci. Ser.,* **63**, 1–33.

658 Raiswell, R., Brimblecombe, P., Dent, D. L. and Liss, P. S. (1980) *Environmental Chemistry.* Arnold, London.

659 Rankin, J. C. and Davenport, J. A. (1981) *Animal Osmoregulation.* Blackie, Glasgow.

660 Raskin, I. and Kende, H. (1985) Mechanism of aeration in rice. *Science,* **228**, 327–9.

661 Raven, J. A. (1970) Exogenous inorganic carbon sources in plant photosynthesis. *Biol. Rev.,* **45**, 167–202.

662 Raven, J. A. and Glidewell S. M. (1975) Photosynthesis, respiration and growth in the shade alga. *Hydrodictyon africanum Photosynthetica,* **9**, 361–71.

663 Reay, P. J. (1979) *Aquaculture.* Arnold, London.

664 Redfern, M. (1976) Revised field key to the invertebrate fauna of stony hill streams. *Field Studies,* **4**, 105–15.

665 Redfield, A. C. (1934) On the proportions of organic derivatives in sea water and their relation to the composition of plankton. *James Johnstone Memorial Volume,* pp. 176–92. Liverpool University Press, Liverpool.

666 Reinertsen, H., Jensen, A., Langeland, A. and Olsen, Y. (1986) Algal competition for phosphorus: the influence of zooplankton and fish. *Can. J. Fish. Aquat. Sci.,* **43**, 1135–41.

667 Renberg, I. and Hellberg, T. (1982) The pH history of lakes in south-western Sweden, as calculated from the subfossil diatom flora of the sediments. *Ambio,* **11**, 30–3.

668 Reynolds, C. S. (1984a) *The Ecology of Freshwater Phytoplankton.* Cambridge University Press, Cambridge.

669 Reynolds, C. S. (1984b) Phytoplankton periodicity: the interactions of form, function and environmental variability. *Freshwat. Biol.,* **14**, 111–42.

670 Reynolds, C. S. and Butterwick, C. (1979) Algal bioassay of unfertilized and artificially fertilized lake water, maintained in Lund tubes. *Arch. Hydrobiol. Suppl.,* **56**, 166–83.

671 Reynolds, C. S., Harris, G. P. and Gouldney, D. N. (1985) Comparison of carbon-specific growth rates and rates of cellular increase of phytoplankton in large

limnetic enclosures. *J. Plankton Res.,* **7**, 791–820.

672 Reynolds, C. S. and Walsby, A. E. (1975) Water-blooms. *Biol. Rev.,* **50**, 437–81.

673 Reynolds, C. S., Wiseman, S. W. and Clarke, M. J. O. (1984) Growth- and loss-rate responses of phytoplankton to intermittent artificial mixing and their potential application to the control of planktonic algal biomass. *J. Appl. Ecol.,* **21**, 11–39.

674 Reynolds, T. B. (1966) The distribution and abundance of lake-dwelling triclads—towards a hypothesis. *Adv. Ecol. Res.,* **3**, 1–71.

675 Reynoldson, T. B. (1983) The population biology of Turbellaria with special reference to the freshwater triclads of the British Isles. *Adv. Ecol. Res.,* **13**, 235–326.

676 Reynoldson, T. B. and Bellamy, L. S. (1970) The establishment of interspecific competition in field populations, with an example of competition in action between *Polycelis nigra* (Mull.) and *P. tenuis* (Ijima) (Turbellaria, Tricladida). *Proceedings Advanced Study Institute on Dynamics, Number and Populations* (Oosterbeek) 282–97.

677 Ribbink, A. J. (1987) African lakes and their fishes: conservation scenarios and suggestions. *Env. Biol. Fishes,* **19**, 3–26.

678 Richardson, J. L. and Richardson, A. E. (1972) History of an African rift lake and its climatic implications. *Ecol. Monogr.,* **42**, 499–534.

679 Richardson, K., Griffiths, H., Reed, M. L., Raven, J. A. and Griffiths, N. M. (1984) Inorganic carbon assimilation in the isoetids, *Isoetes lacustris* L. and *Lobelia dortmanna* L. *Oecologia,* **61**, 115–21.

680 Richey, J. E., Perkins, M. A. and Goldman, C. R. (1975) Effects of kokanee salmon (*Oncorhynchus nerka*) decomposition on the ecology of a sub-alpine stream. *J. Fish. Res. Bd. Can.,* **32**, 817–20.

681 Riessen, H. P. (1984) The other side of cyclomorphosis: why *Daphnia* lose their helmets. *Limnol. Oceanogr.,* **29**, 1123–7.

682 Rigler, F. H. (1964) The phosphorus fractions and the turnover time of inorganic phosphorus in different types of lakes. *Limnol. Oceanogr.,* **9**, 511–18.

683 Rigler, F. H. and Downing, J. A. (1984) The calculation of secondary productivity. *A Manual on Methods for the Assessment of Secondary Productivity in Freshwaters.* pp. 19–58. Blackwell Scientific Publications, Oxford.

684 Ritter, J. A. and Porter, T. R. (1980)

Issues and promises for Atlantic salmon. management in Canada. *Atlantic Salmon: its Future* (Ed. by A. E. J. Went), 108–27. Fishing News Books, Farnham.

685 Robarts, R. (1985) Dam troubles. *Scientiae*, **26**, 17–23.

686 Robertson, W. B. Jr. (1955) *A Survey of the Effects of Fire in Everglades National Park*. U.S. National Park Service, Homestead, Florida, U.S.A.

687 Robertson, W. B. Jr. and Kushlan, J. A. (1974) The South Florida Avifauna. *Environments of South Florida: Present and Past* (Ed. by P. J. Gleason). pp. 414–52. Miami Geological Society Memoir 2, Miami.

688 Rogers, F. E. J., Rogers, K. H. and Buzer, J. S. (1985) *Wetlands for Wastewater Treatment with Special Reference to Municipal Wastewaters*. Witwatersrand University Press, Johannesburg.

689 Rogers, K. H. (1984) *The role of Potamogeton crispus L. in the Pongola River floodplain ecosystem*. Ph.D. Thesis, University of Natal.

690 Rogers, K. H. and Breen, C. M. (1981) Effects of epiphyton on *Potamogeton crispus* L. leaves. *Microbial Ecology*, **7**, 351–63.

691 Round, F. E. (1961) The diatoms of a core from Esthwaite Water. *New Phytol.*, **60**, 43–59.

692 Round, F. E. (1964) The ecology of benthic algae. *Algae and Man* (Ed. by D. F. Jackson) pp. 138–84. Plenum, New York.

693 Round, F. E. (1981) *The Ecology of Algae*. Cambridge University Press, Cambridge.

694 Rounick, J. S. and Winterbourne, M. J. (1983) The formation, structure, and utilization of stone surface organic layers in two New Zealand streams. *Freshwat. Biol.*, **13**, 57–72.

695 Royal Commission on Environmental Pollution (1979) *Seventh Report. Agriculture and Pollution*. H.M.S.O., London.

696 Royal Society (1983) *The Nitrogen Cycle of the United Kingdom*. Royal Society, London.

697 Ruby, S. M. Aczel, J. and Craig, G. R. (1977) The effects of depressed pH on oogenesis in flagfish (*Jordanella floridae*). *Water Res.*, **11**, 757–62.

698 Russell, E. S. (1931) Some theoretical considerations on the 'overfishing problem'. *J. Cons. Int. Explor. Mer*, **6**, 3–20.

699 Ruttner, F. (1953) *Fundamentals of Limnology*. Transl. by D. G. Frey and F. E. J.

Fry. University of Toronto Press, Toronto.

700 Ryder, R. A. (1978) Fish yield assessment of large lakes and reservoirs—a prelude to management. *Ecology of Freshwater Fish Production* (Ed. by S. D. Gerking) pp. 403–23. Blackwell Scientific Publications, Oxford.

701 Ryder, R. A. and Henderson, H. F. (1975) Estimates of potential fish yield for the Nasser Reservoir, Arab Republic of Egypt. *J. Fish. Res. Bd. Canada*, **32**, 2137–51.

702 Ryder, R. A., Kerr, S. R., Loftus, K. H. and Regier, H. A. (1974) The morphoedaphic index, a fish yield estimator-review and evaluation. *J. Fish. Res. Bd. Canada*, **31**, 663–88.

703 Rzoska, J. (Ed.) (1974) *The Nile—Biology of an Ancient River*. Junk, The Hague.

704 Rzoska, J. (1976) A controversy reviewed. *Nature*, **261**, 444–5.

705 Salonen, K., Jones, R. I. and Arvola, L. (1984) Hypolimnetic phosphorus retrieval by diel vertical migrations of lake phytoplankton. *Freshwat. Biol.*, **14**, 431–8.

706 Sand-Jensen, K. (1978) Metabolic adaptation and vertical zonation of *Litorella uniflora* (L.) Ashers and *Isoetes lacustris* L. *Aquat. Bot.*, **4**, 1–10.

707 Sand-Jensen, K. (1983) Photosynthetic carbon sources of submerged stream macrophytes. *J. Exper. Bot.*, **34**, 198–210.

708 Sand-Jensen, K. and Borum, J. (1984) Epiphyte shading and its effect on photosynthesis and diel metabolism of *Lobelia dortmanna* L. during the spring bloom in a Danish lake. *Aquat. Bot.*, **20**, 109–19.

709 Sand-Jensen, K. and Prahl, C. (1982) Oxygen exchange with the lacunae and access leaves and roots of the submerged vascular macrophyte *Lobelia dortmanna* L. *New Phytol.*, **91**, 103–20.

710 Sand-Jensen, K., Prahl, C. and Stockholm, H. (1982) Oxygen release from roots of submerged aquatic macrophytes. *Oikos*, **38**, 349–54.

711 Sand-Jensen, K. and Søndergaard, M. (1981) Phytoplankton and epiphyte development and their shading effect on submerged macrophytes in lakes of different nutrient status. *Int. Rev. ges. Hydrobiol.*, **66**, 529–52.

712 Sanger, J. E. and Gorham, E. (1970) The diversity of pigments in lake sediments and its ecological significance. *Limnol. Oceanogr.*, **15**, 59–69.

713 Scavia, D., Fahnenstiel, G. L., Evans, M.

S., Jude, D. J. and Lehman, J. T. (1986) Influence of salmonid predation and weather on long-term water quality trends in Lake Michigan. *Can. J. Fish. Aquat. Sci.*, **43**, 435–43.

714 Scavia, D., Laird, G. A. and Fahnenstiel, G. L. (1986) Production of planktonic bacteria in Lake Michigan. *Limnol. Oceanogr.*, **31**, 612–26.

715 Scheider, W. and Wallis, P. (1973) An alternate method of calculating the population density of monsters in Loch Ness. *Limnol. Oceanogr.*, **18**, 343.

716 Schierup, H. H. (1978) Biomass and primary production in a *Phragmites communis* Trin. swamp in North Jutland, Denmark. *Verh. int. Verein. theor. angew. Limnol.*, **20**, 93–9.

717 Schindler, D. W. (1974) Eutrophication and recovery in experimental lakes: implications for lake management. *Science*, **184**, 897–8.

718 Schindler, D. W. (1977) The evolution of phosphorus limitation in lakes. *Science*, **195**, 260–2.

719 Schindler, D. W. (1978) Factors regulating phytoplankton production and standing crop in the world's freshwaters. *Limnol. Oceanogr.*, **23**, 478–86.

720 Schindler, D. W. and Fee, E. J. (1974) Experimental lakes area: whole lake experiments in eutrophication. *J. Fish. Res. Bd. Canada*, **31**, 937–53.

721 Schindler, D. W., Welch, H. E., Kalff, J., Brunskill, G. J. and Kritsch, N. (1974) Physical and chemical limnology of Char Lake, Cornwallis Island (75°N lat.) *J. Fish. Res. Bd. Canada*, **31**, 585–607.

722 Schindler, J. E. (1971) Food quality and zooplankton nutrition. *J. Anim. Ecol.*, **40**, 589–95.

723 Schlesinger, W. H. (1978) Community structure, dynamics and nutrient cycling in the Okefenokee cypress swamp-forest. *Ecol. Monogr.*, **48**, 43–65.

724 Schmidt, J. A. and Andren, A. W. (1984) Deposition of airborne metals into the Great Lakes: an evaluation of past and present estimates. *Toxic Contaminants in the Great Lakes* (Ed. by J. O. Nriagu and M. S. Simmons) pp. 81–104. Wiley, New York.

725 Schmidt-Nielsen, K. (1983) *Animal Physiology.* Cambridge University Press, Cambridge.

726 Schmitt, M. R. and Adams, M. S. (1981) Dependence of rates of apparent photosynthesis on tissue phosphorus concentra-tions in *Myriophyllum spicatum* L. *Aquat. Bot.*, **11**, 379–87.

727 Schoenberg, S. A. and Carlson, R. E. (1984) Direct and indirect effects of zooplankton grazing on phytoplankton in a hypereutrophic lake. *Oikos*, **42**, 291–302.

728 Schoenberg, S. A. and Maccubbin, A. E. (1985) Relative feeding rates on free and particle-bound bacteria by freshwater macrozooplankton. *Limnol. Oceanogr.*, **30**, 1084–90.

729 Scourfield, D. J. and Harding, J. P. (1966) *A Key to the British Species of Freshwater Cladocera* 3rd edn. *Sci. Publ. Freshwat. Biol. Assoc.*, **5**, 1–55.

730 Scullion, J., Parish, C. A., Morgan, N. and Edwards, R. W. (1982) Comparison of benthic macroinvertebrate fauna and substratum composition in riffles and pools in the impounded River Elan and the unregulated R. Wye, mid-Wales. *Freshwat. Biol.*, **12**, 579–96.

731 Sculthorpe, C. D. (1967) *The Biology of Aquatic Vascular Plants.* Arnold, London.

732 Setaro, F. V. and Melack, J. M. (1984) Responses of phytoplankton to experimental enrichment in an Amazon floodplain lake. *Limnol. Oceanogr.*, **29**, 972–84.

733 Sevalrud, I. H., Muniz, I. P. and Kalvenes, S. (1980) Loss of fish populations in southern Norway. Dynamics and magnitude of the problem. *Proceedings of an International Conference on the Ecological Impact of Acid Precipitation* (Ed. by D. Drablos and A. Tollan) pp. 350–1. Sandefiord, Norway.

734 Shapiro, J. (1980a) The need for more biology in lake restoration. *Lake Restoration. Proceedings of a National Conference August 1978.* U.S.E.P.A. 444/5-79-001. Minneapolis.

735 Shapiro, J. (1980b) The importance of trophic-level interactions to the abundance and species composition of algae in lakes. *Hypertrophic Ecosystems* (Ed. by J. Barica and L. R. Mur). Junk, The Hague.

736 Shapiro, J., Lamarra, V. and Lynch, M. (1975) Biomanipulation: an ecosystem approach to lake restoration. *Water Quality Management through Biological Control* (Ed. by P. L. Brezonik and J. L. Fox). Report ENV-07-75-1. University of Florida, Gainsville.

737 Shapiro, J. and Wright, D. I. (1984) Lake restoration by biomanipulation: Round Lake, Minnesota, the first two years. *Freshwat. Biol.*, **14**, 371–83.

738 Sheldon, R. W. and Kerr, S. R. (1972) The population density of monsters in Loch Ness. *Limnol. Oceanogr.*, **17**, 796–8.

739 Sheldon, R. W., Prakash, A. and Sutcliffe, W. H. Jr. (1972) The size distribution of particles in the ocean. *Limnol. Oceanogr.*, **17**, 327–40.

740 Shoard, M. (1980) *The Theft of the Countryside.* Temple Smith, London.

741 Sikora, L. J. and Keeney, D. R. (1983) Further aspects of soil chemistry under anaerobic conditions. *Mires: Swamp, Bog, Fen and Moor. General Studies* (Ed. by A. J. P. Gore) pp. 247–56. Elsevier, Amsterdam.

742 Simpson, P. S. and Eaton, J. W. (1986) Comparative studies of the photosynthesis of the submerged macrophyte, *Elodea canadensis* and the filamentous algae *Cladophora glomerata* and *Spirogyra* sp. *Aquat. Bot.*, **14**, 1–12.

743 Singer, A. and Ehrlich, A. (1978) Palaeo-limnology of a late Pleistocene-Holocene crater lake from the Golan Heights, Eastern Mediterranean. *J. Sed. Petr.*, **48**, 1331–40.

744 Sinker, C. A. (1962) The North Shropshire Meres and Mosses: a background for ecologists. *Fld. Stud.*, **1**, 101–7.

744a Sioli, H. (Ed.) (1984) *The Amazon.* Junk, Dordrecht.

745 Small, J. W. Jr., Richard, D. I. and Osborne, J. A. (1985) The effects of vegetation removal by grass carp and herbicides on water chemistry of four Florida lakes. *Freshwat. Biol.*, **15**, 587–96.

746 Smid, P. (1975) Evaporation from a reed swamp. *J. Ecol.*, **63**, 299–309.

747 Smith, A. M. and apRees, T. (1979) Pathways of carbohydrate fermentation in the roots of marsh plants. *Planta*, **146**, 327–34.

748 Smith, D. W. (1985) Biological control of excessive phytoplankton growths and the enhancement of aquacultural production. *Can. J. Fish. Aquat. Sci.*, **42**, 1940–5.

749 Smith, S. I. (1972a) Factors of ecologic succession in oligotrophic fish communities of the Laurentian Great Lakes. *J. Fish. Res. Bd. Canada*, **29**, 717–30.

750 Smith, S. I. (1972b) The future of salmonid communities in the Laurentian Great Lakes. *J. Fish. Res. Bd. Canada*, **29**, 951–7.

751 Smith, V. H. (1983) Low nitrogen to phosphorus ratios favour dominance by blue-green algae in lake phytoplankton. *Science*, **221**, 669–71.

752 Smith, V. H. and Wallsten, M. (1986) Prediction of emergent and floating-leaved macrophyte cover in Central Swedish Lakes. *Can. J. Fish. Aquat. Sci.*, **43**, 2519–23.

753 Smyly, W. J. P. (1979) Population dynamics of *Daphnia hyalina* Leydig (Crustacea: Cladocera) in a productive and an unproductive lake in the English Lake District. *Hydrobiologia*, **64**, 269–78.

754 Sollas, W. J. (1884) On the origin of freshwater faunas: a study in evolution. *Sci. Trans. Roy. Dublin Soc. (Ser. 2)*, **3**, 87–118.

755 Sommer, U. and Gliwicz, Z. M. (1986) Long range vertical migration of *Volvox* in tropical lake Cahora Bassa (Mozambique). *Limnol. Oceanogr.*, **31**, 650–3.

756 Sommer, U. and Stabel, H. H. (1983) Silicon consumption and population density changes of dominant planktonic diatoms in Lake Constance. *J. Ecol.*, **73**, 119–30.

757 Sorokin, Y. I. and Kadota, H. (1972) *Microbial Production and Decomposition in Freshwaters.* IBP Handbook No. 23. 112 pp. Blackwell Scientific Publications, Oxford.

758 Southwood, T. R. E. (1966) *Ecological Methods with Particular Reference to the Study of Insect Populations.* Chapman and Hall, London.

759 Sozska, G. P. (1975) Ecological relations between invertebrates and submerged macrophytes in the lake littoral. *Ekol. Polsk.*, **23**, 593–615.

760 Spence, D. H. N. (1964) The macrophytic vegetation of lochs, swamps and associated fens. *The Vegetation of Scotland* (Ed. by J. H. Burnett). Oliver & Boyd, Edinburgh.

761 Spence, D. H. N. (1976) Light and plant response in freshwater. *Light as an Ecological Factor II* (Ed. by G. C. Evans, R. Bainbridge and O. Rackham). Blackwell Scientific Publications, Oxford.

762 Spence, D. H. N. (1982) The zonation of plants in freshwater lakes. *Adv. Ecol. Res.*, **12**, 37–125.

763 Spence, D. H. N. and Chrystal, J. (1970a) Photosynthesis and zonation of freshwater macrophytes. I. Depth distribution and shade tolerance. *New Phytol.*, **69**, 205–15.

764 Spence, D. H. N. and Chrystal, J. (1970b) Photosynthesis, zonation of freshwater macrophytes. II. Adaptability of species

of deep and shallow waters. *New Phytol.,* **69**, 217–27.

765 Spence, D. H. N., Milburn, T. R., Nalawula-Senyimba, M. and Roberts, E. (1971) Fruit biology and germination of two tropical *Potamogeton* species. *New Phytol.,* **70**, 197–212.

766 Stark, F. and Werner, H. (1976) Natural history and management of Everglades National Park. *Proceedings of the Symposium on the Okavango Delta and its Future Utilization* (Ed. by D. N. Nteta) pp. 263–75. Botswana Society, Gabarone.

767 Starkweather, P. L. and Bogdan, K. G. (1980) Detrital feeding in natural zooplankton communities: discrimination between live and dead algal foods. *Hydrobiologia*, **73**, 83–5.

768 Steel, J. A. (1972) The application of fundamental limnological research in water supply system design and management. *Symp. Zool. Soc. London,* **29**, 41–67.

769 Steeman Nielsen, E. (1952) The use of radioactive carbon (C^{14}) for measuring organic production in the sea. *J. Cons. Int. Explor. Mer.* **18**, 117–40.

770 Steinhorn, I. (1985) The disappearance of the long term meromictic stratification of the Dead Sea. *Limnol. Oceanogr.,* **30**, 451–72.

771 Steinhorn, I., Assaf, G., Gat, J. R., Nishry, A., Nissenbaum, A., Stiller, H., Beyth, M., Neev, D., Garber, R., Friedman, G. M. and Weiss, W. (1979) The Dead Sea: deepening of the mixolimnion signifies the overture to overturn of the water column. *Science,* **206**, 55–7.

772 Stenson, J. A. E. (1985) Biotic structures and relations in the acidified Lake Gordsjön system—a synthesis. *Ecol. Bull.,* **37**, 319–26.

773 Stewart, W. D. P., Haystead, A. and Pearson, H. W. (1969) Nitrogenase activity in heterocysts of filamentous blue-green algae. *Nature,* **224**, 226–8.

774 Stewart, W. D. P. and Lex, M. (1970) Nitrogenase activity in blue-green alga, *Plectonema boryanum* strain 594. *Arch. Mikrobiol.,* **73**, 250–60.

775 Stewart, W. D. P. and Pearson, H. W. (1970) Effects of aerobic and anaerobic conditions on growth and metabolism of blue-green algae. *Proc. R. Soc.,* (B) **175**, 293–311.

776 Stockner, J. G. (1968) Algal growth and primary productivity in a thermal stream. *J. Fish. Res. Bd. Canada,* **25**, 2037–58.

777 Stockner, J. G. and Antia, N. J. (1986) Algal picoplankton from marine and freshwater ecosystems: a multidisciplinary perspective. *Can. J. Fish. Aquat. Sci.,* **43**, 2472–503.

778 Stockner, J. G. and Benson, W. W. (1967) The succession of diatom assemblages in the recent sediment of Lake Washington. *Limnol. Oceanogr.,* **12**, 513–32.

779 Stoner, J. H., Gee, A. S. and Wade, K. R. (1983) The effects of acid precipitation and land use on water quality and ecology in the Upper Tywi catchment in West Wales. Report to the Welsh Water Authority.

780 Strayer, D. (1985) The benthic micrometazoans of Mirror Lake, New Hampshire. *Arch. Hydrobiol. Suppl.,* **72**, 287–426.

781 Strickland, J. D. H. and Parsons, T. R. (1972) *A Practical Handbook of Seawater Analysis. Bull. Fish. Res. Bd. Canada* 167.

782 Stumm, W. and Morgan, J. J. (1981) *Aquatic Chemistry,* 2nd edn. Wiley, New York.

783 Suberkropp, K. and Klug, M. J. (1976) Fungi and bacteria associated with leaves during processing in a woodland stream. *Ecology,* **57**, 707–19.

784 Sumpter, J. P. and Wood, C. R. C. (1981) The trout. *Biologist,* **28**, 219–24.

785 Sutcliffe, D. W. (1978) Water chemistry and osmoregulation in some anthropods, especially Malacostraca. *Ann. Rep. Freshwat. Biol. Assoc.,* **46**, 57–69.

786 Sutcliffe, D. W. (1983) Acid precipitation and its effects on aquatic systems in the English Lake District (Cumbria). *Ann. Rep. Freshwat. Biol. Assoc.,* **51**, 50–62.

787 Sutcliffe, D. W. (1984) Quantitative aspects of oxygen uptake by *Gammarus* (Crustacea, Amphipoda): a critical review. *Freshwat. Biol.,* **14**, 443–90.

788 Sutcliffe, D. W., Carrick, T. R., Heron, J., Rigg, E., Talling, J. F., Woof, C. P. and Lund, J. W. G. (1982) Long term and seasonal changes in the chemical composition of precipitation and surface waters of lakes and tarns in the English Lake District. *Freshwat. Biol.,* **12**, 451–506.

789 Swales, S. (1979) Effects of river improvements on fish populations. Proceedings of the First British Freshwater Fisheries Conference, University of Liverpool, pp. 86–99, Liverpool.

790 Talling, J. F. (1957) The phytoplankton populations as a compound photosynthetic system. *New Phytol.,* **56**, 133–49.

791 Talling, J. F. (1966) The annual cycle of

stratification and phytoplankton growth in Lake Victoria (E. Africa). *Int. Rev. ges. Hydrobiol.,* **51**, 545–621.

792 Talling, J. F. (1969) The incidence of vertical mixing, and some biological and chemical consequences, in tropical African lakes. *Verh. int. Verein. theor. angew. Limnol.,* **17**, 998–1012.

793 Talling, J. F. (1971) The underwater light climate as a controlling factor in the production ecology of freshwater phytoplankton. *Mitt. int. Verein. theor. angew. Limnol.,* **19**, 214–43.

794 Talling, J. F. (1976) The depletion of carbon dioxide from lake water by phytoplankton. *J. Ecol.,* **64**, 79–121.

795 Talling, J. F. (1986) The seasonality of phytoplankton in African lakes. *Hydrobiologia,* **138**, 139–60.

796 Talling, J. F. and Talling, I. B. (1965) The chemical composition of African lake water. *Int. Rev. ges. Hydrobiol.,* **50**, 421–63.

797 Talling, J. F., Wood, R. B., Prosser, M. V. and Baxter, R. M. (1973) The upper limit of photosynthetic productivity by phytoplankton: evidence from Ethiopian soda lakes. *Freshwat. Ecol.,* **3**, 53–76.

797a Tarapchak, S. J. and Nalewajko, C. (1986a) Synopsis: phosphorus–plankton dynamics symposium. *J. Fish. Res. Bd. Canada,* **43**, 416–19.

797b Tarapchak, S. J. and Nalewajko, C. (1986b) Introduction: Phosphorus–plankton dynamics symposium. *Can. J. Fish. Aquat. Sci.,* **43**, 293–301.

798 Teal, J. M. (1980) Primary production of benthic and fringing plant communities. *Fundamentals of Aquatic Ecosystems* (Ed. by R. S. K. Barnes and K. H. Mann), pp. 67–83. Blackwell Scientific Publications, Oxford.

799 Templeton, R. (Ed.) (1984) *Freshwater Fisheries Management.* Fishing News Books, Farnham.

799a Tessier, A. J. (1986) Life history and body size evolution in *Holopedium gibberum* Zadach (Crustacea:Cladocera). *Freshwat. Biol.,* **16**, 279–86.

800 Theis, T. L. and McCabe, P. J. (1978) Phosphorus dynamics in hypereutrophic lake sediments. *Water Res.,* **12**, 677–85.

801 Thesiger, W. (1964) *The Marsh Arabs.* Longmans, London: (1967) Penguin, Harmondsworth.

802 Thesiger, W. (1979) *Desert, Marsh and Mountain. The World of a Nomad.* Collins, London.

803 Thompson, K., Shewry, P. R. and Woolhouse, H. W. (1979) Papyrus swamp development in the Upemba Basin, Zäire: studies of population structure in *Cyperus papyrus* stands. *Bot. J. Linn. Soc.,* **78**, 299–316.

804 Thompson, R. (1973) Palaeolimnology and palaeomagnetism. *Nature,* **242**, 182–4.

805 Threlkeld, S. (1979) Estimating cladoceran birth rates: the importance of egg mortality and the egg age distribution. *Limnol. Oceanogr.,* **24**, 601–12.

806 Tilman, D. (1977) Resource competition between planktonic algae: an experimental and theoretical approach. *Ecology,* **58**, 338–48.

807 Timms, R. M. and Moss, B. (1984) Prevention of growth of potentially dense phytoplankton populations by zooplankton grazing, in the presence of zooplanktivorous fish, in a shallow wetland ecosystem. *Limnol. Oceanogr.,* **29**, 472–86.

808 Tippett, R. (1970) Artificial surfaces as a method of studying populations of benthic micro-algae in freshwaters. *Br. Phycol. J.,* **5**, 187–99.

809 Titman, D. (1976) Ecological competition between algae: experimental confirmation of resource-based competition theory. *Science,* **192**, 463–5.

810 Tokeshi, M. (1986) Population dynamics, life histories and species richness in an epiphytic chironomid community. *Freshwat. Biol.,* **16**, 431–41.

811 Tokeshi, M. and Pinder, L. C. V. (1985) Microhabitats of stream invertebrates on two submersed macrophytes with contrasting leaf morphology. *Hol. Ecol.,* **8**, 313–19.

812 Toth, L. (1972) Reeds control eutrophication of Balaton Lake. *Water Res.,* **6**, 1533–9.

813 Townsend, C. R. (1980) *The Ecology of Streams and Rivers.* Arnold, London.

814 Townsend, C. R. and Hildrew, A. G. (1976) Field experiments on the drifting, colonization, and continuous redistribution of stream benthos. *J. Anim. Ecol.,* **45**, 759–72.

815 Townsend, C. R., Hildrew, A. G. and Francis, J. (1983) Community structure in some southern English streams: the influence of physicochemical factors. *Freshwat. Biol.,* **13**, 521–44.

816 Trewavas, E., Green, J. and Corbet, S. A. (1972) Ecological studies on crater lakes

in West Cameroon. Fishes of Barombi Mbo. *J. Zool.,* **167**, 41–95.

817 Tuite, C. H. (1981) Standing crop densities and distribution of *Spirulina* and benthic diatoms in East African alkaline saline lakes. *Freshwat. Biol.,* **11**, 345–60.

818 Turner, J. L. (1982) Lake flies, water fleas and sardines. *Biological Studies on the Pelagic Ecosystem of Lake Malawi.* Fishery Expansion Project, Malawi. FAO/UNDP FI.DP.MLW/75/019. Technical Report 1, Rome.

819 Turner, S. M. and Liss, P. S. (1985) Measurement of various sulfur gases in a coastal marine environment. *J. Atmos. Chem.,* **2**, 223–32.

820 Tyler, P. A. (1986) Anthropological limnology in the Land of Moinee. *Limnology in Australia* (Ed. by P. De Deckker and W. D. Williams), pp. 523–38. Junk, Dordrecht.

821 United States Environmental Protection Agency (1973) *Measures for the Restoration and Enhancement of Quality of Freshwater Lakes.* USEPA, Washington, DC.

822 United States Environmental Protection Agency (1980) *Lake Restoration.* USEPA, 440/5-79-001.

823 Vallentyne, J. R. (1969) Sedimentary organic matter and palaeolimnology. *Mitt. int. Ver. theor. angew. Limnol.,* **17**, 104–10.

824 Vallentyne, J. R. (1974) *The Algal Bowl Lakes and Man.* Misc. Publ. Dept. Env. Fish. Mar. Service, Ottawa.

825 Vanni, M. J. (1986) Competition in zooplankton communities: suppression of small species by *Daphnia pulex. Limnol. Oceanogr.,* **31**, 1039–56.

826 Vannote, R. L., Minshall, G. W., Cummings, K. W., Sedell, J. R. and Cushing, C. E. (1980) The river continuum concept. *Can. J. Fish. Aquat. Sci.,* **37**, 120–37.

827 Varley, M. E. (1967) *British Freshwater Fishes.* Fishing News (Books) Ltd, London.

828 Van Vierssen, W. and Prins, Th. C. (1985) On the relationship between the growth of algae and aquatic macrophytes in brackish water. *Aquat. Bot.,* **21**, 165–79.

829 Vincent, W. F., Wurtsbaugh, W., Vincent, C. L. and Richerson, P. J. (1984) Seasonal dynamics of nutrient limitation in a tropical high-altitude lake (Lake Titicaca, Peru–Bolivia): application of physiological bioassays. *Limnol. Oceanogr.,* **29**, 540–52.

830 Viner, A. B. (1973) Responses of a mixed phytoplankton population to nutrient enrichments of ammonia and phosphate and some associated ecological implications. *Proc. R. Soc. London,* (B) **183**, 351–70.

831 Viner, A. B. and Smith, I. R. (1973) Geographical, historical and physical aspects of Lake George. *Proc. R. Soc.,* (B) **184**, 235–70.

832 Vollenweider, R. A. (1975) Input–output models with special reference to the phosphorus loading concept in limnology. *Schweiz. Zh. Hydrol.,* **37**, 53–84.

833 Vollenweider, R. A. and Kerekes, J. J. (1981) Background and summary results of the OECD cooperative programme on eutrophication. Appendix 1 in *The OECD Cooperative Programme on Eutrophication* Canadian Contribution (compiled by L. L. Janus and R. A. Vollenweider) Environment Canada. Scientific Series 131.

834 Waddy, B. B. (1966) Medical problems arising from the making of lakes in the Tropics. *Man-Made Lakes* (Ed. by R. H. Lowe-McConnell), 87–94.

835 Walker, K. F. (1973) Studies on a saline lake ecosystem. *Aust. J. Freshwat. Res.,* **24**, 21–71.

836 Walker, K. F. and Likens, G. E. (1975) Meromixis and a reconsidered typology of lake circulation patterns. *Verh. int. Verein. theoret. angew. Limnol.,* **19**, 442–58.

837 Walsby, A. E. (1965) Biochemical studies on the extracellular polypeptides of *Anabaena cylindrica* Lemm. *Br. Phycol. Bull.,* **1**, 514–515.

838 Walshe, B. M. (1950) The function of haemoglobin in *Chironomus plumosus* under natural conditions. *J. Exp. Biol.,* **27**, 73–95.

839 Ward, J. V. and Stanford, J. A. (Eds.) (1979) *The Ecology of Regulated Streams.* Plenum Press, New York.

840 Ward, R. C. (1975) *Principles of Hydrology.* McGraw-Hill, London.

841 Watson, R. A. and Osborne, P. L. (1979) An algal pigment ratio as an indicator of the nitrogen supply to phytoplankton in three Norfolk Broads. *Freshwat. Biol.,* **9**, 585–94.

842 Webster, K. E. and Peters, R. H. (1978) Some size-dependent inhibitions of larger cladoceran filterers in filamentous suspensions. *Limnol. Oceanogr.,* **23**, 1238–45.

843 Welch, E. B. (1980) *Ecological Effects of Waste Water.* Cambridge University Press, Cambridge.

844 Welcomme, R. L. (1976) Some general and theoretical considerations on the fish yield of African rivers. *J. Fish Biol.,* **8**, 351–64.

845 Welcomme, R. L. (1979) *Fisheries Ecology of Floodplain Rivers.* Longman, London.

846 Welcomme, R. L. and Hagborg, D. (1977) Towards a model of a floodplain population and its fishery. *Env. Biol. Fishes,* **2**, 7–24.

847 Went, A. E. J. (Ed.) (1980) *Atlantic Salmon: its Future.* Proceedings 2nd International Atlantic Salmon Association, Edinburgh. Farnham Fishing News Books.

848 Werner, E. E., Hall, D. J., Laughlin, D. R., Wagner, D. T., Wilsmann, L. A. and Funk, F. C. (1977) Habitat partitioning in a freshwater fish community. *J. Fish Res. Bd. Can.,* **34**, 360–70.

849 Westlake, D. F. (1967) Some effects of low velocity currents on the metabolism of aquatic macrophytes. *J. Exp. Bot.,* **18**, 187–205.

850 Westlake, D. F. (1978) Rapid exchange of oxygen between plant and water. *Verh. int. Verein. theor. angew. Limnol.,* **20**, 2363–7.

851 Westlake, D. F. (1982) The primary productivity of water plants. *Studies on Aquatic Vascular Plants* (Ed. by J. J. Symoens, S. S. Hooper and P. Compère), pp. 165–80. Royal Botanical Society of Belgium, Brussels.

852 Westlake, D. F. and Dawson, F. H. (1982) Thirty years of weed cutting on a chalk stream. *Proc. Eur. Weed Res. Soc. Sixth Symp. Aquat. Weeds.* 132–40.

853 Westlake, D. F. *et al.* (1980) Primary production. *The Functioning of Freshwater Ecosystems* (Ed. by E. D. Le Cren and R. H. Lowe-McConnell), pp. 141–246. Cambridge University Press, Cambridge.

854 Wetzel, R. G. (1964) A comparative study of the primary productivity of higher aquatic plants, periphyton and phytoplankton in a large shallow lake. *Int. Rev. ges. Hydrobiol.,* **49**, 1–61.

855 Wetzel, R. G. (1975) *Limnology.* Saunders, Philadelphia.

856 Wetzel, R. G. (1979) The role of the littoral zone and detritus in lake metabolism. *Arch. Hydrobiol. Beih Eraben. Limnol.,* **13**, 145–61.

857 Wetzel, R. G. (Ed.) (1983) *Periphyton of Freshwater Ecosystems.* Junk, The Hague.

858 Wetzel, R. G. and Manny, B. A. (1972) Secretion of dissolved organic carbon and nitrogen by aquatic macrophytes. *Verh. int. Verein. theor. angew. Limnol.,* **18**, 162–70.

859 White, I. D., Mottershead, D. N. and Harrison, S. J. (1984) *Environmental Systems. An Introductory Text.* Allen and Unwin, London.

860 Whitford, L. A. and Schumacher, G. J. (1961) Effect of current on mineral uptake and respiration by a freshwater alga. *Limnol. Oceanogr.,* **6**, 423–5.

861 Whitton, B. A. (1970) Biology of *Cladophora* in freshwater. *Freshwat. Biol.,* **4**, 457–76.

862 Whitton, B. A. (1975) *River Ecology.* Blackwell Scientific Publications, Oxford.

863 Whitton, B. A. (Ed.) (1984) *Ecology of European Rivers.* Blackwell Scientific Publications, Oxford.

864 Whitton, B. A. and Say, P. J. (1975) Heavy metals. *River Ecology* (Ed. by B. A. Whitton), pp. 286–311. Blackwell Scientific Publications, Oxford.

865 Williams, D. D. (1987) *The Ecology of Temporary Waters.* Croom Helm, London.

866 Williams, W. D. (1986) Limnology, the study of inland waters: a comment on perceptions of studies of salt lakes, past and present. *Limnology in Australia* (Ed. by P. De Deckker and W. D. Williams), pp. 471–86. Junk, Dordrecht.

867 Winterbourn, M. J., Hildrew, A. G. and Box, A. (1985) Structure and grazing of stone surface organic layers in some acid streams of southern England. *Freshwat. Biol.,* **15**, 363–74.

868 Winterbourn, M. J., Rounick, J. S. and Cowie, B. (1981) Are New Zealand stream ecosystems really different? *N.Z. J. Mar. Freshwat. Res.,* **15**, 321–8.

869 Wium-Andersen, S. (1971) Photosynthetic uptake of free CO_2 by roots of *Lobelia dortmanna. Physiol. Plant.,* **25**, 245–8.

870 Wium-Andersen, S. and Andersen, J. M. (1972) The influence of vegetation on the redox profile of the sediment of Grane Langsø, a Danish *Lobelia* lake. *Limnol. Oceanogr.,* **17**, 948–52.

871 Wium-Andersen, S., Anthoni, U., Christophersen, C. and Houen, G. (1982) Allelopathic effects on phytoplankton by substances isolated from aquatic macrophytes (Charales). *Oikos,* **39**, 187–90.

872 Woodwell, G. M. (1983) Aquatic systems as part of the biosphere. *Fundamentals of*

Aquatic Ecosystems (Ed. by R. S. K. Barnes and K. H. Mann), pp. 201–15. Blackwell Scientific Publications, Oxford.

873 World Health Organisation (1985) *The Control of Schistosomiasis*. Technical Report 728, WHO, Geneva.

874 Worthington, E. B. (1964) Conservation of water and fisheries in 1970. *Salmon and Trout Assoc. Lond. Conf.*, **1**, 1–7.

875 Worthington, S. and Worthington, E. B. (1933) *Inland Waters of Africa*. Macmillan, London.

876 Wright, J. F., Armitage, P. D., Furse, M. T. and Moss, D. (1985) The classification and prediction of macroinvertebrate communities in British Rivers. *Ann. Rep. Freshwat. Biol. Assoc.*, **53**, 80–93.

877 Wright, J. F., Moss, D., Armitage, P. D. and Furse, M. T. (1984) A preliminary classification of running water sites in Great Britain based on macro-invertebrate species and the prediction of community type using environmental data. *Freshwat. Biol.*, **14**, 221–56.

878 Wright, R. T. (1975) Studies on glycolic acid metabolism by freshwater bacteria. *Limnol. Oceanogr.*, **20**, 626–33.

879 Wright, S. J. L., Redhead, K. and Maudsley, H. (1981) *Acanthamoeba castellanii*, a predator of Cyanobacteria. *J. Gen. Microbiol.*, **125**, 293–300.

880 Wurtsbaugh, W. and Li, H. (1985) Diel migrations of a zooplanktivorous fish (*Menidia beryllina*) in relation to the distribution of its prey in a large eutrophic lake. *Limnol. Oceanogr.*, **30**, 565–76.

881 Wyatt, J. T. and Silvey, J. K. G. (1969) Nitrogen fixation by *Gloeocapsa*. *Science*, **165**, 908–9.

882 Young, G. and Wheeler, N. (1977) *Return to the Marshes*. Coilins, London.

883 Zaret, T. M. (1969) Predation-balanced polymorphism of *Ceriodaphnia cornuta* Sars. *Limnol. Oceanogr.*, **14**, 301–3.

884 Zaret, T. M. and Paine, R. T. (1973) Species introduction in a tropical lake. *Science*, **218**, 444–5.

885 Zaret, T. M. and Suffern, J. S. (1976) Vertical migration in zooplankton as a predator avoidance mechanism. *Limnol. Oceanogr.*, **21**, 804–13.

Index

Abramis brama 297–8
Acer saccharum 72
Acetic acid 119, 222
Achnanthes 264
Achromatium 119
Acid rain 35, 36, 59
Acidification
    detected by diatoms in sediments 342–4
    in lakes 185, 197, 198
    reversal by liming 93
    of streams 90–3
Acidobionts 343
Acidophils 343, 347, 352
Acinetobacter 119
Acipenser fulvescens 317
Acipenseriformes 232
Acridine orange 222
Acroloxus lacustris 267
Activated sludge process 149
Adenosine triphosphate 222
Aedes 137
    A. aegypti 21
Aeration of reservoirs 221
Aerobacter 227
Aeromonas 262
Aeschynomene 130, 319
Aeshnidae 151
Afforestation
    and acidification 91
    effects on streams 94
Agriculture
    Act (1947) 54
    effects on water chemistry 51–5
    lowland 52–5
    policy of EEC 54
    and pollution 154
Agriidae 151
Agrobacterium 262
Air breathing
    invertebrates 24, 123
    vertebrates 123, 324
Air lift sampler 79
Alanine 122, 245
Alcaligenes 262
Alcohols 38

Alewife 317
Algae 17
    hot spring 66
    red 74
    see also Epiphytic algae; Phytoplankton
Algicides 221
Aliphatic acids 39
Alkalibionts 343, 352
Alkaline phosphatase 264
Alkaliphils 343, 347, 352
Allen curve 80, 81, 290, 306
Allerød 347, 348
Alligator 123, 141
Allophycocyanin 162
Alma 123
Alnus 115, 119
    A. rugosa 72
Alonopsis elongata 276
Alosa
    A. aestivalis 236, 237
    A. pseudoharengus 255, 317
Aluminium 14, 15, 40, 59, 92, 93, 197, 198
Amazon streams, chemistry of 19, 49
Ambatch 319
Amia 123
Amino acids 18, 38, 39, 121, 122, 245
Ammocrypta 76
Ammonia 11
Ammonium ion 37, 44, 178, 217, 244
Amphibia 22
Anabaena 217, 219, 248, 300
    A. cylindrica 245
Anabaenopsis 301
Anabantid fish 125
Anabas 123
Anacystis peniocystis 248
Anadromous fish 333
Angling 90
Anguilliformes 232
Ankar Watt 326
Ankistrodesmus 229
Anodonta cygnea 21
Anomoeneis 349
Anopheles 136, 137
Antibiosis against diatoms 250

Antimony 38
Anuradhapura 326
Aphanizomenon 219, 229
  A. flos-aquae 251
Aphanocapsa grevillei 248
Aphanothece 219
Aphelochiridae 151
Arabs, Marsh 125–7
Arapaima 123, 133
Araphidineae 343, 345
Arbor viruses 137
Arenicola 19
Arsenic 38
Artemia salina 202
Artificial substrata
  in epiphyte studies 264–5
  in invertebrate studies 266–7
Asellidae 151
Asellus 93, 147, 198, 272, 274, 276
  A. aquaticus 117, 275
Aspartic acid 122
Asplanchna 237, 239
Astacidae 151
Asterionella 207, 345, 347, 348
  A. formosa 214, 215, 217, 227, 248, 249, 251
Astrocaryum jauri 133
Aswan High Dam 2
  pros and cons 334
Atheriniformes 232
Atmosphere 10
  absorption of radiation 160
  gases of 16, 17, 33, 37
  industrial pollution 58, 59
Audubon, John James 141
Aulonacara nyassae 30
Australia 10
Autochthonous production 69
Azolla 325

Bacillariophyta 205, 209, 219, 221, 247, 264, 278, 282, 341–4
Bacillus 119
Bacteria
  high temperature marine 66
  in iron 97
  measurement of activity 222, 261, 262
  measurement of biomass 222
  planktonic 205, 206, 221–2
  on plants 261, 262
  in streams 71
Baetidae 151
Baetis 64
Bagrus 130
Balanites 128, 130
Baltic Sea 20, 27, 28
Banyoro tribe 319

Barbus
  B. aeneus 231–3
  B. barbus 297
  B. paludinosus 29
  B. sharpeyi 127
Barium 15
Bartrachospermum 62
Base flow 60
Bay of Quinte, benthic studies 288–94
Bayluscide 138
Bear, black 94, 141
Beaver 159
Beetles *see* Coleoptera
Beggiatoa 119, 147
Benthos 257
  in Bay of Quinte 288–94
Beraidae 151
Bertalannfy, von, equation 310
Beryllium 15
Bicarbonate 19, 33, 37, 217
  uptake 104
  users 104
Bichir 123, 125, 232
Big Cypress Swamp, USA 143
Biological oxygen demand 147, 150
Biological score system 150, 151
Biomanipulation 251–6
Biomass change method for plant production 106
Biomass removal as control of eutrophication 194
Biomass of zooplankton related to that of phytoplankton 226
Biomphalaria
  B. pfeifferi 5
  B. sudanica 123
Biotic index 149
Biotin 209
Biraphidineae 343
Birge–Ekman dredge 266
Bittern 258
Bivalve molluscs *see* Lamellibranchia
Black flowers 146
Blackfish 125
Blackfly *see Simulium*
Blanket weed *see Cladophora glomerata*
Blood, roast 129
Blooms, algal 219, 220
Blue-green algae (bacteria) *see* Cyanophyta
Boar, wild 127
Bog 357
Boron 15, 37
Bosmina 223, 236, 277
  B. coregoni 349, 350
  B. longirostris 227, 228, 238, 349, 350
Botryococcus braunii 209, 341
Bottom up control 256
Boundary layer 62

Brachionus 223, 239
Brachycentridae 151
Brachyplatystroma flavicans 133
Brecon Beacons 198
Brine shrimp 202
Bromine 37
Brugia malayi 137
Brycon 133
Bryophyta 69, 74, 97, 180
Bryozoa 341
Buffalo fly *see Simulium*
Bugs, water *see* Hemiptera
Bulinus globosus 5
Bullhead 85, 96–8
Burbot 318
Busk's fluke 138
Butyric acid 118
Bythinia 139
Bythrotrephes 238

Caddis-flies *see* Trichoptera
Cadmium 38
Caenidae 151
Caesium 15
Calanoid copepods 225, 238
Calcium 15, 19, 32, 37, 40, 43, 44, 46, 49, 85,
    92, 119, 217
California 10
Calluna vulgaris 358
Calopharyphus 67
Campylodiscus noricus 347
Canalization 144
Capniidae 151
Capybara 131
Carbon 15
    C-14 dating 336, 338
    C-14 uptake method for plant production
        108
    sources for photosynthesis 104
Carbon dioxide 119, 217
    and algal periodicity 250, 251
    determinant of pH in rain 33–4
    diffusion 104
    solubility 16, 175
Carbonate 15, 33, 42, 67
    equilibria 217
Carboxylic acids 38
Carcass retention in streams 94
Carcinus maenas 21
Carex 110, 266
Caridina 321
Carnivores 72
Catadromic migration 22
Catchments, water composition 36–44, 56
Catfish 76, 232
Catla catla 324–5
Cattail *see Typha*

Cattle keeping 129, 130
Centrales 343
Centrarchid fish 232
Centroptilum luteolum 274
Ceratium 207, 217
    C. hirundirella 219, 251
Ceratophyllum demersum 104, 269, 325, 354
Ceratopogonidae 92
Ceratopogonids *see* Ceratopogonidae
Cercarium 5, 138
Cercosulcifer 66
Ceriodaphnia 236, 237
    C. cornuta 239
Chaetogaster 252
Chaoborus 202, 239, 286, 299, 322, 354
    C. alpinus 286
    C. flavicans 286
Chara 104, 252, 259, 269, 357
Characin fish 131, 132
Charophyta 18, 180
Charr
    Arctic *see Salvelinus alpinus*
    Dolly varden *see Salvelinus malma*
Chelators 246
Chilotilapia rhoadesii
Chinese black roach 324
Chironomidae 64, 72, 92, 117, 146, 151, 252,
    281–5, 350
Chironomus
    C. anthracinus 282–5, 294
    C. imicola 199
    C. luyubris 281
    C. transvaalensis 280
Chlamydomonas 149, 199, 207, 227
Chlorella 149, 228
Chloride 19, 32
Chlorination of water 186
Chlorobium phaeobacteroides 208
Chlorococcales 217
Chloroflexus 66
Chloroperlidae 151
Chlorophyll 162
Chlorophyta 18, 149, 153, 199, 205
Chlorthiamid 113
Cholera 57
Chromium 35
Chroococcus dispersus 248
Chrysophyta 206, 217, 223, 247, 341
Chrysosphaerella longispina 248
Chydoridae, niche differentiation 276
Chydorus 238
    C. sphearicus 252, 287
Chytrids 209, 223
Cichla ocellaris 133, 321
Cichlid fish 232
Ciliates 223
Cirrhina mrigala 324–5
Ciscoe 232, 317

Civet 199
Cladium jamaicensis 115, 139, 140, 141
Cladocera 24, 97, 223, 225–31, 269, 270, 271, 281, 341
Cladophora glomerata 111, 147, 264, 265
Clambidae 151
Clarias 123, 130
    C. mossambicus 29, 281
Clay minerals 42, 49
Climax vegetation and nutrient leaching 50
Cloeon dipterum 267
Clonorchis sinensis 139
Closterium 217
Clostridium pasteurianum 119
Clupeidae 320
Clupeiformes 232
Coarse particulate organic matter 69, 74
Cobalt 37, 38, 49
Cobitis 76
Coccochloris peniocystis 252
Cocconeis 264, 343
Coelocanth 26
Coenagridae 151
Cohort 80
Coleoptera 82, 268
Collectors 72–4
Colossoma macropomum 132
Communities, controversy over causes of composition 82
Community respiration 289
Competition in stream communities 86
Complex anions 13, 14
Conochilus dossuarius 227
Consents to discharge 58, 153
Conservative ions, definition of 37
Control of Pollution Act (1974) 154
Coot 141, 258, 271
Copepoda 25, 223, 225–31
Copepods *see* Copepoda
Copper 37, 49
Copper sulphate 113, 221
Cordulegasteridae 151
Corduliidae 151
Coregonid fish 186, 232, 238
Coregonus clupeiformis 317
Corematodus shiranus 30
Corers, sediment 336–7
Corixidae 151
Corophiidae 151
Corrie lakes 159
Cosmarium 217
Cost–benefit analysis (Pongolo floodplain) 9
Cottus gobio 76
Coypu 116
Crab-eating frog 22
Crane 141
Crangonyx pseudogracilis 274
Crassula helmsii 330

Crassulacean acid metabolism (CAM) 103
Crayfish and acidification 92
Crocodile 141, 258
Crow 199
Crustacea 16, 83, 146, 267, 272, 278
Cryptomonas 206, 223, 248, 250
    C. erosa 227
Cryptophyta 205, 247
Crysomelidae 151
Ctenopharyngodon idella 139, 296–301, 324, 325
Cuffa 133
Culex 137
Cultivation on swamps 145
Currulionidae 151
Cyanocobalamin (vitamin B_{12}) 38, 209
Cyanophyta 17, 18, 44, 205, 208, 209, 217, 219, 220, 221, 223, 228, 245
Cyathochromis obliquidens 30
Cyclomorphosis 239
Cyclopoid copepods 225, 238
Cyclops 223
    C. scutifer 228
Cyclostomes 318
Cyclotella 207, 217, 218, 343, 349
    C. comta 248
    C. meneghiniana 214
    C. michiganiana 248
    C. striata 361
Cygnus olor 271
Cymbella 343, 347, 349, 350
Cynodon dactylon 5–7
Cynotilapia afra 30
Cyperus
    C. papyrus 110, 115, 119, 120, 159, 331
    C. rotundus 127
Cypress *see Taxodium*
Cyprinid fish 76, 187, 232
Cypriniformes 232
Cyprinodont fish, annual species 123
Cyprinus carpio 139, 196, 297, 324
Cytophaga 262

Dalapon 113
Damaliscus korrigonus 129
Dams
    African 327
    Brazilian 327, 335
    Chinese 327, 335
    effects downstream 332–3
    effects of fish migration 95
Daphnia 206, 223, 227, 229, 236, 237, 252, 254, 255
    and biomanipulation 252
    D. galeata mendotae 228, 234, 240
    D. hyalina 227
    D. lumholtzi 239

D. pulex 238, 252
D. rosea 227, 229, 230
Darters 76
Dasyhelea thompsoni 199, 200
Dating of sediments 336, 338–9
Datura 127
DDT 88, 136, 137
Dead Sea, Israel 19
Deforestation
    effects on stream chemistry 47–9
    and salmon 94, 317
Dehydration and land organisms 23
Dendrocoelidae 151
Dendrocoelum lacteum 272
Dendrocygna viduata 5
Denitrification 110, 119, 120, 279
Deoxygenation in new lakes 327
Depth–time diagrams, interpretation of 165–6
Desmids 207, 209, 217–18
Desulphomaculatum 119
Desulphovibrio desulphuricans 119
Detergents
    industrial 57
    non-phosphate 194
    phosphate 55, 193, 194, 319
Detrended correspondence analysis (DCA) 83
Detritus
    feeders in sediments 287
    and light absorption 161
Dialysis bags 220
Diaphanosoma 238
Diaptomus 223
    D. minutus 228
    D. spatulocrenatus 227
Diatoma 218, 343
Diatoms
    in pollution monitoring 150
    in sediments 342–4
    see also Bacillariophyta
Dichlobenil 113
Digitalis 127
Dimethyl sulphide 34
Dinka 128
Dinobryon 207, 247, 250
    D. bavaricum 248
    D. divergens 248
    D. pediforme 248
Dinoflagellates *see* Pyrrophyta
Diptera 72, 82, 84, 123
Diquat 113
Direct stratification 165
Discharge of streams 61
Disparalona 276
Dissolved organic matter 69, 74
Disturbance and aquatic plant colonization 261
Diura bicaudata 274, 275

Diuron 113
Docimodus johnstoni 30
Dolphin 131
Drainage 139–45
Drift 64–5, 92
Duck 4, 141
Duckweed *see* Lemna
Dugesia
    D. lugubris 272–4
    D. polychroa 272–4
Dunaliella salina 202, 217
Dysentery 127

East Africa, arid areas 10
Ecclesiastes 86
Ecdyonurus 64
    E. dispar 274
Echinocloa
    E. pyramidalis 129
    E. stagnina 129
Eels 22, 232
Effluent diversion 191
Effluent stripping 191, 192
Egeria radiata 332
Egg
    fish 297
    guarding 297
    ratio method 234–5
Egregia 330
Egret 141
Eichhornia crassipes 110, 129, 329–30
Elatine 261
Electivity index 301
Electrofishing 80
Electrophorus 123, 124
Elephant snout fish 232
Elephantiasis 137
Elminthidae 151
Elodea canadensis 104, 267, 330
Emys orbicularis 144
Encephalitis 136
Enclosure experiments with zooplankton 228
Endorheic lakes 202
Energy budgets
    Bear Brook 70
    hot spring 68
    swamp 117
English Lake District
    early nutrient concentrations 49, 346–51
    invertebrates 274–6
    palaeolimnology 346–51
    rain composition 32
    rates of sedimentation 340
    water chemistry 41
Engraulicypris sardella 322
Enterobacteriaceae 262

Ephemerellidae 151
Ephemeridae 151
Ephemeroptera 24, 64, 65, 73, 82, 84, 92, 258, 272
Epilimnion 165
Epilithic organisms 62, 65, 74
Epipelic algae 278
Epiphytic algae 106, 108, 258, 262, 263–5, 269
Epipsammic algae 258, 278
Epischura 236–7
Epithemia 349, 350, 352
Eriocheir sinensis 21
Eriophorum vaginatum 358
Erpobdella octoculata 273, 275
Erpobdellidae 151
Esox lucius 281
Ethanol 121, 122
Etheostoma 76
Ethylene diamine tetraacetic acid 246
Euglena 199, 227
Euglenophyta 205, 217, 244
Eukaryotes 66
Eunotia 342, 343, 347, 348, 349, 352
Euphotic zone 163, 213, 220
Eurycercus 238
Eutrophic lakes 191, 192, 217
Eutrophication 185–97
    and animal droppings 199
    artificial 185–6
    and fisheries 318
    natural 185, 359
    perception of 187
    and sediment release 196
    solution 188–90
    symptoms 186
Everglades, Florida 116, 123, 139–43
Evolution of freshwater organisms 17–18
Extinction coefficient 161
    see also Net downward attenuation
        coefficient

Faeces, monkey 132
Fagus grandifolia 72
Fasciolopsis buski 138
Feeding rates, zooplankton 229–30
Fen 357
Fenlands and malaria 136
Fenuron 113
Fertilizers 48, 54
Filariasis 2, 136, 137
Filina 224
Filipendula ulmaria 121
Filter clogging organisms 186
Filter feeding in zooplankton 224–5
Fine particulate organic matter 69, 74
Fire in swamps 116, 143

Fish
    and acidification 198
    African 5
    biology 296–303
    breeding 5, 6, 76, 301–3, 323
    cartilaginous 20, 22
    culture 323–5
    demersal 303
    digestion of blue-green algae 300, 301
    digger 30
    egg production 247–8
    employers of subterfuge 30
    escapes 99
    and eutrophication 186–7, 318
    eye biter 30
    faunas 231
    feeding 298, 299
    fin chopper 30, 277
    floodplain 5, 130–3
    fruit eating 132
    generalist species 27
    growth 306
    growth in River Tees and Maize Beck 98
    growth rings 306
    hatching and acidification 92
    hybridization 99, 323
    introductions 255
    ladders 95
    in Lake Chilwa 27, 29
    in Lake Malawi 27, 30
    leaf chopper 30
    marine 11
    mimic 30
    mollusc crusher 30
    nesting in swamps 123
    open water 231–41
    orders 232
    osmoregulation 20–2
    pelagic 303
    piscivorous 30
    plant scraper 30
    predation on zooplankton 231
    production 295–325
    reproduction 76
    respiration 24
    rock scraper 30, 277
    salmonid 75–8
    scale eater 30, 277
    seed eaters 131
    smoked 321
    spawning 6
    specialist species 27, 30
    stimulation of breeding 324
    stomach contents 29
    stream 75–8
    teleost 20, 22
    toxins 127
    yield 295

zooplanktivorous 30, 230–2
Fisheries
  approximate method of yield assessment
    314–15
  choice of fish 303–4
  commercial 307
  drift net 90
  East African Great Lake 319–22
  exploitation of stock 307
  floodplain 4, 5, 130–3
  forest 137
  game 88–90
  ideal requirements 295
  management of single species 309–12
  mechanized 320–1
  in new lakes 330–1
  North American Great Lakes 316–17
  primitive 127, 129
  production 295–325
  salmon 89, 90
Flagfish 92
Flandrian 348
Flatworms *see* Flukes; Tricladida
Flavobacterium 262
Floating plant problems 327–30
Flood flow 60
Floodplains 2–9, 100, 114–15, 119–122, 125–45
Flow rate and photosynthesis 104
Flukes 136, 138
Fluorine 37
Flushing coefficient 172
Fontana di Papa Leone 355
Fontinalis 62
Forest
  flooded 131
  role in nutrient supply 46–50, 53
Fossil fuel burning and rain chemistry 35
Fragilaria 343, 347 9, 350, 352
  F. crotonensis 248, 251, 353
Frustulia 342
Fulica atra 271
Fungi
  in lakes 223
  in streams 71
Fynbos 51

Gallionella 119
Galloway 90, 198
Gambusia affinis 252
Gammaridae 151
Gammarus 64, 72, 93, 267, 272, 276, 287
  G. pulex 72, 73, 85, 87, 97, 267, 274, 275
Gar 232
Gar vesicles 208, 219, 220
Gasterosteus aculeatus 301–2
Gastropoda 82, 83, 84, 117, 123, 198, 258,
  272, 274

Gastrotrichs 223
Gelbstoffe 39, 161
Genet cat 199
Genychromis mento 30, 277
Geological history of Earth 26
Gerridae 151
Gerris 268
Ghost Mountain 2
Gill nets 127, 305, 320
Gills 23
  effects of acidification 93
  in osmoregulation 19
  rakers 299
Glaciation 26
Gloeocapsa 245
Gloeotrichia 354
Glossina morsitans 129
Glossiphoniidae 151
Glucose 222
Glutamic acid 122
Glyceria 267
  G. maxima 116
Glycollic acid 121, 222
Glyptotendipes 354
Gobies 76
Gobio gobio 297
Goeridae 151
Goldfish ponds, algal problems 240
Gomphidae 151
Gomphonema 264, 349, 350
Graptoleberis testudinaria 276
Grass carp 139, 296–301
Gravel 258
  and salmonid reproduction 76
Grayling 232
Grazing by zooplankton 226–31
Gulf of Bothnia 28
Gull 141
Guppies 232
Gyrinidae 151
Gyttja 347

Haggard, Rider 2
Half-saturation constant 214
Haliplidae 151
Halobacteria 202
Haplochromis 277, 319, 321, 322
  H. burtomi 301–2, 304
  H. compressiceps 30
  H. cyaneus 30
  H. euchilus 30
  H. intermedius 30
  H. livingstonii 30
  H. pardalis 30
  H. placodon 30
  H. polyodon 30
  H. rostratus 30

Haplochromis (*cont.*)
  H. similis 30
Haptophyta 206
Hedriodiscus trusquii 66
Helodidae 151
Heloecius cordiformis 21
Hemiptera 82, 83, 261
Hemitilapia oxyrhynchus 30
Hepatitis 149
Heptagenia lateralis 274, 275
Heptagenidae 151
Herbicides 112–13
Heron 141, 258
Herring 232
Heterocysts 245
Hevea 133
Hildenbrandia 62
Hippeutis 138
Hippopotamus 5, 258
Hippuris 269
  H. vulgaris 101
Hirudidae 151
Hirudinae 73, 83, 268
Homoeostasis in nature 362
Hoplosternum 123
Hot spring streams 65–8
Hubbard Brook experimental site 46–7
Hydra 97, 268, 276
Hydracarina 73
Hydrated cations 14, 15
Hydrilla 325
Hydrobiidae 151
Hydrocynus vittatus 299, 319
Hydrodictyon africanum 261
Hydroelectric power 90
Hydrogen fluoride 11
Hydrogen selenide 11, 12
Hydrogen sulphide 11, 12, 34, 66, 220, 327
Hydrometra 268
Hydrometridae 151
Hydrophilidae 151
Hydropsychidae 151
Hydroptilidae 151
Hygrobiidae 151
Hyparrhenia 128
Hypertrophic lakes 191–2, 217
Hyphomycetes 71, 74, 92
Hypolimnion 165
  accumulation of substances 179
  aeration 194–5
  deoxygenation 175–6, 186
Hypophthalamichthys molitrix 324, 325

Ibis 141, 143
Ice
  caps 10

structure 12
and thermal stratification 167–8
Ilyodrilus 282
  I. hammoniensis 285–6
Index B 343
Indicator species 345–6
Indole 39
Industrial pollutants 57
Industry 57
Inert gases 37
Infectious pancreatic necrosis (IPN) 99
Insecta 22
Inverse stratification 167–8
Ionic charge 14, 15
Ionic potential 14, 15
Ionic radius 14, 15
Iron 14, 37, 38, 40, 49, 85, 97, 119, 178, 246
Irrigation 7
IsiFonya 4
Isoetes 101, 258, 261, 269
  I. lacustris 104
Itshaneni 2, 3

Jaluo tribe 320
Japanese B encephalitis 137
Jellyfish 223
Jordanella floridae 92
Jussiaea diffusa 127

K selection 304
Kelp (in salmon) 77
Keratella
  K. cochlearis 227
  K. quadrata 205–6
  K. tropica 239
Kettle holes 159
Kick sampling 78, 97
Killifish 232
King Solomon's mines 2
Konia dikume 202

Labeo cylindricus 277
Labeotropheus fuelleborni 30, 277
Labidochromis vellicans 30
Lacunae 101, 107
  and pressure 180
Lakes
  Albert, Zaire, Uganda 158, 239, 296, 301, 319
  Baikal, USSR 10, 27, 157
  Bangweulu, Zambia 158
  Baringo, Kenya 204
  Barombi Mbo, Cameroon 202
  Barton Broad, UK 19, 44

Base Line, USA 234
Bassenthwaite, UK 41, 275
Big Momela, Kenya 185
Blaxter Lough, UK 287
Blelham Tarn, UK 177, 212
Bodensee, W. Germany 318
Bogoria, Kenya 204
Borax, USA 181
Brundall Broad, UK 44
Buttermere, UK 227, 275
Caban Coch Reservoir, UK 97
Cahora Bassa, Mozambique 219, 328
Char, Canada 168
Chilwa, Malawi 19, 27, 29, 158, 181, 279, 304
Chiuta, Mozambique 158
Coniston, UK 41, 275
Constance, W. Germany, Switzerland 227
Cow Green Reservoir, UK 96–8
crater 157
Crater, USA 162, 164
Crummock, UK 41, 275
Crystal, USA 236–7
Dead Sea, Israel 19, 201
Derwentwater, UK 41, 275
development 326–63
Edward, Uganda 158
Elmenteita, Kenya 185, 204
endorheic 157
Ennell, Ireland 196
Ennerdale Water, UK 41, 189, 212, 275, 340
Erie, USA, Canada 159, 316–19
Esrom, Denmark 281
essential features 156
Esthwaite, UK 41, 179, 189, 227, 275, 340, 348–51
exorheic 157–9
Eyasi, Kenya 158
filling in 335–6
formation by ice 158, 159
Gatun, Panama 239, 321
Gek-Gel, USSR 208
George, Uganda 157, 229, 241, 247, 301
Gull, USA 165, 167, 176, 248
Harriet, USA 254
Hartbeespoort Dam, S. Africa 221
Heart, Canada 230, 241
Hemmelsdorfersee, W. Germany 201
Hickling Broad, UK 44
Hodsons Tarn, UK 266
Hornindulsvatn, Norway 176
Horsey Mere, UK 44
Huron, Canada, USA 159, 316–19
Kainji, Nigeria 327, 332, 333
Kariba, Zaire, Zimbabwe 2, 322, 327–31, 333
Kiev, USSR 181

Kioga, Uganda 158
Kivu, Zaire, Rwanda 157, 158, 201
Latnajaure, Sweden 181
Lawrence, USA 181, 263
Le Roux, S. Africa 231–3
Leven, UK 213
Lilla Galtsjon, Sweden 198
Lillesjon, Sweden 197
Magadi, Kenya 185, 204
Malar, Sweden 28
Malawi, Malawi 10, 27, 201, 231, 277, 304, 321, 322
Manyara, Kenya 158, 185, 204
Marion, Canada 181
Martham Broad, UK 44, 189
Memphremagog, USA, Canada 264–5
Mendota 245
Michigan, USA 159, 215, 254, 316–19
Mikolajskie, Poland 181
Mjosa, Norway 28
Monoumi, Cameroon 202
Mont Bold Reservoir, Australia 220
Monterosi, di, Italy 353–6
Mweru, Zambia 158
Myastro, USSR 181
Naivasha, Kenya 158
Nakuru, Kenya 158, 185, 202–4
Nasser–Nubia, Egypt, Sudan 315, 327, 332, 333, 334
Natron, Kenya 158
Neagh, N. Ireland 157, 213
Ness, UK 315
Nios, Cameroon 202
Okeechobee, USA 139–42
Oresjon, Sweden 28
Pickerel, USA 351–3
production, general models 182–5
Ranu Lamongan, Indonesia 202
Reshitani, Kenya 185
Rockland Broad, UK 44
Roseires Reservoir, Sudan 328
Round, USA 255
Rudolf *see* Lake, Turkana
Rukwa, Kenya 158
Sagami, Japan 246
Sennar Reservoir, Sudan 328
Sommen, Sweden 28
South Walsham Broad, UK 44
Superior, Canada, USA 159, 316–19
Tahoe, USA 259
Tanganyika, Tanzania 10, 27, 157, 158, 201, 277, 302, 320, 322, 330, 356
Tokke, Norway 201
Trummen, Sweden 195, 204
Turkana, Kenya 27, 158, 204, 296
Ullswater, UK 41, 275
Upton Broad, UK 189
Valencia, Venezuela 359–61

Lakes (*cont.*)
　Vennes, Sweden 28
　Victoria, Kenya, Uganda, Tanzania 130,
　　157, 158, 165, 166, 249, 277, 309, 320,
　　321, 322
　Volta, Ghana 327, 330-4
　Washington, USA 191, 193, 227, 254
　Wastwater, UK 41, 176, 189, 340
　Windermere, UK 41, 165, 167, 179, 189,
　　212, 213, 245, 249, 273-5, 340
Lamellibranchia 82, 146
Lampreys 20, 317
Larvae, definition 24
　colonization of fresh waters 25
Laterite 40
Lates 320
　L. niloticus 319, 322
　　biology 296-301
　　introduction 321-2
Latimeria chalumnae 26
Latitude, effects on production 183
Lead 271, 272
Lead-210 dating 336, 338
Leaf packs 65
Lechwe 125, 129
Leeches *see* Hirudinae
Lemna 110, 325
　L. trisulca 104
Lemnanea 62
Lepidoptera 64
Lepidosira 123
Lepidosireniformes 232
Lepidostomatidae 151
Lepisosteus 123
Lepomis
　L. cyanellus 255
　L. macrochirus 252, 255
Leptoceridae 151
Leptodora 223, 238
Leptophlebia 267
　L. vespertina 267
Leptophlebiidae 151
Lesser flamingo 202
Lestidae 151
Lethal concentration (LC_{50}) 112, 153
Lethrinops brevis 30
Leucichthys 317
　L. artedi 317
Leuctra nigra 65, 85
Leuctridae 151
Levees 3
Libellulidae 151
Light
　absolute levels in river water 103, 104
　absorption in water 103, 104, 159, 163
　and dark bottle method for plant
　　production 107, 108
　saturation 211, 212

Limnephilidae 151
Limnephilus 109
Limnocalanus macrurus 27, 28
Limnothrissa miodon 320, 322, 330
Limpets 62
Lion, mountain 141
Liquids, structure of 11, 12
Lithium 15
Litter 65
Littoral zone 257-9
　open water relationships 279
Littorella uniflora 101, 102, 105, 258, 266, 269
Liver fluke 139
Loading 172
Lobelia dortmanna 103, 104, 106, 263
Loch Ness monster 315
Logistic equation 304
Lota lota 318
Lucius Cassius Longinius 355
Lugworm 19
Lund Tubes 177, 271
Lung 23
Lung fluke 138
Lungfish 123-5
Lymnaea peregra 267
Lymnaeidae 151
Lyngbya 301
　L. taylori 248

Mackereth corer 336-7
Macrobrachium niloticum 299
Macrophytes *see* Plants
Madan 125-7
Magnesium 15, 19, 32, 40, 41, 44, 46, 49, 92
Major ions 37, 40, 41
Makatini Flats 3
Makumba *see* Fish, in Lake Chilwa
Malaria 2, 136, 137, 321
Maleic hydrazide 113
Malic acid 121
Manatee 131
Manganese 14, 15, 37, 40, 49, 119, 178, 246
Man-made lakes 159, 327-35
Mansonia 137
Marduk, the Great God 127
Mark, release, recapture method 305
Marl 263
Marsh, definition 114
Mastigloia 353
Mastigocladus laminosus 66
Matemba *see* Fish, in Lake Chilwa
Maximum economic yield 308-9
Maximum sustainable yield 308-9
Mayflies *see* Ephemeroptera
Meanders 61
Megaloptera 268

Melanin 39
Melanoidin 39
Melosira 217, 218, 249, 301, 343, 347, 348
  M. ambigua 353
  M. arenaria 349
  M. granulata 342, 352, 353, 355
  M. italica 227
    M. italica var *subarctica* 250
    M. italica var *tenuissima* 228
Mercaptans 39
Mercury 38
Meridion 264
Meromictic lakes 201–2
Mesh size of nets 312
Mesocyclops 236, 237
Mesopotamia 2, 125
Mesotrophic lakes 191–2, 217
Mesovelidae 151
Metacercariae 138, 139
Metalimnion 165
  algae in 250
Metallogenium 208
Metals, heavy 152–3
Methane 11, 39, 148
Methanobacterium 119
Methanomonas 119, 148
Metrifonate 138
Microbial complexes in plankton 223
Microcystis 207, 219, 229, 300, 301
  M. aeruginosa 251
Microfilariae 137
Micropterus salmoides 255
Miller's thumb *see Cottus gobio*
Millet 129–30
Miocene 27
Mineral lodes 38
Miracidium 5
Mites 223
Mixing and phytoplankton 218
Mixing depth 213
Mkuzi 2
Mlamba *see* Fish, in Lake Chilwa
Mollanidae 151
Molybdenum 37, 49, 246
Monoraphidineae 343
Monuron 113
Moraine 159
Moraxella 262
Mormyridae 299, 330
Mormyriformes 232
Morpho-edaphic index 185
Mosquitoes 136
Mosses *see* Bryophyta
Mougeotia 198
Mouth brooding 302
Mrigal 324–5
Mudhif 127
Multiple regression analysis 85

Muskrats 116, 117, 251
Mylopharyngodon piceus 324
Mylossoma duriventris 132, 133
Myosotis 111
Myriophyllum 269, 354
  M. spicatum 101, 104, 105, 270, 330
Mysids 226, 238
Mysis 226

Naididae 97
Najas (*Naiar*) 325
  N. flexilis 263, 353
  N. marina 269, 353
Nanoplankton 210
Nasturtium officinale 262
Natal 2
Naucoridae 151
Nauplii 25
Navicula 264, 343
  N. elkab 342
Nebulon 113
Nemacheilus 76
Nematoda 186, 278
Nematode diseases 137
Nemoura
  N. avicularis 274
  N. cinerea 268
Nemourella picteti 65
Nemouridae 151
Neocertatodus 123
Neolithic Invasion 351
Neomysis 226
  N. mercedis 254
Nepidae 151
Neritidae 151
Net downward attenuation coefficient 161,
  163, 164, 252
Net plankton 210
New England 90
Nile Perch *see Lates niloticus*
Nitella 259
Nitrate 15, 37, 44
  concentrations in British rivers 54
  leaching from catchments 48, 52, 53
  removal by plants 109
  in streams 85
Nitrilotriacetic acid 246
Nitrite 44
Nitrobacter 119
Nitrogen 15, 16, 17, 37, 56, 135
  fixation 44, 217, 245
  as limiting nutrient 170
  oxidides in atmosphere 35
  in plants 105
  solubility in water 17
  in streams 71
Nitrogenase 245

Nitrosomonas 119
Nitzschia 264
  N. amphibia 360
  N. frustulum 342
NO_2 35
NO_x 35, 40, 59, 90
Non-conservative ions 37
Norfolk Broadland 44, 218
North Buvuma Island fishery 310–14
Norway 90
Nothofagus 75
Notonecta 268
Notonectidae 151
Nova Scotia 90
Nuer 127–30
Nutrients
  budget for River Bure 56
  effects of agriculture 51–5
  in effluent 55
  gains and losses from natural vegetation
    47–50
  key 37, 168–70
  limitation, discussion of 170–1, 183–4, 214
  pump hypothesis 280
  supply to demand ratio 49
  tissue contents in aquatic plants 105
  trace 37
  uptake in phytoplankton 214–16
Nymphaea alba 101, 267, 354
  N. lotus 130
Nymphs, definition 24
Nyssa 115

Ocean
  length of existence 10
  phyla in 11
  salinity 10, 18
  size 10
  volume 10
Odonata 268
Odontoceridae 151
Ohanapecosh hot springs 67–8
Oligochaeta 82, 83, 97, 123, 146, 151, 278, 345
Oligochaete worms *see* Oligochaeta
Oligomictic lakes, rare overturn and disaster
  202
Oligotrophic lakes 191, 192, 217
Oligotrophication 197
Onchocerca volvulus 67
Oncorhynchus
  O. kisutch 75, 94, 254
  O. nerka 75
  O. tshwawytscha 75, 89, 254
Ontario lakes, phosphorus levels in 189
Ophiocephalid fish 125
Opisthorcis felinus 139
Opossum 141

Oreochromis 297
  O. andersonii 323
  O. aureus 323
  O. esculentus 309, 312
  O. macrochir 323
  O. mossambicus 323
  O. niloticus 323, 330
  O. urolepis hornorum 323
Organic matter
  aliphatic 16
  analytical problems 38
  aromatic 16
  charged 15
  dissolved 51
  films on stones 62–3
  input to streams 50, 68–70
  labile 36, 51, 222
  nature in water 39
  origins in water 39
  pollution 46–7
  processing in streams 70–1
  refractory 36, 51
  in sediment 340–1
Orthocladinae 97
Oryza 128
  O. barthis 129
Oscillatoria 207, 217, 218, 219
  O. agardhii 341
  O. rubescens 19, 341
Oscillaxanthin 341
Osmoconcentration 18
Osmoconformers 19–21
Osmolality 18
Osmoregulation 18–21
  fish 20, 21
  invertebrates 19, 21
  vertebrates 20, 21
Osteoglossid fish 125
Ostracoda 66
Otter 141, 258
Overfishing 295, 308
Oxamniquine 138
Oxidation ponds 149
Oxidized microzone 178, 196, 197
Oxygen 16, 17, 38
  concentration with depth 176
  solubility 17

P:R ratio 74–5
Pacific Ocean 212
Pacutoba 132
Palaeolimnology
  of Blea Tarn 347
  of Esthwaite 348–51
  of Lago di Monterosi 353–6
  of Lake Valencia 359–61
  problems in interpretation 344–6

of Tarn Moss, Malham 356–9
Palaeomoneter varians 21
Papyrus *see Cyperus papyrus*
Paragonimus 138
Parr 77
Particle size hypothesis 315–16
Pasture reclamation in uplands 93
Peat formation 117, 279
Peaty waters 51
Pediastrum 207, 217, 341, 355
Pelican 141, 320
Peptides 38, 39, 222
Perca
    P. flavescens 299
    P. fluviatilis 271
Perch 232, 271
Percids 76
Perciformes 232
Percina 76
Peredinium 217
Periodic table 11, 15
Periphyton 261, 267
Perlidae 151
Perlodidae 151
Pesticides
    and loss of submerged plants 271
    and rice culture 324
    safety precaution scheme 112
Petromyzon marinus 317
Petrotilapia tridentiger 277
pH
    and carbon availability 104, 119, 217, 218
    definition 34
    in fish gut 300
    problems in determination 36
    rock weathering 40
    in streams 85, 90
    see also Acidification
Phacus 207
Phaeophyta 17
Phagotrophy in algae 223
Phalaris arundinacea 121
Phenols 39, 51
Philopotamidae 151
Phoeniconaias minor 202
Phosphorus 15, 49, 56, 135
    chlorophyll relationship 171, 190, 191
    cycle in plankton 241–4
    establishment of concentration 171–4
    leaching from catchment areas 47, 48
    loss to sediments 177
    relative scarcity 169–70
    sources to lakes 190
    in streams 71
    stripping 191–2
    upland pastures 93–4
Photon flux density *see* Light
Photosynthesis curve with depth 211–12

Photosynthetically active radiation 159
Phoxinus phoxinus 297
Phragmites 110, 116, 117, 135
    P. australis 115
    P. mauritianus 129
Phryganea
Phryganeidae 151
Physa fontinalis 275
Physidae 151
Phytane 39
Phytoplankton 159, 205–21
    communities 217
    competition with plants 269
    distribution 216
    grazing 216
    gross photosynthesis 213
    growth equation 210
    layering 208
    light absorption 163
    net production 213–14
    nutrient uptake 209
    parasites 210
    seasonal changes 246–51
    and sex 210
    sinking 209
    size 216
    vertical movements 219, 220
Picarucu 133
Picoplankton 210
Pigments in sediments 341
Pike 232, 281
Pinnularia 347
Pinus caribaea 139
Piranha preta 132
Piscicolidae 151
Pisidium 282
    P. casertanum 285, 286
Pister, E. P., change in view 99
Pistia stratiotes 137
Planariidae 151
Plankton
    absence from streams 62
    community structure 205
    compartment 241–2
    nitrogen cycle 244–5
    phosphorus cycle 241–4
    scaled up model 205, 206, 210
    seasonal changes 246–51
Planorbidae 151
Planorbis vortex 267
Plantago lanceolata 351
Plants, emergent 115, 116
    aeration by mass flow 121
    flood tolerance 120–2
    in lakes 180
    see also Swamps; Wetlands
Plants, floating
    problems with 327–30

Plants, submerged
  active or inert substrata 264-5
  alkaloid production 264
  carbon uptake 104
  competition with phytoplankton 269
  depth distribution 180, 259, 260, 261
  general biology 101-3
  grazing of 264
  growth 103-5
  invertebrate communities of 265-9
  in lakes 159, 180-2, 259-61
  management 111-14
  nutrient uptake 102
  organic secretions 263-5
  photosynthesis 102-3, 103-5, 106, 107, 108
  productivity 105-9, 110, 181
  refuges for Cladocera 255
  root: shoot ratios 106
  seed dispersal 261
Plasmodium 136
Plastron 25
Platycnemididae 151
Plecoptera 64, 65, 82, 84, 85, 272
Plecostomus 123
Plectrocnemia conspersa 65
Pleidae 151
Plover 141
Podocarp hardwoods 75
Podostemonaceae 62
Podostemonads 62
Poliomyelitis 149
Pollen 344
Pollen zones 346
Pollution management 154
Pollution monitoring 149-52
Polycelis 272-4
Polycentropidae 151, 268
Polycentropus flavomaculatus 273, 275
Polychlorinated biphenyls 57, 58, 153
Polyculture 324
Polygonum senegalensis 127
Polypedilum vanderplanki 199, 200
Polyphemus 223, 238
Polypteriformes 232
Polypterus 123
Polyzoa 258
Pomoxis nigromaculatus 255
Pond culture 323-5
Pongolapoort dam 7, 8
Pontoporeia 282, 318
Pools
  experimental 251, 282
  temporary 199
Porphyrins 39
Porto Velho 131
Potamanthidae 151
Potamobius fluviatilis 21

Potamogeton 260, 269
  P. crispus 5, 6, 104
  P. filiformis 260
  P. lucens 127
  P. natans 101
  P. obtusifolius 260
  P. pectinatus 353
  P. polygonifolius 260
  P. praelongus 260
  P. Richardsonii 264
  P. x Zizii 260
Potamophylax latipennis 79
Potassium 15, 18, 19, 32, 40, 41, 44, 46, 49, 92
  leaching from catchment 47, 48
Povilla adusta 331
Prairie development 351-3
Prawns, in Dominican streams 73
Predator avoidance by zooplankton 239-40
Predators, lurking and hunting 268
Pressure, effect on plants 180
Pristane 39
Prochlorophyta 205
Procladius pectinatus 286
Productivity
  algal, measurement methods 105-9
  animal, measurement methods 78-86
  distribution among communities in lakes 181
  in endorheic lakes 203-4
  fish 304-25
  oxygen exchange method 67, 68
  plant, measurement methods 105-9
  swamps 110, 115, 116
Profundal zone 257, 281-8
Prokaryotes 66
Proline 122
Proteins 38
Protista 17
Protopterus 123, 130
Protozoa 17, 22, 221-3, 265, 278
Proziquantel 138
Prymnesium 207
Psephenus 64
Pseudomonas 119, 262
Pseudotropheus
  P. livingstonii 277
  P. tropheops 277
  P. zebra 277
Psychomyiidae 151
Ptychadena anchiaetae 200
Punic Wars 355
Purse seine 305, 320
Pyrrophyta 205, 247
Pyruvic acid 121

Quercus alba 72

r selection 304
Raccoon 94, 141
Radionuclides 57
Rain
  atmospheric pollution 35
  composition 32
  determination of pH 33–4, 36
  dust contribution 35
  sea spray contribution 34–5, 43
  sulphur compounds in 32
Ramphochromis macrophthalmus 30, 231
Ranunculus penicillatus var *calcareus* 105
  bacteria on 262
  production in rivers 109
Raphe 264
Raphidioidinae 343
Redd 77
Redfield ratio 215
Redox potential
  definition 117
  values in sediments and soils 118, 120, 178,
    179
Reed *see Phragmites*
Reed mace *see Typha*
Reflection and light loss 160
Renewal time 10
Residence time 10, 192
Respiration 67
  measurement in algae 213
Rhizosolenia 217, 345
  R. eriensis 248
Rhodomonas 207
  R. minutus var *nannoplantica* 248
Rhodotorula glutinis 229
Rhopalodia gibba 348
Rhyacophilidae 151
Ribulose diphosphate carboxylase 261
Rice 2, 127, 324
Rift Valley, East African 27
Rinderpest 130
Riplox process 197
Rithrogena 64
River
  Amazon, Brazil 131
  Ant, UK 44, 56
  Bahr el Ghazal, Sudan 127
  Bahr el Jebel, Sudan 127
  Bear Brook, USA 68–70
  blindness 87
  Bure, UK 44, 56
  Caloosahatchee, USA 141
  Charles, USA 134
  classification 150
  continuum concept 73, 75
  Dorset Frome, UK 54
  Elan, UK 97
  Euphrates, Iraq 2, 125
  Great Ouse, UK 54, 111

Hubbard Brook, USA 46–9
Hudson, USA 318
Indus, Pakistan 2
Lee, UK 54
Machado, Brazil 132–3
Madeira, Brazil 131
Maize Beck, UK 96–8
management 111–14, 144
Medway, UK 85
Mersey, UK 57
Muck Fleet, UK 44
in New Zealand 73
Nile, Egypt, Sudan 2, 27, 130, 297
Norris Brook, USA 90
Orange, South Africa 50, 232
Orinoco, Venezuela 361
Ouse, UK 85
Owens, USA 99
Pongolo, South Africa 2–9
regulation 96, 97
Rhine, W. Germany, Netherlands 144
Rutshuru, Zaire, Uganda 157
St. Lawrence, USA, Canada 317–18
Sobat, Sudan 127
Spixworth Beck, UK 44
Stour, UK 54
Tees, UK 96–8
Thames, UK 54
Thurne, UK 44, 56
Tigris, Iraq 2, 125
Tywi, UK 93
Usutu, South Africa, Mozambique 3
Volta, Ghana 2
Whangamata Stream, New Zealand
  110
White Nile, Sudan 127, 212
Wye, UK 52, 97
Yare, UK 44
Zambesi, Zimbabwe, Mozambique,
  Zambia 2
Road building, effects on lakes 355
Rock fishes in Lake Malawi 277–8,
  304
Rocks
  continuous oceanic sequence 26
  of English Lake District 41
  igneous 40
  metamorphic 42
  sedimentary 26
  weathering 39–44, 50
Rocky shores in lakes 272–8
Roots, flood tolerance 121
Rorippa nasturtium aquaticum 105
Rotenone 255
Rotifera 24, 223, 300
Ruppia 202
Russell equation for fisheries 307
Rutilus rutilus 231, 297

Salamander 237
Saline lakes 202, 217
Salix 115
Salmo
  S. gairdneri 76, 99, 254
  S. salar 75, 76, 77–8, 89, 90, 316, 317
  S. trutta 75, 76, 85, 88, 89, 96–8, 99, 254, 296–301
  S. tshawytscha 75
Salmon
  Atlantic *see Salmo salar*
  chinook *see Salmo tshawytscha*
  coho *see Oncorhynchus kisutch*
  effects of afforestation 94
  farming 90, 323
  fishing 89, 90
  poaching 90
Salmonid fish
  effects of acidification 92
  effects of eutrophication 186
Salmoniformes 232
Salt lakes 202, 203
Salt pans 202
Salvelinus
  S. alpinus 89
  S. fontinalis 76, 99
  S. malma 89
  S. namaycush 254, 317
Salvinia
  S. auriculata 328
  S. biloba 328
  S. molesta 328–30
Samplers
  air lift 78, 79
  artificial substrata 78
  kick 78
  in plant beds 265–6
  Surber 78, 79
Sandy shores 278
Sarotherodon 297
  S. galilaeus 330
  S. niloticus 296–301
Scalar irradiance 161
Scandium 15
Scenedesmus 149, 206, 217
Schistosoma
  S. haematobium 5
  S. mansoni 5
Schistosomiasis 2, 5, 8, 127, 136–8, 333
Scirpus 115, 116
  S. cubensis 328
  S. subterminalis, secretion of organic matter 263
Scrapers 72–4, 80
Sculpins 76
Sea water composition 32
Seattle 191
Secci disc 164

Sediment
  chemistry 178, 179, 196, 339
  formation 335–6
  fossils 341–4
  removal for lake restoration 195–6
  traps 288
Sedimentation coefficient 172
Segmentina 138
Seiches 174
Seine net 305
Self-shading 213
Semionotiformes 232
Serenoa serrulata 191
Sericostomatidae 151
Serine 122, 245
Serrasalmus thombeus 132
Sesarma erythrodactyla 21
Setaria 128
Sewage
  effects on invertebrate communities 274
  effects on stream chemistry 55
  effluent 55
  fungus *see Sphaerotilus*
  mains 55
  nutrients in 55
  oxidation ponds 209, 210
  septic tanks 55
  treatment 147–9
  treatment in wetlands 135
Shad 232
Shikimic acid 121
Shoreline development 181
Shredders 71, 72, 74
Shropshire Meres 159, 189
Si:P ratio 215
Sialidae 151
Sialis 268
Siderocapsa 208
Silicate 37, 249
Silicic acid 40
Silicon 14, 15, 49
  lattices and minerals 39, 40
Silt 50
Siluriformes 232
Siluroid fish 125, 232
Silver 38
Silver carp 324
Simuliidae 151
Simulium 64, 73, 85, 87, 88, 144
Siphlonuridae 151
Sitatunga 125
Size–efficiency hypothesis 236
Size-selective predation 237
Small mouth yellowfish *see Barbus aeneus*
Smelt 232
Smolt 77
Snails *see* Gastropoda
Snakes, sea 25

Snow 32
  acid 36
Sodium 15, 17, 18, 19, 32, 40, 41, 44, 46, 49
Solvents, ionic and polar 14
Sorghum 129
Southey, R. 351
Specificity of invertebrate communities on
  plants 267
Sphaeriidae 151
Sphaerium 288
Sphaerocystis 217
  S. schroeteri 228
Sphaerotilus 119, 147, 149
Sphagnum 198, 354, 357, 358
Spiders, water 25
Spirodela 101, 325
Spirogyra 252, 263
Spirulina 217
  S. platensis 202
Sponges 186
Spoonbill 141
Sporobolus 128
Staurastrum 207, 217
Staurodesmus 217
Stephanodiscus 217, 218
  S. hantzchii 227, 250
  S. niagarae 248, 353
Stickleback 301, 302, 304
Stizostedion
  S. lucioperca 255, 296–301
  S. vitreum 255, 296–301
Stock farming, nutrient effects 52–3
Stolothrissa tanganyikae 320, 330
Stoneflies *see* Plecoptera
Stone loach 85
Stork 141
Stream
  acidification of 90–3
  adaptation to 64–5
  basic hydrology 60–1
  communities 62, 81–7
  discharge 61
  drift 64
  food relationships 74
  ideal 60–1
  organic input 65, 68–70
  physical alteration 95–6
  upland, effects of man 87, 88–99
  see also River
Stromatolites 360–1
Sturgeon 232, 317
Sudd 127–8
Sugars 38–9
Sulfolobus acidocaldarius 65
Sulphate 15, 19
Sulphide 119
Sulphur 15
Sulphur bacteria 206

Sulphur dioxide 34, 35, 40, 59, 90
Surber sampler 78–9
Suspension feeding in rotifers 223
Swamps 114–45
  cultivation 145
  cypress *see Taxodium*
  definition 114
  diseases in 127, 136–9
  drainage 137, 139–45
  fish 123–5
  grazing in 116
  nutrient retention 134–5
  productivity 110, 115–16
  relationships with open water 279–81
  soils 116–22
  tree 110
Swan 258
  poisoning by lead shot 271–2
Sweden 90
Symbranchus marmoratus 124
Synechococcus 65
Synedra 218, 310, 345
  S. acus 248

Tabellaria 217, 349
Tadpoles, Savannah ridge frog 200
Taeniopterygidae 151
Tambaqui 132
Tanypus nebulosus 153
Tanytarsus 282
Tapir 131
Tardigrades 278
Tarn Moss, Malham 356–9
Tarns 159
Taxodium 115, 140
Telphura fluviatilis 21
Tench 271
Teotonio Rapids 131
Terbutyrene 113
Tern 141
Thembe-Tonga 4, 7
Thermal stratification 164–8, 174–7, 218, 219
Thermocline 165
Thermocyclops hyalinus 229, 245, 247
Thiamine 209
Thiobacillus thiooxidans 66
Thymidine 222
Tiang 129
Tilapia 296–301, 309, 320, 323
  T. shirana chilwae 29
  T. zilii 330
  see also Oreochromis; Sarotherodon
Tin 38
Tinca tinca 271
Tipulidae 92, 151
Titanium 15
Toic 128–30

Top down control 256
Trace elements 31
    analytical problems 38
Trapa natans 138
Travertine 67
Trawl net 305
Tree swamps 110
Trematoda 137
Trichoptera 79, 82, 84, 109, 258, 272
Trickling filters 148
Tricladida 24, 83, 223, 272–6
Tropocyclops 236–7
Trout
    Brook *see Salvelinus fontinalis*
    Brown *see Salmo trutta*
    farming 99
Trypanosomiasis 129, 136
Tsessabe 129
Tsetse fly 129
Tuberculosis 127
Tubifex 282, 289
Turions 5, 9, 101
Turtles 123
Two-way indicator species analysis 83–4
Typha 110, 115, 116, 127, 361
    T. australis 129
    T. domingensis 250

Ulmus americana 72
Ultra-oligotrophic lakes 191–2
Ultraplankton 210
Underfishing 308
Unionidae 151
Unit stock of fish 309
Upstream–downstream method for
        production 106
Urea 22, 39, 245
Utricularia vulgaris 117

Vahlkampfia 66
Vanadium 15, 37
Varves 339, 347
Vascular plants 22
Vegetables 7
Vegetation, effect on water chemistry 45–50
Veronica beccabunga 111, 262
Via Cassia 355
Viral haemorrhagic scepticaemia 99
Virunga volcanoes 157
Vitamins
    B$_{12}$ 38, 209, 246
    biotin 209, 246
    concentration in water 246
    requirement by algae 209, 250
    thiamine 209, 246
Volcanic lava 38

Vollenweider model 172–4, 190
Volta dam, pros and cons 333
Volvox 219
Vossia cuspidata 129

Wading birds 237
Washout 211, 218
Water
    absorption of light 161–2
    athallasic 19
    Authorities 85, 144, 188
    bugs *see* Hemiptera
    buffalo 127
    chemistry 37–9
    chestnut 138
    cress 139
    cycle 1, 2, 10
    density 13
    distribution pipes, clogging of 186
    filtration 186
    fleas *see* Cladocera
    gas diffusion in 23
    hyacinth *see Eichhornia*
    hypersaline 19
    lily, aeration of rhizome 121
    meadows 111
    molecular structure 12
    physical properties 11–14, 23
    physiological problems of living in 18–19
    pure 32
    respiration in 23–4
    saline 4, 10, 19
    snakes 123
    supply 57, 221
    temperature 13
    treatment 221
Weichselian 348
Wetlands
    evaporation from 134
    flood control 134
    values 134
White bighead 324
Whitefish
    floodplain 125
    Great Lake 317
Wolffia columbiana 101
Wordsworth, William 351
Wuchereria bancrofti 137
Wurm 355

Xanthomonas 262
Xyloborus torquatus 331

Yaws 127
Yellow fever 136

Yellow substances 39, 279
  see also Gelbstoffe

Zinc 35, 37, 49
Zooplankton 223–31
  and biomanipulation 253
  community composition 236–9
  development in rivers 100
  diet 227–8

feeding rate 229–31
food limitation 227
grazing rates 229–31
and phosphorus cycling 244, 254
population dynamics 234
predator avoidance 239
seasonal changes 247
vertical migration 239
Zulus 4
Zygnema 263